BF
455
.C6747
1978

Conference on
Cognitive Process
Instruction,
University of
Massachusetts at
Amherst, 1978.

Cognitive process
instruction

DATE		

EDUCATION & PHILOSOPHY

Social Sciences and History Division

COGNITIVE PROCESS INSTRUCTION

RESEARCH ON TEACHING THINKING SKILLS

Jack Lochhead and John Clement, editors

THE FRANKLIN INSTITUTE PRESS℠

© 1979
THE FRANKLIN INSTITUTE PRESS
Philadelphia, Pennsylvania.

Current printing (last digit):
5 4 3
THE FRANKLIN INSTITUTE PRESS NUMBER: I/1-078
ISBN Number: 0-89168-014-4
Library of Congress Catalog Card Number: 78-22122
Printed in the United States of America.

Preface

They should have a course to teach you *how* to learn
. . . all they have is courses on *what* to learn.

—David Kocot (college student)

Cognitive process instruction is an approach to teaching which emphasizes understanding, learning, and reasoning skills as opposed to emphasizing rote memorization of factual knowledge. This book describes some of the most recent and innovative approaches to cognitive process instruction and describes some recent research studies on thinking skills that have direct implications for instruction of this kind.

In June 1978 a group of approximately 50 faculty from various American universities met at the University of Massachusetts to discuss issues related to cognitive process instruction. They were joined by representatives from eight private and federal agencies. All the participants held a common belief that it is possible to conduct serious applied research in education, and that through such research ways can be found to train students to learn productively and think more clearly.

The conference brought together, for the first time, representatives of a small but rapidly growing movement. The movement consists of people who are reawakening the 19th century belief that education can improve the functioning of the mind through training and that the role of the University is not just to pass on information, but to teach students how to become knowledgeable learners and problem solvers. Added to this revival of practical interest are some very recent theoretical developments, both in the science of cognition and in the epistemology of science. The application of these theoretical developments to instructional practice is a challenging goal and one with great potential benefits for education.

This book represents only a sampling of the viewpoints expressed at the conference. Even so it serves as an introduction to an exciting and expanding new field. It is a diverse collection: Papers range from the theoretical to the applied, the conjectural to the carefully researched, and from brief comments to exhaustive investigations. This diversity of approaches is in one sense entirely appropriate; the field is very young and a subject as complex as human thought is probably best researched from a variety of perspectives.

Cognitive process instruction cannot be defined by stating a universal point of view or a single research methodology. But investigators in the field do share an interest in the following sorts of questions:

1. How can general learning and problem-solving skills be taught?
2. How can instruction encourage lasting understanding rather than short-term recall based on rote learning?

3. How can creativity and inventiveness be encouraged without sacrificing disciplined inquiry, systematic methodology and attention to detail?

4. What learning and reasoning skills do students actually employ and how are these affected by instruction?

Certainly, definitive answers to these questions would have revolutionary implications for education. All of these questions touch on some aspect of cognition, but there is considerable disagreement as to what cognitive theories, if any, can be usefully applied to instruction. At one extreme there are those who are profoundly skeptical of all cognitive theory. Others base their approach on information-processing psychology, developmental (Piagetian) psychology, or artificial (computer) intelligence. Finally, there are a few who have attempted to integrate these theories into a consistent whole. The papers in this volume sample each of these points of view.

This book is intended for college and high school teachers as well as those doing research on cognitive processes relevant to higher education; its focus is primarily, but not exclusively, on science and mathematics. It is an introduction—not a handbook or text—and we hope it will encourage the reader to pursue further reading in one or more of the areas covered. It might also serve as a book of readings for a graduate seminar on cognitive process instruction. We have yet to hear of the existence of such a seminar but hope that some of our readers will perhaps be inspired to start one.

Jack Lochhead
John Clement
Amherst, Massachusetts, 1978

ACKNOWLEDGEMENTS

Many people have helped in the preparation of this book. We would especially like to thank Lois B. Greenfield, Julia S. Hough and Joseph W. Martin for their editorial assistance. We are grateful to the Exxon Education Foundation for its support of this project and the Fund for Improvement of Post Secondary Education for its sponsorship of the Conference from which these papers were selected. The Fund and Exxon have been strong supporters of Cognitive Process Instruction and their vision has been critical to the early development of this field.

Finally we want to acknowledge the help of Frederick W. Byron Jr., Belinda Cajka, Robert L. Gray, Jane Luzbetak, Jim Nix, Ron Narode and Susan White.

CONTRIBUTORS

Arnold Arons
Department of Physics & Education, University of Washington
Seattle, WA 98195

Robert Bauman
Department of Physics, University of Alabama at Birmingham
Birmingham, AL 35294

Thomas Wdowiak
University of Alabama at Birmingham
Birmingham, AL 35294

Irene Loomis
Department of Mathematics, Memphis State University
Memphis, TN 38152

John J. Clement
Department of Physics and Astronomy, University of Massachusetts
Amherst, MA 01003

Gene D'Amour
Directorate for Science Education, National Science Foundation
Washington, DC 20550

Andrea diSessa
M.I.T. Artificial Intelligence Laboratory, 545 Technology Square
Cambridge, MA 02139

John A. Easley, Jr.
College of Education, University of Illinois
Champaign, IL 61820

Ira Goldstein
Xerox Corporation, Palo Alto Research Center
3333 Cayote Hill Road, Palo Alto, CA 94304

John Seely Brown
Xerox Corporation, Palo Alto Research Center
3333 Cayote Hill Road, Palo Alto, CA 94304

Robert L. Gray
Department of Physics and Astronomy, University of Massachusetts
Amherst, MA 01003

Lois Greenfield
University of Wisconsin-Madison, Engineering Freshman Office
Madison, WI 53706

James J. Kaput
Mathematics Department, Southeastern Massachusetts University
North Dartmouth, MA 02747

Robert Karplus
SESAME, Lawrence Hall of Science, University of California
Berkeley, CA 94720

Elizabeth Karplus
Campolindo High School
Moraga, CA

Marina Formisano
University of Rome
Rome, Italy

Albert-Christian Paulsen
Royal Danish College of Educational Studies
Copenhagen, Denmark

Herb Koplowitz
R.R. 1, Carrying Place
Ontario, KOK 1L0, Canada

Jill H. Larkin
Department of Psychology, Carnegie-Mellon University, Schenley Park
Pittsburgh, PA 15213

Herbert Lin
Physics Department, 20 B-145, MIT
Cambridge, MA 02139

Jack Lochhead
Department of Physics and Astronomy, University of Massachusetts
Amherst, MA 01003

Alan Schoenfeld
Department of Mathematics, Hamilton College
Clinton, NY 13323

Dorothea P. Simon
Department of Psychology, Carnegie-Mellon University, Schenley Park
Pittsburgh, PA 15213

Herbert Simon
Department of Psychology, Carnegie-Mellon University, Schenley Park
Pittsburgh, PA 15213

Robert E. Sparks
Biological Transport Laboratory, School of Engineering
And Applied Science, Washington University
St. Louis, MO 63130

Ruth Von Blum
Directorate for Science Education, National Science Foundation
Washington, DC 20550

Richard Wertime
Department of English, Beaver College
Glenside, PA 19038

Arthur Whimbey
Mathematics Department, Xavier University
New Orleans, LA 70125

CONTENTS

An Introduction to
Cognitive Process Instruction

Jack Lochhead

We should be teaching students how to think; instead we are primarily teaching them what to think. This misdirection of effort in education is the inevitable
consequence of an overemphasis on objectively measurable outcomes. In brief,
we are more concerned with what answers are given than with how they are
produced. But recent developments in cognitive studies have made it possible to
conduct systematic scientific research on high-order human cognitive processes.
Cognitive process instruction applies these research techniques to problems in
education in order to develop instructional programs which teach students how
to think. The long-range goal is to improve the level of reasoning typically used
by students; some even predict that it will raise the level of human intelligence
in general. At the very least, research on human thought processes should be
able to help us eliminate some of the more stultifying aspects of education and
bring thinking back into the classroom without sacrificing content or substance.

Interest in cognitive process instruction has grown steadily during the past
decade. To a certain extent this reflects nothing more than a reaction to the behaviorist excesses with which psychology has burdened both itself and education over the past 70 years. But there are also substantive reasons for the
renewed interest in the human thought process. Major developments in the
fields of artificial intelligence, cognitive psychology and genetic epistemology
now make it possible to study mental processes in a more systematic and sophisticated manner than was possible in the time of William James and John
Dewey. Computers have provided a mechanism for simulating thought and a
framework for conceptualizing mental processes. A new approach to Cognitive
Psychology has overcome old taboos against the analysis of verbalized thought
processes and created a wider recognition of the importance of Jean Piaget's
work. Although they are very different in style, computer simulation and
clinical interview techniques have produced strikingly similar conclusions on
the nature of intelligence. One is the important role existing knowledge plays in
determining how experience is perceived, and hence in how new knowledge is
constructed. This constructivist epistemology suggests that much of our effort
in education has been seriously misguided.

One hundred years ago schools strove to develop good habits of mind. Since
little was known concerning the detailed structure of these habits they could
only be taught indirectly. It was widely believed that subjects such as Latin and
mathematics required, and therefore developed, both discipline and logical
analysis. Although this model of implicit instruction lost favor after it was "dis-
credited" by Thorndike's transfer experiments, it is still used to justify many
practices in higher education. However the impact of this approach has been
weakened because students are no longer asked to consider how they think or to

discipline their habits of mind. Furthermore in the primary and secondary schools, where behaviorist educational psychology has had its biggest impact, there is now almost no stress on disciplining the mind. In some schools the emphasis on objectively measurable behavioral outcomes has meant that multiple-choice test have virtually eliminated the need to write (and grade) essays. The ultimate irony is that by overemphasizing objectively measurable products (and ignoring the process by which they are obtained) some schools have managed to decrease their effectiveness as measured by the same "objective" standards (SAT exams) they stress.

Advocates of cognitive process instruction do not claim that their approach can solve all problems in education. But they do believe it will provide a framework for systematic research, or as Frederick Reif (1976) put it, an applied science of education, with the promise of gradual but highly significant improvements for our methods of instruction. Cognitive process instruction is based on a simple premise: cognitive processes can be studied and students can benefit from the knowledge gained through such studies. The concept differs from the approach of the last century in the level of specificity possible. One hundred years ago logic was conveyed through Latin. Today it is becoming possible to isolate specific cognitive skills and to design instructional material appropriate for each skill. Since the instructional components have well-defined objectives, one can, in principle, determine how well each is working and modify those that are ineffective.

Cognitive process instruction is more than a shift of emphasis towards basic skills; it implies a radical change in our current conception of learning. The origins of this "new" view are quite old; Galileo once said, "You cannot teach a man anything; you can only help him to find it within himself." This is the essence of constructivist epistemology; it implies that students can only learn when they are actively involved in piecing together their own ideas.

Constructivism threatens many traditional educational maxims, especially those of established fields. Yet it is consistent with some instructional practices in skill-oriented disciplines such as music and athletics. Whenever the objective of instruction is an observable skill one cannot avoid recognizing the need for students to act. In academic areas action does not necessarily mean motion; it means that students must be actively involved in putting ideas together, testing and modifying them. The teacher is most effective as a tutor/coach, helping students develop their own knowledge systems; he is least effective when he tries to pass on his own system directly. From the constructivist perspective it is not surprising if students fail to comprehend a brilliant lecture; what is surprising is that they can learn anything at all from a lecture.

During the past five years a variety of cognitive programs has been started in colleges across the country. Many are based on the developmental theory of Piaget and are oriented to preparing students for the kinds of abstract reasoning expected in college. There are science courses for students majoring in education such as the physics courses at the University of Oklahoma (Renner, 1972) and the University of Washington (Arons, 1972) which train students to teach new science curricula such as Science Curriculum Improvement Study and Elementary Science Study. A more comprehensive approach is taken at the University of Nebraska (Fuller, 1977); in the special freshman program all courses are taught in accord with Piagetian principles of intellectual development. This program is now being adopted by several colleges across the country.

While Piagetian programs are appropriate for students who are poorly prepared, courses in problem solving are often oriented to the superior student. Schools of engineering are increasingly interested in such courses; the program at McMaster University (Woods, 1975) and the Guided Design course at the University of West Virginia (Wales, 1977) are of particular interest because they involve students in a detailed analysis of their own problem-solving strategies. Awareness of one's own mental processing is stressed in many cognitive projects; in 1977 a conference on "Loud Thinking" was held at the Massachusetts Institute of Technology. An equivalent approach forms the basis for a workbook on analytic thinking skills written by Arthur Whimbey (Whimbey, 1978). The Whimbey method has been used in several courses including Spanish and mathematics at Bowling Green State University, and in physics, mathematics and writing at the University of Massachusetts. Finally, at a few universities— University of California at Berkeley, Carnegie-Mellon University, Massachusetts Institute of Technology, and University of Massachusetts—a continuous program of course development is being conducted in close cooperation with active research into human cognitive processing.

Most educational innovations have turned out to be little more than fads. What reasons are there for believing that the interest in cognitive process instruction will last? This question cannot be answered without making serious criticisms of recent work in psychology and education. But before doing that I would like to explain that these criticisms concern the application of certain theories to inappropriate contexts. They are not meant to be absolute rejections of the entire theory nor condemnations of every application. Behaviorism and the open-classroom movement both have a place in education; the Skinner box and the skate board each have significant roles. But neither can address more than a small aspect of the total picture. Cognitive process instruction is likewise only one part of a complete education and although there are reasons for believing it to be an important part, any program that ignores other aspects is probably doomed to failure.

Perhaps the most important reason for believing that cognitive process instruction may be an innovation of substance is the nature of the people who are developing it. Almost all have strong backgrounds in traditional quantitative sciences and most are from fields such as physics, engineering, mathematics and computer science. This background is desirable not only because it insures that the proponents of cognitive process instruction can deal with highly complex analytical problems, but also because it provides them with an appreciation of the limitations of quantitative theories. One of the reasons that educational psychology has made so little progress is that it has tried to build mathematical models of learning without first investigating basic qualitative relationships. The observational techniques and insights of biology and ethology have been largely ignored.

Most researchers in cognitive process instruction, perhaps because of their experience with the wave-particle duality, accept the need for employing a variety of theories which may be incompatible with each other. They investigate psychological theories in terms of usefulness to education and they expect each theory to have a limited domain of application. Finally these applications include some of the most difficult conceptual courses in the University. It is no accident that interest in high order cognitive processes is greatest in those disciplines for which they are most important.

The origin of research on cognitive process instruction is another factor that distinguishes it from other approaches. As an applied science it is strongly anchored to observation. Most educational innovations have been derived from psychological or social theories which have been developed outside the educational context. Such innovations are limited by the domain of the theory. Much of the research into cognitive process instruction has placed observation ahead of theory. Since this work is based on research traditions from the biological sciences, we propose to call it Human Cognitive Ethology. Ethology, the study of the behavior of animals in their natural environment, stresses naturalistic observation; it is more concerned with the contextual validity of the situation than with the extent to which all factors can be controlled. At this stage in its development it is a "messy" science in which one can draw tentative hypotheses rather than firm conclusions. It strives to make simplified theories for complex phenomena rather than complex theories of simplified phenomena. Human Cognitive Ethology as applied to instruction is the study of thought processes of students in realistic educational contexts. Although simple in concept, it is unfortunately difficult in practice and strongly depends on skilled observers. Nevertheless, it is hard to imagine how any serious student of education can expect to make significant progress without first making careful observations of the manner in which people think and learn.

REFERENCES

Fuller, R. "The ADAPT Book," ADAPT Program, University of Nebraska, Lincoln, 1977.

Arons, A. and Smith, J.P. "A Coordinated Program of Instruction in Physical Science Teaching for Pre-Service Elementary School Teachers," Progress Report to NSF. University of Washington, March 1972.

Fuller, R. "The ADAPT Book." ADAPT Program, University of Nebraska, 1977.

Reif, F. "Toward an Applied Science of Education." Presented at AAAS Symposium, Feb 1976.

Renner, J.W. and Stafford, D.G. *Teaching Science in the Secondary School.* New York: Harper and Row, 1972.

Wales, C.E. and Stager, R.A. *Guided Design.* West Virginia University, 1977.

Whimbey, A. and Lochhead, J. *Problem Solving and Comprehension, A Short Course in Analytical Reasoning.* Philadelphia: The Franklin Institute Press, in press.

Woods, D.R., Wright, J.D., Hoffman, T.W., Swartman, R.K., and Doig, I.D. "Teaching Problem Solving Skills." *Engineering Education,* 1 (1) p 238, 1975.

SECTION I

RESEARCH

Related to

Cognitive
Process
Instruction

Introduction: Research

The hope that researchers are in a position to help practitioners in the classroom is certainly not a new hope, but there is some reason to believe that significant advances have recently been made in that direction at the college level. A number of factors lead us to be optimistic about this. First, researchers are using tasks that are directly related to college–level academic work, such as solving word problems in mathematics and physics and organizing written essays in rhetoric. Second, researchers are looking at cognitive tasks that are at a much higher level of complexity—tasks that tap the student's deeper understandings of a situation rather than tasks that measure a student's ability to perform correctly using a straightforward algorithm or retrieval operation: this can for example involve modeling the student's use of several levels of representation in problem solving. Third, we seem to now have a group of researchers who combine an awareness of the extremely subtle problems of fostering understanding in the classroom with a concern for doing some serious scientific work in cognitive process research. All of the researchers represented in this section, for example, are trained in science, and at the same time all have been involved in educational reform projects.

Fourth, a growing group of enlightened teachers are realizing that clarity of lecture presentation is not enough. They are recognizing the need for dealing with the tacit knowledge and process skills used in their own field. This requires input from researchers on the cognitive processes used by experts in the field. Furthermore, these teachers are recognizing the need to encourage an active role in knowledge construction on the part of the student by tapping into his or her preconceptions—anchoring new concepts to common-sense notions such as physical or social intuition. This requires input from researchers on the preconceptions and intuitive reasoning processes possessed by naive students.

The papers in this section provide evidence that both researchers and practitioners can in the future benefit from each other's work. The first two papers deal with the methodological aspects of cognitive process research. Herbert Lin compares several of the research methodologies now used to study issues relevant to cognitive process instruction. This is not an easy task because many of these methodologies are still in a state of development.

Jack Easley has directed a number of studies of cognitive processes occurring in students and teachers using case study and clinical interviewing approaches. His paper defines a methodology that can help investigators generate qualitative models for the cognitive structures of individual subjects. After arguing that there are two basic paradigms for research activity in science, the quantitative measurement paradigm and the qualitative structural analysis paradigm, he points out that the structural paradigm is often the more appropriate one for progress in a field such as cognitive science which is at an early stage in its development. He also describes some intriguing techniques for diagramming models of human knowledge structures. These diagramming techniques can be used to map cognitive structures and explicitly show their relations to observed behavior.

Papers in the second part of this section describe actual research studies related to cognitive process instruction. The papers by Robert Karplus, et al. deal with tasks for assessing formal reasoning skills. The finding that many college students are not able to give immediate correct responses to questions on written tests designed to tap formal reasoning skills has prompted a growing number of attempts at interpretation. The Karplus papers provide an excellent introduction to these tests and the issues surrounding their interpretation. They are the only papers in the book which deal directly with secondary students, and they are included because they suggest certain limits on the type of preparation we can expect college freshmen to have. The first study compares the performance of students from seven western countries on tasks involving proportional reasoning and the control of variables and raises some serious questions about the level of attainment in certain reasoning skills.

Karplus' second paper is a pilot study using one of these tasks and was conducted in the People's Republic of China. His preliminary findings show that the performance of Chinese students may be significantly higher than that of western students on this task.

Jill Larkin provides a concise overview of recent important work in information-processing psychology relevant to cognitive process instruction. She describes a number of ways that this work can be applied to science instruction.

The last three papers in this section by Simon and Simon, Clement, and Lochhead give the reader a close-up view of individual students thinking out loud while solving problems. These papers are empirical in emphasis—they include verbatim sections of interview transcripts and concentrate on identifying new phenomena to be observed during the process of problem solving. In this respect they are poised at the frontier of our knowledge about complex cognitive processes—they exhibit behavior that can be explained currently only in broad theoretical terms, with many subtle aspects of the behavior being as yet unexplained.

In the past, most research on problem solving has emphasized relatively simple tasks in which the problem is unambiguously defined and in which the required knowledge base is small and specific. Dorothea and Herbert Simon in their paper, however, examine the protocols of two individuals working on a very open-ended engineering design problem. They describe striking differences in their solution approaches to the same problem. The individual differences highlight some important features of problem solving in open-ended tasks and raise some important questions concerning the relevance of the engineering curriculum to real-world problems.

John Clement's paper analyzes a student's responses to an open-ended question of a different kind. The student is asked a series of questions about the role of friction and momentum in a simple mechanics experiment. This amounts to a request not so much to decide on an action which will solve a problem, as to give an explanation or theory of why the apparatus will behave in a certain way. Explanation protocols of this kind are particularly useful for analyzing the structure of the conceptions a student is using to interact with a particular situation. The causal structures described in this paper are of interest because they appear to be related to physical intuitions that expert scientists and engineers use to organize problem solutions before applying detailed quantitative methods.

Jack Lochhead's paper is a field study of two students engaged in a self-directed instructional experience. Since no interviewer or instructor was present during most of their taped, one-hour session in the physics laboratory, the transcript presents a particularly realistic view of the repetitive cycle of triumphs and setbacks that can occur in cognitive process instruction. The students had previously failed a number of tasks designed to detect patterns of reasoning at the formal operational level. The protocol is of interest because it gives a close-up view of two transition-level students in their struggle to organize and use these reasoning patterns successfully. The extensive and largely uncondensed sections of transcript show how the students work on the problem using several different types of representation and how they help each other by complementing each other's strengths and weaknesses.

One of the initial goals for this volume was that it be a place where extended sections of unedited transcripts of problem-solving behavior could appear. The last three papers just discussed include data of this kind. We feel that such raw transcripts are valuable in published form for several reasons. First, the inclusion of verbatim sections of transcript reflects the growing use of a new research method which may be called *cognitive microanalysis*. In this approach the transcript must be examined in great detail because one is interested in describing the sequence of cognitive processes occurring in the subject during the solution of a problem, rather than just describing the correctness or incorrectness of the solution method.

Second, *transcripts allow the reader access to the researcher's data base.* This accessibility allows the reader to make his own interpretations of the data, and compare these interpretations with those of the author. It is true that the author may have access to an even richer data base in the form of tape recordings which include the intonations, pauses, and changes in pacing, which are sometimes very helpful in suggesting the nature of the student's thought processes. But at the same time, written transcripts provide a great deal more information to the reader than would a mere summary of a student's approach.

Herbert Lin points out in his paper that descriptive models of cognitive processes are ideally verified formally by a panel of judges who are "trained observers." But in the beginning stages of a science, such as the study of complex cognitive processes, this standard may be difficult to attain. A weaker but useful alternative is to have the researcher receive feedback on the plausibility of his interpretations of a protocol from readers who have access to the same protocol. An alternative theory proposed by a reader for the same protocol could then replace the author's theory if it were able to account for more of the details present in the transcript or provide a more plausible theory. This kind of constructive reaction and criticism is only possible when relatively complete sections of the transcript are included by the original author.

Finally, transcripts allow the reader to build up a *non-formal but important set of intuitions about the nature of a student's thought processes* in the situation being studied. That is, the reader can acquire a different, less articulated kind of knowledge from transcripts in addition to the knowledge he acquires from the formal analysis section of a report.

"Raw" protocol data sometimes comes as a shock to the uninitiated because the statements of the subject (and sometimes those of the interviewer!) are often ungrammatical, discontinuous, broken, and punctuated by spontaneous "aha's" or "uh-oh's" or silent frowns which interrupt the flow of speech. In addition, some of a student's comments may at first appear to be contradictory or illogical, and often it is only after some intensive analysis that enough insights are achieved concerning the form of the subject's knowledge structures so that it becomes clear how the student's comments fit together. But these "imperfections" in transcripts are what make them valuable and interesting in their own right. They open up a more realistic view of the human mind: acting sometimes as a precise, hierarchical, logic machine, but more often as a conjecturing, analogy-seeking, multi-leveled, inconsistent, and—crucially—self-correcting biological system.

Approaches to Clinical Research
In Cognitive Process Instruction

Herbert Lin

INTRODUCTION

Cognitive process instruction (CPI) is a view of instructional design which emphasizes thinking skills over factual knowledge. Research related to CPI usually makes use of a clinical method which seeks to illuminate the processes underlying behavior in complex intellectual domains. However at least two different perspectives on CPI research exist: *deterministic* and *descriptive*. The first verifies models by comparing the behavior of the model with that of the subject. The second verifies models through the consensus of a community of experts in either a psychological paradigm such as developmental stage theory (if the emphasis is on models of cognitive process) or a disciplinary paradigm such as physics (if the emphasis is on models for instruction). Certain issues arise in the choice of methodology: the diversity in the cognitive processes people use, the complexity of a model, the role of human judgement, the generality and utility of a particular approach.

This paper is intended as a brief introduction to these issues, with an emphasis on research methodology. It is directed primarily at the college instructor who has more than a passing interest in questions of pedagogy and who wants to know how to begin CPI research in his own subject area, but who does not qualify as an "expert" in cognitive psychology.

It is always risky to discuss the methodology of a field in which the methodology is far from established, so this paper must be taken as provisional. The last section of this paper presents factors that have influenced a personal choice of methodology. Of course, in no sense is this paper a summary judgement of the pros and cons of different approaches.

Cognitive process instruction is a view of instructional design which emphasizes cognitive skills relevant to understanding and thinking. It is different in scope from traditional cognitive psychology, and different in character from other approaches to educational improvement. Standard cognitive psychology deals with the entire range of human cognitive phenomena: perception, language, thinking, memory. Typical tasks include paired-associate learning (memorizing unrelated pairs of words), recall of nonsense syllables, problem solving (involving relatively simple problems), and perception of optical illusions. In contrast, CPI research is concerned almost exclusively with the relatively complex problems and situations that are characteristic of higher education, for example, a physics or a mathematics problem, or an essay assignment.

CPI research tends not to emphasize mathematical models, time-interval measurements, eye movements or statistics. Rather, it makes substantial use of introspection and protocol analysis as windows into the cognitive processes of the subject. The subject may think out loud using a tape recorder, keep a written record of his train of thought, or interact with an interviewer who pursues interesting comments, not adhering to a fixed framework of inquiry.

These techniques lead to richer insights into the nature of thought processes, and retain the essential human qualities of the information sought. They offer windows into the "why" and "how" of what a subject does. Such techniques have their roots in the clinical tradition of the anthropologist or psychiatrist. This clinical tradition seeks to illuminate the underlying processes and offer plausible interpretations and organizations of the observed behavior.[1]

Standard approaches to educational improvement usually emphasize content, coverage, or format. For example, a quick look through any issue of the *American Journal of Physics,* a journal of physics education, will indicate that the majority of its articles concern simplifications of various derivations, changes in course format, new applications of familiar content, new physics labs, and so on.

However, an emphasis on content is only one component of education (though a very necessary and important one). Education must also emphasize the skills necessary to process knowledge of content. It is the primary goal of CPI research to identify and articulate these more general intellectual skills, and devise ways of teaching these skills to students so that they can use them in a reliable and flexible manner. Thus, CPI complements an educational delivery centered around content.

This emphasis on intellectual skills leads to concern with how the novice and the expert think. We try to understand the novice's thoughts so that we may teach him more effectively, asking questions such as: "Why is the student the way he is?"; "What makes him different from another student?"; "What is the student thinking at the time of instruction?"; "What specifically are the errors he makes?"

We try to understand the expert's thoughts so that we may know what to teach the student, asking questions such as: "What makes the expert different from the novice?"; "What specifically does the expert know that facilitates his work?"; "How does he think when solving a problem?"; "How does he catch his mistakes?"

Good models of problem solving or understanding in experts or novices are not easy to construct; in fact, the notion of explicit models of experts or novices is relatively new to current teaching practice. In the most standard case, the most important ways in which experts think remain tacit and for the most part unobservable to the student, except by example. A student with the question "But what should I be *doing*?" should receive a better answer than "Just practice; you'll get better." An articulation of exactly what the expert does and precisely how he does it helps to answer this student's question. Similarly, an articulation of precisely what the novice does and why he does it helps the expert see specifically what leads the student into difficulty.

---

1. Parlett (1977) discusses the notion of an anthropological approach to social science in which human interpretative skills and judgement are essential components of the research effort. He takes an anthropological approach to the evaluation of educational innovation, but his discussion can easily be adapted to the investigation of high-level cognitive phenomena.

In addition, the attempt to articulate the tacit components of a disciplinary paradigm and/or the cognitive structures underlying the novice's behavior often results in a better understanding of the tremendous gap between students and teachers. Teachers often act as though students were simply ignorant professionals capable of thinking and reasoning as experts do if only they had enough factual knowledge. Of course, this is not the case, and critical self-examination often reveals the wealth of *nonfactual* knowledge essential to the professional: values, procedures, heuristics, and organizational strategies (DiSessa, this volume).

Finally, by making expert skills explicit, teachers have more control over the student's implementation of these skills, especially when the student is just learning them. The student may otherwise develop his own (probably inefficient and ineffective) ways of doing things.

There are at least two major dichotomies in CPI research. While in practice most work lies between the extremities, it is useful to polarize them for the sake of discussion and clarity.

The first dichotomy is that between a *deterministic* and a *descriptive* perspective. Both are observationally based, but a deterministic approach requires a strongly falsifiable operating model of some cognitive phenomena, whereas a descriptive approach depends on expert consensus to validate its models. The second dichotomy is characterized by a cognitive science versus educational engineering perspective; the first asks "What's really going on in people's heads?" and the second asks "What should we be doing in the classroom right now?" This paper reviews these perspectives and discusses their specific application to education. Many examples in this paper come from physics, since that is my background, but its main ideas apply more generally.

Finally, the word "model" has different meanings to different people. My use of the word "model" by itself will refer to an abstract representation of behavior or action which bears significant resemblance to that behavior or action in certain *unspecified* ways. If I wish to specify these resemblances explicitly, I will use appropriate adjectives.

DETERMINISTIC MODELS

By tradition, researchers (especially in the hard sciences) have strong preferences for deterministic models of phenomena. A deterministic model is one in which the model's behavior follows deductively with no uncertainty (or at least rigorously calculable probabilities) from the starting assumptions and initial parameter settings. As such, these models have overtones of predicting behavior, and I will use the word "prediction" to refer to a rigorously deduced behavior of the model, even though most work in the field does not make a priori predictions in the strict sense of the word.

By varying the initial parameters, one hopes that the model's behavior will change in a way that corresponds to the subject's behavior; in this way, deterministic models achieve some type of generality. They are also strongly falsifiable; they can be proved wrong or incomplete by virtue of an incorrect prediction—a mismatch between subject and model behavior.

In the domain of complex cognitive phenomena (for example, learning physics, as opposed to learning nonsense syllables), all deterministic models of which I am aware are formulated in information processing or computer science terms, quite ofter as computer programs.

Feigenbaum and Feldman (1963, pp 269-270) describe this common form of deterministic model:

> The computer program is a model which represents the researcher's hypotheses about the information process underlying the behavior. The program is run on a computer to generate the predictions of the model. These predictions are compared with actual human behavior. There are usually some discrepancies between prediction and behavior, and the model is revised to reduce these discrepancies. Then the entire process is repeated. Eventually, the researcher hopes to obtain a model which will be a good predictor of the relevant behavior. As he continues to test his model and to improve it, the researcher gains confidence in the belief that this model represents the processes underlying the behavioral phenomena he is studying. . . .

Proponents of a deterministic approach argue that to the extent that the behavior of the deterministic model matches the human behavior under study, they understand the human behavior (Simon, 1962, pp 101-103).

These researchers believe this approach has a number of major advantages. For example, they argue that the requirement that a model's behavior follow deductively from the premises implies that all assumptions must have been made explicit. For example, Newell and Simon (1974) write that:

> . . . formal rigor assures that all assumptions are made explicit. If certain mechanisms are being postulated to explain some phenomena, formalization reveals whether the mechanisms are indeed sufficient to produce the phenomena. . .

In addition, they believe that deterministic models are unambiguous—the model either does or does not do X. Feigenbaum and Feldman (1963, pp 270-271) argue that

> The researcher may represent a model of human behavior in any of a number of different ways. Perhaps the most common representation is natural language . . [which is inappropriate because of] the difficulty of rigorously determining the predictions of a [natural language] model . . . [This arises] from the use of ambiguous words . . . .
>
> The computer program has several advantages as a medium for model construction. The only real constraint on the model builder is that his statements must be unambiguous and complete . . . Despite the freedom given to the model builder in constructing computer models, he retains the ability to make a rigorous determination of the implications of the model. For the computer can execute the program and determine the behavior of the program in particular situations.

They also believe that they can account for a broad range of phenomena in terms of a limited number of explicitly stated basic concepts. For example, Newell and Simon (1974) write that

> The enhanced inferential power provided by formalization contributes to the parsimony of theories, for it permits relatively large sets of phenomena to be derived from relatively small sets of assumptions . . .

Finally, computer science provides researchers with a very powerful vocabulary with which to design deterministic models. For example, computer science gives the notions of *procedure* and *subroutine* concrete meanings and implementations. Among other things, these ideas focus attention on *action*—how the subject does what he does.

Perhaps the best summary of the advantages of a deterministic theory is offered by Boden (1977, p 402):

> A program provides an explicit theory of the epistemological processes by which the concepts contributing to cues and schemata are manipulated in the mind . . . Detailed questions about the epistemological structure and function of memory cannot be avoided by the programmer, who has a rich set of precisely definable concepts in terms of which to frame hypotheses concerning them.

The generality of a deterministic model is determined by the capability of the model to simulate accurately a large variety of human cognitive processes *without* major structural (but perhaps parametric) changes: larger variety, more general.

Deterministic models may also have pedagogic value; a good deterministic model, say of a novice, might be altered piece by piece in order to make it behave more like an expert, or vice versa. This might give us insight into the specific processes responsible for particular shortcomings or difficulties, or ways of improving novice (or expert!) behavior.

Finally, deterministic models may offer us concrete prescriptions of pedagogic value. A good deterministic model may reveal, for example, that if a student replaces one cognitive process behavioral set (characteristic of poor problem solvers) with another (characteristic of experts), he will solve problems more efficiently.

DESCRIPTIVE MODELS

Description offers an alternative to determinism. In the discussion which follows, I draw a sharp distinction between a descriptive model (which refers to a specific set of data) and the model-building framework from which the model emerges. This framework is essentially a theoretical perspective which provides the basic concepts, relations, mechanisms and principles from which individual descriptive models may be constructed.

An example will help here. Consider a student and a teacher in a room, talking about physics, being observed by a behaviorist, a Piagetian, and a psychoanalyst. When they record a significant event in the interaction, they ring a bell. Under these circumstances, they probably will not ring their bells in synchronization; each person has his own view of what constitutes a significant interaction and what is irrelevant.

In short, people with different perspectives will identify and interpret the same behavior in different ways. Each has his own background and training in psychology on which he draws when he makes new observations. His analysis of a situation is his descriptive model of that situation. The background and training on which he draws to construct this analysis provide his model-building framework.

Descriptive frameworks are typically characterized by one (or a combination) of the following labels:

1. A *micro-analytic* framework provides tools and building materials for modeling cognitive structures. These models of cognitive structure specify the manner in which knowledge is organized, the causal mechanisms responsible for

processing that knowledge, and the mechanisms by which the structure evolves in time to incorporate new knowledge and processes.

Micro-analytic models tend to be quite detailed, in the sense that they provide enough mechanisms to account for a large amount of data. In some sense, they have the most theoretical flavor of all descriptive frameworks described here. For example, Clement (1978 this volume) analyzes a beginning student's conception of momentum and force in terms of a semi-quantitative network of causal chains. Larkin (1976) outlines a qualitative model of an expert and a novice solving standard physics textbook problems. For a more detailed look at the micro-analytic approach, see Easley (this volume).

2. A *developmental* framework provides developmental scales along one or a number of intellectual dimensions onto which the judges may map a particular student. This ordering may take the form "Student is at stage three of our scale." For example, Perry (1970) traces in college students the development of a view of knowledge along a dimension of dualism relativism commitment. A major part of Piaget's work includes such developmental progressions, in particular, his work on the growth of logical thought (Inhelder and Piaget, 1958).

The time evolution of developmental (and some micro-analytic) models provides a basis for a comparison of model and subject behavior. The "movie" provided by the model's time evolution can be judged against the videotape of the subject.

3. A *diagnostic* framework provides descriptors which can be used to diagnose the behavior of a particular student in a given situation. A diagnostic model provides "snapshots" of a subject's behavior, usually ignoring causal mechanisms or possible time evolution. For example, Bloom and Broder (1950) study the problem-solving behavior of college students to identify a number of difficulties which commonly appear. Polya (1945) describes many heuristics commonly employed by expert mathematicians.

4. An *instructional* or *prescriptive* framework is, in contrast to the others, not as sharply distinct from the models it creates. Its emphasis is on "educational engineering" rather than "cognitive science." It asks, "What do we need to know in order to guide the instructional delivery?"

This framework might include principles such as "All students should be able to do X, Y and Z in certain situations." The model would direct a particular teacher to do specific things in his classroom which would enable students to do X, Y and Z.

Instructional models must be sufficiently fine-grained and explicit to tell the student or teacher what to *do*. For example, Reif et al. (1976) describe explicit strategies for understanding equations in physics. They articulate a number of abilities which constitute "understanding" and instruct the student to do certain things in order to acquire those abilities. Arons (1977) describes a physical science course which attempts to cultivate general thinking skills relating to concept formation, inferential processes, quantitative reasoning, and articulation of personal insights. Woods (1975) relates an investigation of engineering problem-solving processes. This ultimately results in a specific classroom implementation of a strategy for solving engineering problems (Leibold et al., 1976).

Frameworks apply to classes of problems, in the sense that a framework can be used to generate a model to deal with each class of problem, each model being structurally different. A framework makes predictions in the sense that it claims that its basic building blocks and principles are sufficient to account for the most essential features of a subject's behavior. However, it does not make predictions concerning the specific organization of and relationships between these items. Descriptive frameworks also assist in the coherent organization of data, facilitating the discovery of principles with explanatory power.

Descriptive models apply to specific sets of data such as interviews, protocols, and exam papers, and need not necessarily make specific predictions concerning subsequent interviews, or protocols. From his model-building framework, an investigator selects and arranges the appropriate building blocks as he constructs a model of any interesting behavior. Consequently, the framework must be sufficiently explicit to guide his selection and arrangement procedures.

On the other hand, deterministic models may be general in that they may make probabilistic predictions through the generation of profiles of one individual over a set of tasks or many individuals over a few tasks. For example, a profile may say that a student did X in seven out of 10 protocols; hence we might guess (or "predict") that he tends to do X in most such situations.

While conceptually distinct, framework and model may have a great influence on each other's development. Obviously, particular models emerge from a given framework, but just as importantly, frameworks evolve as the model builder discovers additional common features in the models he creates. As an example of the latter, Bloom and Broder's 1950 framework emerged from an analysis of many students solving problems. The elements common to many problem-solving sessions became the framework, which they put forth as one applicable to new and different situations.

Incidentally, frameworks can differ in scale. A large-scale framework might be exemplified by Piagetian theory or information-processing psychology. Bloom and Broder (1950) would typify a smaller framework.

Since descriptive models cannot be falsified in a predictive manner (since they apply only to specific sets of data), some other method must ensure that the models which result are in fact reasonable representations of the phenomena in question.

One reasonable procedure for verification is to make use of independent judges trained in the use of a given framework. We would ask these judges to observe the behavior of a subject within this general framework, and then to construct a model from these basic building blocks. To the extent that the judges construct similar models among themselves from a variety of observations, we assume that the framework constitutes a reasonable theory of the observed behaviors.

A good example of this verification process is offered by Perry (1970, p 10):

> Any . . . inferential construct, drawn from . . . varied data, faces the question of being no more than the observer's way of imposing an order where it does not exist. We therefore endeavored to reduce our [original] scheme from its broadly discursive expression into a representation sufficiently condensed, rigorous and denotative to be susceptible to a test of reliable use by independent observers . . . . The extent to which the several observers' placements or "ratings" might agree with one another, beyond the level of chance, would then be the measure of the validity or "existence" of our scheme.

The preceding example, though illustrative, is perhaps a bit extreme, in the sense that it represents a degree of verification usually not present in psychological studies. The reason, I believe, is inherent in the nature of Perry's work: judges were asked to agree on a determination of a student's placement along a single dimension in the developmental scheme. It would have been a quite different and much more difficult task to require such agreement among judges on a detailed model of an individual student involving dozens of elements and relations between elements.

Still, to the extent that such agreement is possible, Perry's verification scheme is a reasonable one to emulate. However, a somewhat less demanding procedure operates all the time. People generally do not work alone, and the agreement of a team on a specific model is a verification of that model in a certain sense. In addition, the circulation of that work (usually published) often results in further refinement of any framework or model, and to the extent that responsible professionals criticize flawed work and do not criticize the particular case in question, the model is verified still further.

However, for the descriptive researcher, a practical question remains: how can he ensure that the models or frameworks he personally creates are representative of more than his own idiosyncratic views of the world? How should judges be selected? What should they have in common?

To answer these questions, we must first answer the following question: are we more directly interested in the cognitive process aspects or the instructional aspects of the problem? The choice we make determines how we choose judges.

DESCRIPTIVE COGNITIVE PROCESS MODELS

If we consider the cognitive processes aspects (and generate descriptive cognitive process models as described above), judges need "only" share an acceptance of the model-building framework. In principle, this is all we mean by "training judges in use of the framework." In practice, however, this training takes places over years, not days or weeks. It is in fact this extensive training that defines various schools of thought such as learning theorist, Piagetian, and structuralist, and also sets the stage for major disagreements among them. In short, a number of paradigms[2] exist, and believers in one paradigm often misunderstand believers in another.

This view parallels that of Easley (1977):

> One has to be trained in methods of investigation that are sensitive to the particular phenomena [of psychological interest] in order to collect evidence relevant to the evaluation of models . . . A psychometrician can no more readily test Piaget's theory by his methods than a geologist can test the theory of nuclear structure of the atom by geological methods. Investigators who work on different levels of structure require special conceptual systems to permit communication [among each other] . . . . [For example, in analyzing protocols,] the recognition phenomenon will not occur to two persons trained with the videotape at the same time, and they may need to have lengthy discussion and more videotape recordings before they can even agree substantially. But years of

---

2. I use the word "paradigm" in the Kuhnian sense; a paradigm refers to the entire set of beliefs, values, techniques, and knowledge shared by the members of a scientific community and passed on to their students in the course of a scientific education. For further discussion, see Kuhn (1970, pp 174-191) and Easley (1977).

practice do bring a state of communication that achieves much greater agreement among the trained than between the trained and the untrained.

DESCRIPTIVE INSTRUCTIONAL MODELS

Alternatively, we may choose to focus on the instructional (or "educational engineering") aspects, and generate descriptive instructional models as described above. In this case we make the assumption that expertise in any scientific field is defined by the consensus of the community of expert professionals in that field. Therefore, judges should be members of this community. Membership in this community usually means that they share largely similar educations, share many of the same values, use largely similar techniques, and make largely similar though independent professional judgments (Kuhn, 1970, p 177).

Since these community members usually have a dual role as researcher and teacher (at least at the college level), their professional judgement necessarily includes a pedagogical as well as a research component; they are therefore the judges of the state of a student's knowledge and abilities.

Such a procedure is appropriate since these experts are the ones who answer questions such as: "What is the student doing incorrectly?" "What does the student know?" "What should the student be doing?"

Indeed, it is only by comparison to their own shared background as recognized experts that they are able to answer such questions at all; if the student does something that experts would not do, then the student is "in error." The task of the expert/teacher then becomes one of transforming this judgement into an appropriate instructional strategy which will bring the student's actions more into line with what the expert does.[3]

An instructional framework or model must be explicit enough to be *prescriptive*. It must offer answers to the question "But what should I as a student *do*?" Consider for example a framework designed to model the problem solving of beginning students. Such a framework might include a descriptor such as the following:

> Student does not use symbols consistently in doing problem. He assigns the same symbol to different physical quantities, appearing to change his definition in the middle of the problem. When discussing the problem, he refers to the symbol or variable without specifying precisely what he means.

This is an accurate descriptor in many situations, and thus might be included in a model for a beginning student. On the other hand, it does not specify what a student should do if he has this difficulty. Some things which an expert problem solver might do include:

- explicitly define all symbols in writing or in a picture as he introduces them;
- check for possible confusion or ambiguity in the mapping of the physical situation onto the equation or symbols in the equation. For example, if the symbol is m, is there more than one "m-like" thing in the problem?

Any instructional model-building framework must be accompanied by a com-

---

3. This is of course not to say that the student is a mold into which expert knowledge can be poured. It may be that the only way to bring a student's action into line is to start with what the student has available and work from there.

mentary which articulates the expert's perspective. If we say that "a student does X," and we as teachers feel that X is inappropriate, we must also make clear what is appropriate, since that perspective is ultimately what we wish to pass on to our students.

Thus, the instructional model-building framework consists of two parts: membership in a community which almost unanimously shares a subject-discipline paradigm, and a plausible articulation of novice or expert behavior, pedagogical strategy, or any other aspect relevant to instruction.

The time scales relevant to these two parts differ substantially. An individual adopts and internalizes the subject-discipline paradigm over many years. The plausible articulation is another matter—if after substantial discussion (empirically speaking, days or weeks, maybe months, certainly not years), the articulation seems implausible or unreasonable to a large number of other experts, it must be discarded or substantially reworked.

An individual working within this framework to build descriptive models would have confidence in his work if the framework were verified against the judgments of a number of representative experts, since he could then expect most experts (including himself) to construct similar models of the same phenomena.

It is important to emphasize that these behavioral specifications go far beyond the typical objectives of self-study materials. For example, here is one set of objectives taken from one of my own units on electrostatics:

- to be able to use Coulomb's law to calculate the forces acting between point charges
- to be able to apply the concept of electric field to the calculations above
- to be able to use Gauss' law to calculate electric fields due to various charge distributions

Note that these objectives *describe* what a student should do. They serve an essential function in that they define the content of the material under study, set the appropriate level of presentation, and provide specific points of accountability. They are not explicit enough in specifying what "being able to do" or "applying the concept" means. They leave many aspects of "understanding" tacit.

More generally, descriptive instructional models seek to articulate the tacit components of what experts know and often assume (incorrectly at times) their students also know. This articulation is often quite difficult, because it involves knowledge that experts have internalized, and often use unconsciously. It requires that the expert/teacher spell out in detail what he means by "understand" or "being able to"; it means the teacher cannot say "I know understanding when I see it" any more than the student can say "I know it but I can't explain it."

DESCRIPTIVE MODELS:
COGNITIVE PROCESS AND INSTRUCTIONAL

The descriptive approach is based on a consensus among psychological investigators or disciplinary professionals such as physicists, using the following research procedure for creating descriptive models:

1. Observe behavior (for example, students solving physics problems).

2. Model the observed behavior.

3. Present model to other recognized experts.

4a. If after substantial discussion the panel of experts cannot agree that the model describes the subject reasonably well (either in a "movie" or a "snapshot" sense), reject the model, concluding that original model reflects the personal biases of the original investigator. Rework the model in question. Then iterate.

4b. If experts do agree, conclude that the model is a good representation for the observed processes.

To create a reliable model-building framework which individual experts can use in constructing descriptive models for themselves:

1. Abstract the features common to many models of various behavioral phenomena.

2. Instruct experts in the use of these abstractions.

3. Ask experts to model subject behavior.

4. To the extent that their models of this behavior are similar (perhaps after discussion), the framework is also reliable. Repeat process as outlined above until agreement is substantial.

Descriptive models apply to quite limited sets of data. They may emphasize cognitive science or educational engineering. Cognitive process models require the background of experts who share a paradigm of psychological investigation. These experts judge the match between the model and significant aspects of the subject's behavior. Descriptive instructional models require the background provided by the parts of the disciplinary paradigm (especially the tacit components) which professionals share (values, exemplars, techniques, heuristics, procedures) relevant to the area of instructional concern.

DISCUSSION

This section sketches a schematic map of possible pitfalls involved in the choice of research methodology as I see them: the variability of cognitive data, the significance of matches between model and subject, the complexity of a model, the role of human judgement, the separation of framework and model.

Human Cognitive Data is Quite Variable

The first issue relates to the nature of the data with which we are dealing. I have heard some educators argue that human learning styles are so diverse that no systematic approach is possible. This is true in one sense. While the detailed behavior of individual people is in fact quite variable, the probabilistic behavior of many people is relatively constant from year to year. Any attempt at understanding human cognition should reflect this reality. Ignoring variability can be risky, as Newell (1978) points out:

> One sees some data, and it has an obvious interpretation, and that's how we all take it. Over the course of the next 20 variations on that experiment, complexities always arise, and consequently, things are never quite what they seem. That means that this game [of modeling human cognition] is a sort of Russian Roulette, in which one decides that some phenomena is really a right phenomena to work on, and then tries to base a model on it, and you get it all

fixed up so that it explains that data and a year later it turns out that the data wasn't anything like what it seemed and you've done all this work to commit your model to exactly what wasn't the case . . . so a key design decision—which I can't justify very well—is what phenomena is worth the risk of making a model to explain.

These points are not new. Consider, for example, the behavior of an individual student doing a set of problems out loud. Certain characteristics may appear in each of his protocols, in which case we may be able to say certain things about his problem-solving processes. On the other hand, he may do certain things which we find significant in one problem and not in another, similar problem. However, the overall behavior of our students is relatively constant from year to year—students make the same mistakes, suffer from the same misconceptions and shortcomings, have the same difficulties, and come in with the same preconceptions. In my view, it is this constancy that admits the possibility of systematic research in education.

Replication Is Not Necessarily Explanation

A second issue is that "understanding" is not necessarily coextensive with accurate matches between the behaviors of model and subject. Consider the following claim:

> A system consisting of a charged ball and a metal sheet is a predictive and hence explanatory model of a monkey learning to fetch a banana from behind a transparent screen. The first time, the monkey sees the banana, goes straight ahead, bumps into the screen, and then goes around the screen to the banana. The second time, the monkey, having discovered the existence of the screen which blocks his way, goes directly around the screen to the banana.
>
> In a similar way, a charged metal ball is hung from a string above the banana, and then held at an angle so the screen separates the ball and the banana. The first time the ball is released, the ball swings toward the screen, and then touches it, transferring part of its charge to the screen. The similar charge on the screen and ball now repel, and the ball swings around the screen. The second time the ball is released, the ball sees a similarly charged screen, and goes around the screen directly.
>
> Conclusion: This model is a reasonable representation of the monkey's learning experience, because the ball behaves as the monkey does in its overt behavior.

This is clearly absurd. The metal ball and screen in no way helps me to explain or understand learning, even though it replicates the most salient feature of the monkey's learning most consistently: both the ball and the monkey dodge the screen on the second attempt. In short, simple prediction (even if accurate) is not enough. We need to know what parts of the model we should take seriously. For example, Dennett (1978) notes that:

> We cannot gauge the psychological reality of a model until we are given the commentary on the model that tells us which features of the model are intended to mirror real saliencies in nature and which are either backstage expediters of the modeling or sheer inadvertent detail.

To put it somewhat differently, we must keep in mind the difference between *performance* and *simulation* modes of model operation (Weizenbaum, 1976, p164). In performance mode, we judge a model by how well it performs a given task. (For example, how well does it reproduce a subject's behavior?) In simulation mode, we judge it by how closely it matches the essential features of the subject. (For example, how well does it conform to the cognitive processes actually used by the subject?)

Obviously, simulation criteria are more restrictive than performance criteria, since simulation requires that *both* behavior and underlying process should match that of the subject. Therefore, we must keep these two criteria for judgement conceptually distinct.

For example, I could easily write a computer program that would mimic any protocol: it would consist of input/output and pause statements in which the output would be exactly the same as the input. I would give the programs as input a transcript of a person's problem-solving protocol which it would duplicate exactly. Can I then put forth this program as a broadly applicable "model" of human problem solving?

Cast in these extreme terms, it is obvious that this "model" does not aid understanding. However, cast in more subtle terms, it is much easier for ad hoc fixes to become part of an increasingly complex model as the researcher attempts to enhance the model's performance at the task of mimicking the human subject.

On the other hand, assuming these criteria are actually kept separate, a powerful test of a simulation model is its ability to replicate the essentials of a subject's behavior. For example, Newell and Simon (1972, p 11) write that

> [Information-processing] theories explain behavior in a task by describing the manipulation of information down to a level where a simple interpreter (such as a digital computer) can turn the description into an effective process for performing the task . . .

Complexity Obscures the Essence of Understanding

The third issue is that of a model's complexity, because as a general rule, models in which essence is not separated from detail yield little insight. Reitman (1965, p 39) points out that

> In principle, a program may be a theory. In practice . . . it is not the program . . . [that is important], but the abstractions—the statements about the key concepts and their interrelations.

These abstractions are central—the high-level features we should take seriously. If anything, it is these abstractions which constitute the basic principles on which a genuine theory of human cognition should be based.

"To understand" means "to understand within the framework of a theory in which essence is extracted from irrelevant complexity," and a logically rigorous deduction or prediction of a fact is not necessarily an explanation of that fact. Putnam (1974) makes a compelling case for this point of view:

> Suppose I deduce a fact *F* from *G* and *I* where *G* is a genuine explanation and *I* is something irrelevant. Is *G* & *I* an explanation of *F*? Normally we would answer, "No. Only the part *G* is an explanation." Now, suppose I subject the statement *G* & *I* to logical transformations so as to produce a statement H which is mathematically equivalent to *G* & *I* (possibly in a complicated way), but such

that the information G is, practically speaking, virtually impossible to recover from H. Then on any reasonable standard, the resulting statement H is not an explanation of F; but F is deducible from H.

While the line between performance and simulation modes is often legitimately vague in the initial, exploratory, research stages, it is deceptive to apply performance criteria to simulation models and then to put forth these models as embodiments of a general theory of a particular cognitive task. Ad hoc patches may be acceptable in a performance model, but not in a simulation model.

Insightful understanding rests on meeting the simulation criteria: how well does the model match the *essential* features of the subject?

Human Judgement Always Counts

The fourth issue is that human judgement enters into all forms of methodology, and hence no model is completely unambiguous. As we have characterized descriptive models, it should be obvious how human judgement enters. However, in deterministic models as well, the question "What is an *essential* feature of the subject's behavior?" comes to the fore. Only people can make this judgement. In addition, the behavior of the model must be interpreted. When is a claimed match really a match? Ultimately, people must decide the existence of "actual" correspondences between model and subject.

Framework and Model Must Be Kept Separate

Of crucial importance is the fact that in both the deterministic and descriptive domains, we must separate framework from model. This separation is obvious from my account of the descriptive approach, but it does not emerge clearly from the writings of those who construct deterministic models. Nevertheless, there too, a theoretical framework shapes the nature of all models which emerge. For example, Newell and Simon (1972) offer a framework of objects and operators, difference reduction and means-end analysis, which are present in a large number of problem-solving processes. Within this general framework, we can model many aspects of human problem solving.

It is relatively easy to say that a framework (deterministic or descriptive) has broad applicability; by construction, it must apply to more than a single instance, since it consists of the essential abstractions common to a multitude of different situations.

It is a much more difficult thing to say that a model which emerges from any framework has the same kind of broad applicability. Any model which claims applicability to a large number of problems must exhibit minimal *structural* changes in its various applications. Parameter changes are allowable, but major structural changes alter the essence of the model.

An example: consider a computer language such as FORTRAN. One can write a number of FORTRAN programs ($A, B, C, D \ldots$), each of which have input statements. FORTRAN is the framework, the programs $A, B, C, D \ldots$ are the models which emerge. Program (or model) A may do different things depending on the particular value of a parameter (call it X). Thus, we say that program A can handle one *class* of problems parameterized by X, but in no reasonable sense can we say that *A* can handle the variety of different *classes* of problems for which programs *B, C, D* . . . were written to handle.[4]

Models Conceal Assumption

A model emerges from a framework. A model is a particular arrangement of framework building blocks. However, the assumptions of the human model-builder are by construction not part of this model: nowhere in the model do they appear explicitly. These assumptions may include the fact that the model can handle only certain types of problems, or the fact that the phenomena being modeled can in fact be handled by the particular framework in use. This may be obvious, but it has tangible consequences.

For example, consider a formal system such as General Problem Solver (GPS), a computer-oriented attempt at understanding human problem solving (Newell and Simon, 1972, pp 414-416):

> GPS is a problem-solving program . . . [that] separates in a clean way a task-independent part of the system containing general problem-solving mechanisms from a part of the system containing knowledge of the task environment . . . [It] operates on problems that can be formulated in terms of objects and operators . . . [it] embodies the [general] heuristic of means-end analysis . . .-classifying things in terms of the functions they serve, and oscillating between ends, functions required, and means that perform them.

This system is limited to considerations of means-end analysis, and is, for example, incapable of accounting for the invocation of specialized knowledge upon recognition of a particular problem. This restriction can be removed easily (Bhaskar and Simon 1977), but the point is that *all* systems operate within a given context.

On the other hand, frameworks contain their basic assumptions. To the extent that one can articulate the framework entirely, one can ensure that all assumptions are made explicit. These assumptions may then be included in a commentary on the model in question, but they are not part of the model itself.

PERSONAL COMMENTARY

In this section I discuss reasons for my choice of methodology. Of course each person must judge for himself the relative advantages and disadvantages, and I don't presume to do that for others.

The deterministic and descriptive cognitive process approaches are unusable for my purposes; consequently I work within a descriptive instructional framework. I am quite uneasy with the deterministic approach, primarily as a result of my own experiences with computers. To the best of my knowledge, all

---

4. To the purist who might insist on self-modifying code or automatic programming, let me point out that the phrase "the model" or "the program" must have a well-defined meaning. If it does not, I am not entitled to talk about "the model for phenomena *Z*" because I am no longer sure what "the model" means. In other words, if I start with something definite, and it changes in the course of events, is the model what I started with or what it changed into? Are the essential features the same or different? I should be able to identify important data structures and processes. If these are the same, then nothing has really changed. If different, then what's "*the* model"? This is true even if I can deduce every step of the transformation.

genuinely deterministic models are cast as computer programs, and as a practical matter, I find it too difficult to keep performance criteria from influencing simulation efforts. However, I find that the majority of those working with computer modeling do not believe this distinction to be significant. For example, Minsky (1967) writes that:

> [A programmer faced with an improperly functioning heuristic program] *would have the options 1.* of thoroughly understanding the existing program and "really fixing" the trouble, or 2. of entering a new advice statement describing what he imagines to be the defective situation and [advising the program accordingly]. When a program grows in power by an evolution of partially understood patches and fixes, the programmer begins to lose track of internal details, loses his ability to predict what will happen, begins to hope instead of know, and watches the results as though the program were an individual whose range of behavior is uncertain.
>
> This is already true in some big programs ... it will soon be much more acute ...

However, despite this awareness, Minsky is nevertheless one of the most prominent advocates of an information-processing approach to psychology (Minsky, 1970).

In performance mode, Minsky's approach may be fine. However, in simulation mode, this is unacceptable, since it is hard to see how partially understood programs consisting of "partially understood patches and fixes" can contribute to well-understand theory.

A second difficulty for me is that computer models require inordinate amounts of detailed specification. While this is often considered an advantage, it cuts both ways. Dennett (1978) notes that

> [Artificial intelligence] programs typically model relatively high-level cognitive features ... and it does indeed often take millions of basic computer events to simulate just a handful of these intended model features. The psychological reality of an [artificial intelligence] program is thus usually only intended to be skin deep; only the highest level of program description and commentary is to be counted as the model.

Consequently I would be forced to expend much effort and time with explicit detailed specifications which are irrelevant to the important top-level structure of useful cognitive process models.

Finally, I feel it is too easy to overlook hidden assumptions in the creation of deterministic models. The deterministic framework makes the assumption that all significant behavior can be formulated in, say, information-processing terms. However, this is not always the case. Consider the problem of modeling an unmotivated student who cares not about his classwork, but who has exhibited advanced reasoning strategies in other domains. I cannot see how information-processing terms yield insight of pedagogical relevance into this behavior.

Descriptive cognitive-process models offer a significant alternative to deterministic computer models. In my view, they do a much better job of separating essence from detail. Descriptive cognitive process models do not rely on an intermediary such as the computer, and hence structures and relations are much more easily identified.

Furthermore, users of descriptive frameworks usually make much more realistic claims regarding generality. Because of the extensive interplay between framework design and model design, I find myself much more conscious of the assumptions (and hence the limitations) of a given framework. Thus, framework and model are both related and distinct. Descriptive work often emphasizes case studies, and hence does not usually claim broad generality.

Descriptive models do generalize in a nonformal manner through the experience of the user (Easley, 1977). The model builder increases his personal capabilities for dealing with a variety of new situations, since he gains insight and knowledge about a particular situation which he can combine with his past experience to apply to new situations.

This is the most common form of generality. For example, an instructional model need not predict that the student will make the same errors each time he works a similar problem—it is enough that he errs once. The teacher therefore corrects student errors which he believes may occur again if he does not intervene; his experience leads him to make the "prediction" of subsequent error.

Finally, descriptive models and frameworks explicitly acknowledge the role of human judgement, and make it an important component of the research effort. By contrast, the deterministic approach attempts to minimize the human component. However, as I have mentioned before, no work is free from those judgements.

Still, despite these comments, I feel that descriptive cognitive–process models suffer from at least one major disadvantage. The existence of competing psychological paradigms implies that agreement among large numbers of workers on the interpretation of a given set of data is difficult if not impossible. White (1977, p 78) offers the following account:

> We are interested in what is [*really*] factual for children. So we subject the manifest fact to analysis to try to detect the latent fact. To picture the analytic process most clearly, imagine that you do not like .. [a particular] journal report, do not want to believe it, are committed to a very different belief . . . So you seek to cast doubt on the report. You might want to see if there were decent controls, if the situation was a plausible one for children, if there was anything funny about the recording of responses . . .
>
> Very few journal articles could survive so merciless an attack. Most psychological research is simply not that airtight . . . . [The research report] has told you something, but you do not know whether that something is . . about children or about the game that the experimenter played with the children or . . . about the . . . game that investigators play [when they publish].
>
> [More generally,] one cannot directly observe the behaviors of children but rather must observe the properties of games reflecting a mixture of children's behavior and one's own behavior.

In short, the investigator's perspective influences significantly what he considers important. Furthermore, his own behavior is interwoven into his account of his observations. Therefore, verification becomes quite problematical. If the model makes no predictions that can be confirmed or denied, and it is likely that large groups of people may disagree substantially, then how can a given descriptive cognitive-process model be said to mirror reality in any reasonable manner?

I personally find this question unanswerable. Because of all the competing claims and counter-claims, I find I cannot make a genuine commitment to any particular psychological paradigm. Hence, I personally am in favor of *educational engineering,* what I have described as a descriptive instructional approach. I therefore propose a different question: "What must we know about the student in order to guide our *instruction*?"

This approach quite naturally brings its models close to its ultimate users, by explicitly accounting for human judgements, since it operates in the natural environment of the classroom. I am willing to sidestep the question of understanding a deep theory of cognitive processes. I leave to others the attempt to develop a rigorous cognitive science. Instead, I take an engineering approach, making use of empirical descriptions of specific situations to guide teaching activities related to these situations. Where abstractions of process arise to help organize these descriptions, I accept them, but these abstractions are my working tools and rules of thumb and not the fundamental cornerstones of a rigorous theory.

Furthermore, since I take an instructional approach, I join the community of disciplinary professionals who *do* share a disciplinary paradigm such as physics. Ultimately, it is this community that will verify my work, and while the existence of a shared paradigm does not guarantee consensus, it certainly makes consensus considerably more possible.

Finally, I believe the "expected payoff" (potential payoff times probability of success) is bigger with an engineering view as contrasted to a scientific view. In some sense, this is reminiscent of a "Salk vaccine" versus "iron lung" debate. Basic research people argue that for a really dramatic payoff, we should put our time and effort into understanding what is really going on in people's heads, and that we are tantilizingly close to achieving a real breakthrough. They argue further that this time the technology/science/vocabulary/conceptual structures are really new and different. If all this is true, then the payoff will be truly enormous.

I am somewhat skeptical of these views. I believe that people have been saying these things for 3000 years, and there's no way of knowing now how things are really different. As for the new technology, I am told that when the telephone first came on the scene, people tried to model humans engaged in intelligent behavior as telephone exchanges. So, I'm hedging my bets: though the potential payoff is huge, I feel that the probability of a dramatic breakthrough is rather small at this point in time. I would rather work on things which I feel are likely to have positive impact on education over a time scale which is short compared to my lifetime, even if that impact is small.

On the other hand, maybe this time we really are on the verge of a breakthrough. Ultimately, of course, there is no rational way of judging the probabilities of success, and so each person must work on whatever that person believes has the highest expected return.

My final reasons for preferring an instructional approach are quite cynical, and I note them with sadness. There is a community of individuals concerned with the substantial improvement of college-level education, hereafter referred to as "we." However, while our personal roles as individual teachers in formulating good models and techniques of teaching and learning are essential, education as a whole progresses only to the extend that these models and techniques are employed by this community. We face skeptical colleagues. Thus, we must concern ourselves with a political problem as well as an intellectual one. It is a sad fact that education is often far down on a college professor's list of priorities, but one with which we must live. Hence, I believe we cannot treat the question of education in exactly the same manner in which we have approached our scientific research.

The contrast between the university's concern for education (even when called education research) and its concern for "legitimate" research is one which most of us have felt. For example, Reif (1974) writes that

> The present educational role of the university seems incongruous with its expected role of intellectual leadership . . . [Naively], one might expect that the university would be a spearhead in educational innovation. . . . In actuality, . . the university's norm in the area of education is reasonable adequacy, rather than excellence or innovative leadership. . . . educational endeavors . . . are usually regarded as being of dubious legitimacy compared to more prestigious activities. . . . Educational innovations are few in number and often marginal in their impact.

Therefore, I believe that not only must we carry out our own investigations, we should also provide our scientific colleagues with tools which they find readily accessible. If we are to affect such a person's teaching in a concrete manner, we should give him tools which he can use to discover for himself what we claim to be true about students.

Consequently I believe the difficulties of the deterministic approach—the need to specify unnecessary detail, the incomprehensibility of complex models, the existence of hidden assumptions—argue against its widespread adoption. Their complexity implies a long time to completion with little immediate reward. The multiplicity of psychological paradigms lead many (especially those in the hard sciences) to believe that cognitive process models (either deterministic or descriptive) are either garbage or gold, depending on who you happen to believe. In addition, many experts in math or science have little expertise with computer (or even psychological) modeling techniques, and more to the point, would be unwilling to adopt such approaches. It is in fact hard to integrate these efforts into one's day-to-day teaching.

I have heard it claimed that cognitive process instruction presents the wave of the future in education. Perhaps, I'm not sure. However, if it really is, we must not forget or ignore our roots. I believe that our best hope for educational change lies with people who are actually engaged in the real world of education. Our approach to education will survive and prosper to the extent that these people find our work credible. Thus, our task is to move the thoughtful middle (those professionally concerned with but methodologically unsure about education), give them tools they can use for themselves, and attempt to change against those aspects of current educational practice that we find objectionable.

We must remember that significant change is possible only if it actively involves large segments of the affected community. I believe an instructional approach is the most natural for most teachers. Pedagogical insights occur informally all the time—students seem to suffer from the same shortcomings and misconceptions year after year. The working tools of the instructional approach are pencil and paper, a tape recorder to allow the taking of transcripts, and some time from one's colleagues. One can even suggest that the student transcribe the tape himself, arguing reasonably that he will learn something from listening to what he did himself. Many teachers are good at listening to students and identifying what difficulties might be present, and all teachers value this ability. Plugging into the already existing network of teaching professionals also facilitates the personal discussions so crucial to any sort of progress; without it, one runs a very real risk of unproductive isolation.

In summary, the instructional approach is close to what teachers are (or think they should be) doing. It offers many advantages for most teachers, and as such is unlikely to generate much resistance in the academic community beyond the resistance to pedagogical issues in general.

CONCLUSION

This attempt to discuss methodology has helped to make clear the reasons underlying my preferences. Nevertheless, it is very important to leave our collective options open. I don't believe any approach offers such clear advantages that alternative approaches should not be pursued. At this stage, insightful tolerance of different approaches is absolutely necessary. It means acknowledging the difficulties of one's own approach and dealing with them as best as one can, and trying to understand and learn about other points of view.

We should be constantly aware that any framework (deterministic or descriptive) imposes limitations on the kinds of phenomena which can be modeled, and we should be willing to borrow effective tips and techniques from anyone. An old saying says that if the only tool you have is a hammer, you tend to treat everything as though it were a nail. If we use our entire tool chest, it should be interesting to see what we build in future years.

ACKNOWLEDGEMENTS

Many people have helped me greatly, mostly by tearing this paper apart and forcing greater clarity and precision on me. These wonderful souls include Andy Adler, Andy DiSessa and Joe Weizenbaum of Massachusetts Institute of Technology, John Clement and Jack Lochhead of the University of Massachusetts at Amherst, Dan Dennett of Tufts University, and Dana Roberts of Wellesley College. Of course, any remaining inadequacies are mine alone.

REFERENCES

Arons, A. "Cultivating the Capacity for Formal Reasoning: Objectives and Procedures in an Introductory Physical Science Course." *American Journal of Physics*, 44(9), p 834, 1976.

Bhaskar, R. and Simon, H. "Problem Solving in Semantically Rich Domains: An Example from Engineering Thermodynamics." *Cognitive Science*, 1, p 193, 1977.

Bloom, B. and Broder, L. *Problem Solving Processes of College Students.* University of Chicago, 1950.

Boden, M. *Artificial Intelligence and Natural Man.* New York: Basic Books, Inc. 1977.

Clement, J. "Catalogue of Students' Conceptual Models in Physics, Section 1: Movement and Force," Technical Report. University of Massachusetts at Amherst, 1977.

Clement, J. "Mapping a Student's Causal Conceptions from a Problem Solving Protocol." This volume.

Dennett, D.C. "Artificial Intelligence as Philosophy and as Psychology." In *Brainstorms; Philosophical Essays on Mind and Psychology,* Montgomery, VT: Bradford Books, in press.

diSessa, A. "On 'Learnable' Representations of Knowledge: A Meaning for the Computational Metaphor." This volume.

Easley, J. "Seven Modeling Perspectives on Teaching and Learning—Some Interrelations and Cognitive Effects." *Instructional Science,* (6), pp 319-367, 1977.

Easley, J. "The Structural Paradigm in Protocol Analysis." *Journal of Research in Science Teaching,* 11(3), pp 281-290, 1974 (reprinted in this volume).

Feigenbaum, E. and Feldman, J., editors. *Computers and Thought.* New York: McGraw-Hill, Inc., 1963.

Inhelder, B. and Piaget, J. *Growth of Logical Thinking from Childhood to Adolescence.* New York: Basic Books, Inc., 1958.

Kuhn, T. *The Structure of Scientific Revolutions,* Second edition. Chicago: University of Chicago Press, 1970.

Larkin, J. "Cognitive Structures and Problem-Solving Ability." SESAME working paper, University of California, 1976.

Leibold, B., et al. "Problem-Solving: A Freshman Experience." *Engineering Education,* Nov 1976, p 172.

Minsky, M. "Why Programming is a Good Medium for Expressing Poorly Understood and Sloppily Formulated Ideas." In *Design and Planning II,* edited by M. Krampers and P. Seitz, New York: Hastings House, 1967, p 121.

Newel, A. and Simon, H. *Human Problem Solving.* Englewood Cliffs, NJ: Prentice-Hall, Inc., 1972.

Parlett, M. *Introduction to Illuminative Evaluation.* Berkeley, CA: Pacific Soundings Press, 1977.

Perry, W. *Forms of Intellectual and Ethical Development in the College Years: A Scheme.* New York: Holt, Rinehart & Winston, 1970.

Polya, G. *How to Solve it.* New York: Doubleday & Co., Inc., 1973.

Putnam, H. "Reductionism and the Nature of Intelligence." *Cognition,* 2(1) pp 131-146, 1974.

Reif, F. "Educational Challenges for the University." *Science,* May 3 1974, p 537.

Reif, F., et al. "Teaching General Learning and Problem Solving Skills." *American Journal of Physics,* 44(3), p 212, 1976.

Reitman, W. *Cognition and Thought.* New York: John Wiley & Sons, Inc., 1965.

Simon, H. "Simulation of Human Thinking." In *Computers and the World of the Future,* edited by M. Greenberger. Cambridge, MA: MIT Press, 1962.

Simon, H. and Newell, A. "Thinking Processes." In *Contemporary Developments in Mathematical Psychology,* Vol 1, edited by D.H. Krantz, San Francisco: W.H. Freeman & Company Publishers, 1974.

Weizenbaum, J. *Computer Power and Human Reason.* San Francisco: W.H. Freeman & Company Publishers, 1976.

White, S. "Social Proof Structure: The Dialectic of Method and Theory in the Work of Psychology." In *Life-Span Developmental Psychology,* New York: Academic Press, Inc., 1977, chapter 3.

Woods, D., et al. "Teaching Problem Solving Skills." Engineering Education, Dec 1975, p 238.

The Structural Paradigm
in Protocol Analysis[1]

J.A. Easley, Jr.

Piaget's objections to tests as ways of identifying cognitive structures and processes appear to have been largely ignored in most of the replication studies of his work conducted by English and American psychologists. Piaget (1929) pointed out that tests cannot provide enough information to decide what structures are involved in a child's thinking, and Piaget and Inhelder later characterized tests as giving only the "results of efficiency of mental activity without grasping the psychological operations in themselves." (Piaget, 1947) What we wish to argue here is that there is a basic conflict between the general scientific paradigms for structural analysis and for measurement and that the identification of cognitive structure requires the former. To replicate the determination of a population parameter like the mean age of appearance of conservation of substance, a clear conception of the underlying structure is required. It is our claim that cognitive structures in Piaget's sense, and indeed structures underlying phenomena in many other branches of science (e.g. electron, cell, tectonic plate), cannot be treated like measurable quantities or given so-called operational definitions in the sense in which Bridgman defined length and mass by specifying the operations of measurement. Because of the interest which Piaget's studies of children's thinking have for science and mathematics educators, understanding the appropriate paradigm for analysis of cognitive structures and disengaging structural analysis from the confusions of the measurement paradigm is a task that could be pursued fruitfully by researchers in science and mathematics teaching, who have training in the conceptual theories of the natural sciences in which these paradigms are clearly discernable.

The point has been suggested by Smedslund (1969) who argued that "the constructs involved (in Piaget's tasks) are anchored neither to distal or proximal physical stimuli, nor to physical response categories, but to the *meanings* of the subjects' acts." (emphasis added) The problem then is to discover a structural theory of meaning rather than of phenomena. It has not often been noted in the literature on scientific thought that measurement and structural analysis paradigms in science are quite distinct and that they may be appropriate in quite different circumstances, but those familiar with the appropriate sciences will testify that electrons, benzine rings, fault zones, the jet stream, and DNA are concepts that are not defined operationally in Bridgman's sense. That is, they are not initially "framed in terms of operations which can be unequivocally performed." (Bridgman, 1938, p 114) Furthermore, they do not function as in-

1. ©1974 by the National Association for Research in Science Teaching. Published by John Wiley & Sons, Inc.

tervening variables in multidimensional mathematical models. Rather they function as constructs which are mechanisms that help scientists conceive of subtle and complex phenomena. The history of the development of fundamental scientific structures as intellectual constructs has been documented by Toulmin and Goodfield (1962), Einstein and Infeld (1954), Hanson (1958), Kuhn (1957), Jammer (1957), and Van Melsen (1952).

In psychology, cognitive structures correspond methodologically to such structures as atoms and the cosmos in physical science. But what needs to be made clear is how such structures *relate* to observable events if they cannot be *defined* in terms of procedures of observation and measurement. What constrains the creative imagination of the theoretician? An important clue lies in the property of underlying structures pointed out by Witz (1971) that, because of the complexity-reducing role they play in scientific thought, structural units stand in many-many relations with observable molar events and patterns of events. That is, they cannot have a one-one or one-many relation with these events, as operational definitions of measurable quantities would require. (While definitions are, strictly speaking, one-one relations, logical explications of so-called operational definitions, e.g. Carnap's reduction sentences (1937), take the form of one-many relations. That is, the quantity defined can be identified only under the conditions necessary to perform the operation and when certain results are obtained, but under other conditions it is still assumed. For example, we assume that familiar objects have temperatures within normal ranges even though the conditions for temperature measurement are not often met, i.e. when no thermometer or other thermal-sensitive instrument is in contact with them). Direct demonstration of the existence or the nature of underlying structures is therefore much more difficult, but a many-many relation does provide essential linkage between theory and observation though it be a very complex system of links. Yet, despite the difficulty of direct demonstration of precisely when or where a given structure is present or absent (demonstration often destroys or blocks the normal function of a structure), their indirect capacity to help explain most natural phenomena has made structures the backbone of scientific advance. That is, when scientists learn to see the world in terms of structures, measurements and other applications are suggested which often yield surprising (and therefore convincing) results.

Einstein, in the forward to Jammer's book, *Concepts of Space,* describes the paradox of structural concepts like the nature of space which we cannot do without and which must nevertheless be treated with skepticism.

> The eyes of the scientist are directed upon those phenomena which are accessible to observation, upon their apperception and conceptual formulation. In the attempt to achieve a conceptual formulation of the confusingly immense body of observational data, the scientist makes use of a whole arsenal of concepts which he imbibed practically with his mother's milk; and seldom if ever is he aware of the eternally problematic character of his concepts. He uses this conceptual material, or speaking more exactly, these conceptual tools of thought, as something obviously, immutably given; something having an objective value of truth which is hardly ever, and in any case not seriously, to be doubted. How could he do otherwise? How would the ascent of a mountain be possible, if the use of hands, legs, and tools had to be sanctioned step-by-step on the basis of the science of mechanics? And yet in the interests of science it is necessary over and over again to engage in the critique of these fundamental concepts, in order that we may not unconsciously be ruled by them. This be-

comes evident especially in those situations involving development of ideas in which the consistent use of the traditional fundamental concepts leads us to paradoxes difficult to resolve. (Jammer, 1957, p xi).

Thomas Kuhn (1961) argues that, in the history of physical science, measurement has only rarely led to theory but on the contrary, qualitative theory development has usually preceded and often guided and inspired measurement. It would appear reasonable, with regard to these views, that some insight into structures underlying observed phenomena is needed before we can reasonably hope to define measurable quantities appropriate for further illuminating and systematizing our psychological observations.

Psychometric approaches to psychology and education today take the opposite tack, seeking to guess which quantities to measure in hopes of discovering general laws (Cronbach, 1960). While a few psychometricians have recently been looking to Piaget's tasks as a more unified domain for test construction and to his theory for general orientation to individual differences (Tuddenham, 1971), we may require much more precise description of Piaget's postulated cognitive structures and their dynamic functioning before we can hope that task-related phenomena measurable in tests will be usefully identified.

As an example of the many-many relations by which structures are linked to observable phenomena, we may consider how John Dalton succeeded in discovering the key relations which connected hypothetical molecular formulas for compounds (e.g. H_2O) and hypothetical atomic weights (e.g. one unit hydrogen to 16 units oxygen) with observable weight ratios (e.g. one part hydrogen to eight parts oxygen). Because each hypothetical molecular structure depended on two or more equally hypothetical quantities (atomic weights) to relate to a given observation, it was necessary to construct systems of relations involving more than just a few elements and several compounds before the ambiguities between even the simplest hypotheses could be eliminated (Easley and Tatsuoka, 1968). But working with larger and larger groups of compounds whose composition had been measured, Dalton was able to produce a table of atomic weights for 20 elements which has remained substantially unchanged. Modern structural chemistry not only started from this somewhat playful exploration of loose relations between structural models and observations, but has continued to depend on such many-many relations between structures and observations in its development. It is this wholistic process (the antithesis of the piecemeal procedure of most psychological studies) which has made possible the description of such complex structures as DNA, RNA, and proteins. A similar story can be told in biology where learning to think in terms of cells, the circulation of the blood, other organ systems, ecosystems, and recently, molecules, has created a series of major advances leading to modern biological science.

In contrast, the psychology of cognitive structures is in a very primitive state as compared with chemistry, physics, or biology. Still, scholars in science education can study the structural paradigm in the history of the physical and biological sciences to learn its guiding principles and should therefore feel more challenged than most other educational specialists in studying structural psychology. Recognizing that many different structures can give rise to and be involved in the same response in an experimental situation, and that many different responses in that situation can come from or involve the same structure, should help researchers relinquish their debilitating dependence on operational definitions and testing. Recognizing that no single structure acts alone, but always in concert with other structures, which are present and brought into action by the situation, directly or indirectly, should provide a balance for the oversimplicity of most psychological theory.

When a psychologist interviews a subject, he may be tempted to think of the conversation in terms of the common meanings he shares with the subject (see Smedslund's suggestion, quoted above). In the processes of interpretation, he imposes his own categories on the protocol. This may be necessary in starting a new theoretical approach. However, Piaget's concept of cognitive structure (especially of "schemes") now opens the possibility of interpreting regularities in the subject's behavior in terms of structural relationships which account for these regularities by their own dynamic tendencies. Sometimes a brief part of the interview will strikingly suggest a key structure, but to make it more than an intelligent guess, other parts of the interview, involving major variations in situations and possibly counter suggestions, must also be shown to be accounted for by that same structure. As Witz (1971) and Knifong (1971) have shown, almost every interaction in the entire protocol of an interview may be accounted for by a small number of structures. This kind of documentation can greatly add to our confidence and also refine our understanding of cognitive structures (Witz and Easley, 1972).

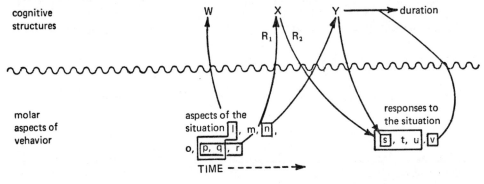

Figure 1

The relation of cognitive structures to observable phenomena is represented in Figure 1 (after Witz, 1973). A particular structure X is said to assimilate certain aspects of the immediate situation, say p, q, r, for which the structure provides an interpretation or a response; another structure W may assimilate 1, p, q; Y assimilates only n; and none of the structures presently active will assimilate others, m and o. That is a domain of structures (W, X, Y, . . .) is related to a domain of situational aspects (l, m, n, . . .) in a many-many relation R_1. No structure is uniquely defined by a known set of situational aspects (because in new situations new aspects will be assimilated to old structures), and no aspect observed in a situation can be defined by a given set of structures (because later, new structures will assimilate and thus redefine old aspects). Similarly, the overt responses, s, t, u, which the subject makes as a consequence of the activity of structure X, and which thereby give evidence of the involvement of X, form part of another many-many relation R_2 with cognitive structures

(Witz, 1971). This is the relation which Piaget and most Piagetians have failed to develop—or, if some Piagetians have tried to explicate it (Tuddenham, 1971), they have treated it as a one-many or even a one-one relation, forgetting that structures cannot act alone. Structures which are in evidence through R_2 over a substantial period of time, e.g. Y, may be indicated as being active for a specified interval in the diagram. Relative occurrences in time can thus be preserved as needed in these diagrams.

The accommodation of structures to given aspects of the situation (which, as Piaget stresses, is a process going on simultaneously with the assimilation of aspects of the situation to the structures) is not represented in Figure 1 but it can be added in different ways, depending on whether one wants to represent explicitly or implicitly the changes in structures resulting from their accommodation to objects or events of the situation. An explicit version, Figure 2 represents two different structures before and after accommodation, and the differing aspects assimilated to them, as displaced in the time dimension.

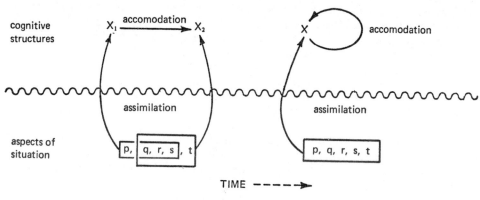

Figure 2

This method, however, would interfere with the representation of duration of activity of a given structure. More implicitly, one may represent accommodation by the looped arrow in the right-hand part of Figure 2 (Furth, 1969). This is an instantaneous representation of the process of accommodation, which process must be understood as going on simultaneously with the process of assimilation that is represented by the upward arrow. The need for explicitness in representing accommodation, however, is dependent on the availability of data to give evidence of structural changes (Witz and Goodwin, 1971). In this paper, we shall follow Witz's even more implicit convention of leaving out arrows representing accommodation altogether.

To understand how this paradigm can be applied formally to the protocols of conservation interviews, we consider an excerpt from one of Kamara's interviews, translated into English from Themne (Kamara, 1971). Kamara investigated the question whether, when interviews are conducted in the child's native tongue, Themne children perform as well as children in advanced cultures of the West. The subject in this interview is Amadu, a nine-year-old, unschooled Themne boy, who after several other discussions, was shown two balls of plasticine and then questioned in the manner described by Piaget and Inhelder (1962) with such adaptations to Themne culture and language as letting the plasticine represent cakes of rice bread.

Experimenter:	Which of you would have more bread to eat?
Subject:	They are equal, I had shared it equally.
Experimenter:	If Aruna should come, and you tell him that they are both equal, and he denies it, what would you do to convince him?
Subject:	(Pauses) . . . He will think that this is more (pancake).
Experimenter:	What would you do to show him that they are equal?
Subject:	I will tell him that this looks bigger because it has been flattened, but they are equal.
Experimenter:	What would you do to convince him?
Subject:	(Pauses) . . . Then I will gather (ball) this (pancake) and make a ball with it . . .

The primary question for those convinced of the reality or general usefulness of operational schemes and other structures described by Piaget, is whether the judgment of equality Amadu makes is due to a general conservation structure (Figure 3) supported by concrete operations or whether this judgment is due to low-level perceptual structures such as would be involved in figurative recall, designated as "pseudoconservation" (Piaget and Inhelder, 1971.) In terms of Witz's paradigm, the question is whether the first interaction of experimenter and subject (i.e. the first entry for each under "observable behavior" in the figure below) should be interpreted as involving the high-level structure or low-level structure. This cannot be answered on the basis of the first interaction alone.

In the second interaction, we have evidence of another structure which appears to be primarily perceptual; we designate it here as a comparison of the surfaces of the ball and the pancake. (Subjects who say that the pancake provides more bread to eat than does the ball often appear to be centering on surface areas, but this is not necessarily "area" in the geometrical sense, for the diameter or some other aspect of the object's appearance may account just as well for the observed behavior.) Amadu attributes some such comparison to his friend, Aruna, implying that he, himself, has not changed his belief in conservation. This means that the structure accounting for conservation is stronger than the comparison structure, which is allowed to function only vicariously. Note that only the first part of the experimenter's question is assimilated by Amadu. An assimilation "overload" may, in fact, be the reason for the pause preceding his response. The last part of the question, "what would do to convince him?" has to be asked twice more before it is fully assimilated, i.e. before a system of structures is activated that can assimilate "what would you do?" as meaning "what motor act will you perform?"

The next response gives supporting evidence for two structures already mentioned and evidence of one additional structure. (We omit reference to linguistic structures such as would be indicated by the opening phrase, "Tell him that.") The second phrase, "this looks bigger," provides additional evidence of the comparison structure, but the third phrase, "because it has flattened," gives evidence of an object-manipulation structure (like a sensory-motor scheme) concerned with flattening the ball into a pancake. This has evidently been activated by the persistency of the questioner and by the perception of the pancake,

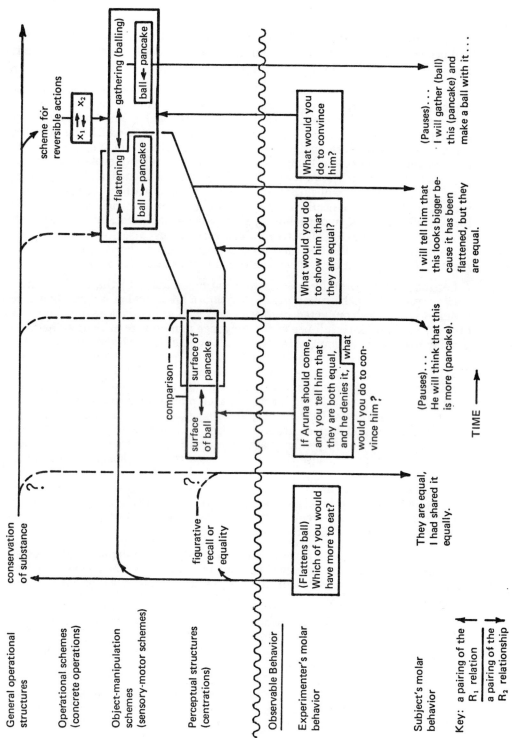

Figure 3

recalling the act of flattening. The final phrase, "but they are equal," shows the still persistent conservation of substance, suggesting now by its persistency that it is something other than figurative recall.

Clear evidence of a concrete operation relating to the conservation of substance does not emerge, however, until the next repetition of the probe, "What would you do to convince him?" Now the response, which comes after another pause (suggesting this time a search for a new structure to which the question can be assimilated), is recognized by Piagetians as evidence of one of the reversible concrete operational schemes, a scheme that generates or evokes the reverse of an action in the subject's mind. If one accepts the existence of such operations, the response, "I will gather this and make a ball with it," in this context, is evidence that the connections between the conservation structure, the scheme for reversible actions, and a related pair of object-manipulation schemes (flattening and gathering) are all active. Of course, one must also presume that some perceptual and kinesthetic structures associated with gathering plasticine-like material (mud balls or balls of rice flour bread) have also been at work.

Language and culturally-specific associations enter into the clinical interviews required in such an important way that a great deal of weight should be given to interviews conducted by researchers who are natives of the children's own culture and make full use of the language and culture of their subjects in probing to remove ambiguities in responses as well as in the interpretation of the responses. Kamara's data on schooled children from Sierra Leone (Kamara, 1971) indicate no significant retardation in comparison with Geneva norms (Kamara and Easley). It thus supports the possibility that the earlier evidence of a retardation of the rate of intellectual development in nonwestern cultures, which most researchers (Dasen, 1972, and Piaget, 1964), have accepted may be suspect. The methodological situation of this area of research is essentially no better today than in 1966, when Piaget commented, "cross-cultural studies are difficult to carry out because they presuppose a good psychological training in the techniques of operational testing, namely with free conversation and not standardization in the manner of tests, and all psychologists do not have this training, a sufficient ethnological sophistication, and a complete knowledge of the language are also prerequisites. We know of only a few attempts of this quality." (translated by Dasen, 1972)

The structural analysis paradigm of Witz helps make explicit the complex system of relationships that must be kept in mind as one interviews children and classifies their responses. But the above discussion is only illustrative of the idea of structural analysis.

We shall examine now a somewhat different application, the incorporation of adult, theoretical conceptions into a child's own conceptions. The following excerpts are from an exploratory interview conducted by George Triplett with David, aged nine years and eight months, to discover his conceptions of heat and temperature (Witz and Easley, 1972). In the course of a long discussion of a variety of physical systems to which heat could be supplied by an alcohol lamp, David invoked three times the concept of molecular motion about which he had evidently heard in school. Of several dozen children from eight to 14 interviewed, the three excerpts below from David's interview show the most functional incorporation of this concept into novel situations.

In the first selection he explains thermal expansion and in the second he explains the heating of a block of metal and in the third the sizzling of drops of water on a hot metal stand. The first two exhibit nothing very creative, and he receives some help from the experimenter in formulation. But in the third selection the expression is clearly David's, and the idea exhibits a power of its own.

Experimenter:	Can you explain to me—I think you sort of did before—but just exactly why does a thermometer go up?
Subject:	Well, uh, it starts getting heated, whenever it's surrounded by water, air or anything, uh, it causes the—I don't know what the liquid's called that's inside.
Experimenter:	I think that's alcohol in there.
Subject:	Well it causes it to expand, and when it expands it just has to take up more room, so it starts rising.
Experimenter:	Um hum. And, uh, why does it expand, uh, do you have any ideas about that?
Subject:	Yeah. When it gets heated it causes the molecules to get spread apart.
Experimenter:	Oh, uh, huh.
Subject:	And when things get hot the molecules start spreading apart, and so, it causes it to expand, and that causes it to go up in the tube.

The next reference occurs a few minutes later when the experimenter has asked David to explain what would happen when a metal block is heated by the flame.

Subject:	Well it's pretty hard to explain molecules but they're just little microscopic things that make up—everything. You can't really say what molecules are made up of—but they make up things.
Experimenter:	Yeah, uh huh. OK. Now, what would happen to the molecules in that iron then? Do you know? When we lit the burner?
Subject:	Well, the burner would start heating it up and the molecules inside the iron, uh, when they're not heated they just sit there. When they heat up they start moving around, and they bump into each other, and when molecules start going real fast it starts getting hot, like a sidewalk on a summer day, if you ever walked on it barefooted.
Experimenter:	Oh, how is that?
Subject:	The sun heats it up.
Experimenter:	And?
Subject:	And the molecules inside the sidewalk start bumping into each other and they go faster and faster.
Experimenter:	You mean they hit your feet?
Subject:	No. (laughs) No, inside the sidewalk, some molecules bounce into other molecules and they bounce off each other, and they keep—it's like a crash course, they just bounce into other molecules, they keep going like little tiny super balls, and they just bounce real faster and faster until it gets hot.

The final excerpt concerns water splattering out of a flask on to the hot screen of the support below where it makes a sizzling noise.

Experimenter:	Yeah, where does the noise come from?
Subject:	When the water hits the hot—the hot part? When it sizzles?
Experimenter:	Yeah, why does it sizzle?

Subject:	Well, when the water touches something hot it just—(laughs)
Experimenter:	Yeah, (laughs) yeah, it what?
Subject:	Well, it always does that. I haven't really sat down and thought about it.
Experimenter:	Well, you're sitting down right now.
Subject:	Yeah.
Experimenter:	Let's think about it.
Subject:	Well, when it touches something hot, it heats up the water, and there's little water there so it just—heat it up and everything starts—since the thing is so hot—the screen is so hot that it just—sends the molecules all over and it just—in a split second it just—makes—the molecules are in there, going around real fast right now, and when it hits that things real fast—hits the hot thing—it just makes them start going so fast it just—sort of explodes. (laughs)

In Figure 4 the protocol selections are merely indicated by segments in order to encompass a more substantial development. The cognitive structures suggested are named by words chosen from David's speech; we simply have little else to go on. What Figure 4 mainly represents, however, is the system of relations between structures by means of which one assimilates an activity in another and thus becomes activated or reactivated.

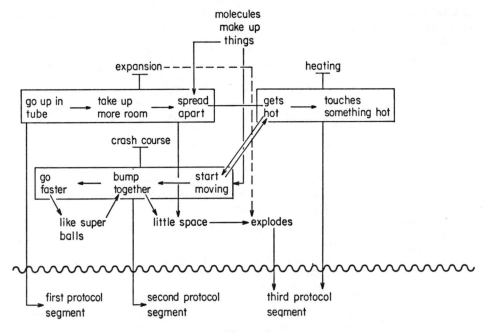

Figure 4

Analysis of Figure 4 permits us to raise several interesting questions.

1. How is the "expansion" structure related to the "crash course" structure? e.g. is (was) movement and bumping involved in "expansion?" We may presume that "explodes," which clearly partakes of "crash course" also involves "expansion," but more evidence is needed for that conclusion. The best connec-

tion we have for that is through his reference to the small space inside the droplets. What we would expect is a more direct connection like that suggested by the broken arrow.

2. Why is it that, in the second segment, David does not assimilate the idea of the molecules in a hot sidewalk hitting his feet? Why is it that his conceptual system doesn't use molecular collisions as the means of heat transfer between two objects in contact? Is it that the touch perception of a hot pavement necessarily involves a change of quality from temperature to discomfort not explainable by treating high temperature as movement? Or is it, as the diagram suggests, that there is simply no structural place within "heating" for the "crash course" structure?

The overriding quest for reliability, which appears to be the dominant concern in the test paradigm, is doomed to generate many errors in the identification of cognitive structures. Analysis of protocols, on the structuralist paradigm, is necessarily a slow and nonmechanical procedure. It begins in subjective, but hopefully educated, judgments and moves toward objectivity as it attains completeness in accounting for the total protocol. Witz, in explaining his axioms for cognitive structure representations (CSRs), says, "In our formulation, we emphasize that it is sufficient merely to account for the observed behavior—the short-term dynamics are vague enough to permit many possible accounts, it limits rather than predicts, and an unambiguous picture emerges only when many items of information are considered and more 'overlapping' CSRs come into play." (Witz, 1971)

In the light of the fragments of interviews we have analyzed, it seems absurd to pretend that one knows how to measure cognitive competences by administering standardized lists of questions when no validating clinical interviews—and certainly no explicit structural analyses—have been published. To advance in public objectivity beyond the highly informal and private discipline of the clinical interview will require the publication of many deeper analyses of rich behavioral records than have yet been published. Fortunately, at least one new journal, *The Journal of Children's Mathematical Behavior,* has been founded for this purpose, and *Cognitive Psychology* has opened the possibility by placing no size limits on articles.

REFERENCES

Bridgeman, P. W. "Operational Analysis." *Philosophy of Science,* 5, pp 114-131, 1938.

Carnap, R. "Testability and Meaning." *Philosophy of Science,* 3, pp 419-471, 1936; and 4, pp 1-40, 1937.

Cronbach, L. J. *Essentials of Psychological Testing.* Second edition, New York: Harper and Bros., 1960, pp 23, 25. (Third edition, New York: Harper & Row Publishers. Inc., 1970.)

Dasen, P. R. "Cross-Cultural Piagetian Research: a Summary." *Journal of Cross-Cultural Psychology,* 3, pp 23-29, 1972.

Easley, J. A., Jr. and Tatsuoka, M. M. *Scientific Thought.* Boston: Allyn and Bacon, Inc., 1968, chapter 4.

Einstein, A. and Infeld, L. *The Evolution of Physics.* New York: Simon & Schuster, Inc., 1954.

Furth, H. *Piaget and Knowledge.* Englewood Cliffs, NJ: Prentice-Hall, Inc., 1969, p 75.

Hanson, N. R. *Patterns of Discovery.* New York: Cambridge University Press, 1958.

Jammer, M. *Concepts of Space.* Cambridge: Harvard University Press, 1957.

Kamara, A. I. "Cognitive Development among School-Age Themne Children of Sierra Leone." PhD dissertation, University of Illinois, Urbana, 1971.

Kamara, A. and Easley, J. A., Jr. "Is the Rate of Cognitive Development Uniform Across Cultures?" In *Culture, Child and School,* edited by M. S. Maehr and W. M. Stalling, Monterey, CA: Brooks/Cole Publishers, in press.

Knifong, J. D. "The Representation of Cognitive Structure of Four and a Half Year Old Children." PhD dissertation, University of Illinois, Urbana, 1971.

Kuhn, T. S. *The Copernican Revolution.* New York: Random House, Inc., 1957.

Kuhn, T. S. "The Function of Measurement in Modern Physical Science." In *Quantification,* edited by H. Woolf, New York: Bobbs-Merrill Co., Inc., 1961.

Piaget, J. *The Child's Conception of the World.* London: Routledge and Kegan Paul, Ltd., 1929.

Piaget, J. "The Development of Mental Imagery." In *Piaget Rediscovered.* edited by Ripple and Rockcastle, Ithaca: Cornell University, 1964, pp 21-32.

Piaget, J. and Inhelder, B. *Le Developpement des Quantites Physique chez l'Enfant* (Deuxième édition augmentée). Neuchâtel: Delachaux et Niestlé, 1962.

Piaget, J. and Inhelder, B. "Diagnosis of Mental Operations and Theory of the Intelligence." *American Journal of Mental Deficiency,* 51, pp 401-406, 1947.

Piaget, J. and Inhelder, B. *Mental Imagery in the Child,* New York: Basic Books, Inc., 1971, p 269.

Smedslund, J. "Psychological Diagnostics." *Psychological Bulletin,* 71, pp 237-248, 1969.

Toulman, S. and Goodfield, J. *The Architecture of Matter.* New York: Harper & Row Publishers, Inc., 1962.

Tuddenham, R. D. "Theoretical Regularities and Individual Idiosyncrasies." In *Measurement and Piaget,* edited by D. R. Green, M. P. Ford and G. B. Flamer, New York: McGraw-Hill Book Co., 1971.

Van Melsen, A. G. *From Atomos to Atom.* Pittsburgh: Duquesne University Press, 1952.

Witz, K. G. "Analysis of 'Frameworks' in Young Children." *The Journal of Children's Mathematical Behavior,* 1(2), pp 44-66, 1973.

Witz, K. G. "Representation of Cognitive Processes and Cognitive Structure in Children I." *Archives de Psychologie,* 15, pp 61-95, 1971.

Witz, K. G. and Easley, J. A., Jr. "Analysis of Cognitive Behavior in Children." Final Report, Grant No. OEC-0-70-2142(508), USDHEW, NCERD, 1972, Appendix 13.

Witz, K. G. and Goodwin, D. "Structural Changes in 4-5 Year Olds." Presented at the Second Annual Interdisciplinary Meeting on Structural Learning, University of Pennsylvania, Philadelphia, April 2-3, 1971.

Proportional Reasoning And Control of Variables In Seven Countries[1]

Robert Karplus, Elizabeth Karplus, Marina Formisano, and Albert-Christian Paulsen

The wide diversity of interests and abilities encountered among secondary school students is well known to teachers. The present study was intended to illuminate this range in regards to logical mathematical reasoning, or what Jean Piaget has called formal thought. We developed two tasks to assess proportional reasoning and control of variables. On these tasks, which could be administered to classroom groups in several countries, the students were asked to solve certain problems and then explain or justify their answers. The logical mathematical reasoning was exhibited by the form and content of the explanations.

Both areas of thought we selected for investigation are vitally important in science instruction. Proportions play a key role in quantitative relations in science. The idea of a controlled experiment, "keeping all other things constant" to isolate the effect of one variable, is essential for determining unambiguous cause-and-effect relationships in science or other subjects. In view of the general ferment in science education, with new course materials and new teaching methods being introduced into secondary schools on an increasing scale, it seemed worthwhile to investigate students' logical-mathematical reasoning.

Traditionally, reasoning has been investigated by means of clinical interviews in which a subject confronts a phenomenon or a problem and is asked to make a prediction, explain what he observes, or find a solution(1). The subject is then asked to justify his reasoning, and is pressed by the interviewer to provide additional reasons or alternate theories regardless of his first response. The resulting conversation is eventually analyzed for the types of relationships, "convincing" arguments, and resolution of ambiguities employed by the subject.

Written responses to a group test clearly do not provide the same depth of information about a single individual. There are many reasons why an individual's first response—and that is what he is likely to write down—may not tap all his intellectual resources. Copying from other students and lack of interest also influence the results. Nevertheless, the written task does have an advantage also, in that it permits rapid surveys of large numbers of subjects interna-

1. This material is based upon research supported by the National Science Foundation under Grant No. SED74-18950. Any opinions, findings, and conclusion or recommendations expressed in this publication are those of the author(s) and do not necessarily reflect the views of the National Science Foundation.

tionally. At the same time, much of a student's school work is more closely similar to the written task situation than to the clinical interview. Hence, student performance on the written task may help a teacher diagnose learning problems or recognize able students that need additional challenges.

Both areas we chose have been investigated through clinical interviews and through group tests in the last ten years (2-13), though no studies that included several national groups have come to our attention. The results of these investigations showed that few average subjects used proportional reasoning unaccompanied by physical actions before age 15. Proportional reasoning has therefore been regarded as one of the later components of formal thought to be acquired by adolescents. Control of variables was found to have roots in children's concept of a "fair" experiment that has been noted during the elementary school years (9,14), but the careful analysis and design of experiments has been a source of difficulty for many high school students and adults (12,13).

A large-scale international study of science education sponsored by the International Education Association (IEA) was completed recently (15). The tests that were used focused on traditional areas of science at various educational levels and used large numbers of multiple-choice items for the assessment. Some of these involve controlling variables or proportional reasoning (16), and we considered gleaning some data on student reasoning from the results. Unfortunately, individual multiple-choice items without student justification can give conclusions about the respondent's reasoning only very indirectly, through a comparison of incorrect choices. We were not able to learn anything beyond the fact that these items had relatively low success rates. We therefore consider our study to complement the IEA study and not to duplicate it.

In this article, we shall first describe the subjects and then our general method of presentation of the tasks. There follows a detailed account of each task and the scoring procedure used to evaluate the responses. Next, we present an overview of the results in seven countries and comparisons among the populations. We conclude with a more general discussion of our observations and some implications.

THE SUBJECTS

This report presents an overview of the results of our study of 13- to 15-year-old students in seven countries: Denmark (Copenhagen area), Sweden (Gothenburg area), Italy (Rome), United States (northeast and north central states), Austria (Vienna), Germany (Gottingen), and Great Britain (London). Almost 1800 boys and an equal number of girls participated. The testing usually occurred in a science or mathematics class, but sometimes a scheduling problem led to the use of other classes. As much as possible, the groups in each school were chosen so as to reduce the effects of any tracking by ability within the school.

To have subjects of closely similar ages in all countries in spite of the differences in school-entrance criteria, we worked with eighth graders in the United States, Italy, Austria, and Great Britain, seventh graders with a few eighth graders in Denmark and Sweden, and ninth graders in Germany.[2] In Austria

2. A special change in the school calendar in 1966 eliminated entering German first graders in the fall of that year. Children who entered school in spring of 1966 were designated as ninth graders at the time of our visit.

and Italy, many children leave school after the eighth grade, so that this age group constituted the oldest group we could sample reliably through school visits.

School organization in the seven countries varies considerably, and we tried to choose the populations for testing to get a fair survey appropriate to each school organization. We do not claim to have a representative sample for any country or even any town, since the duration of our visits, our access to various schools, and our knowledge of the composition of the entire population were completely inadequate for that. Except for Denmark, where the data were quite uniform from community to community, we report results only for the specific groups we tested and not for the entire country or region. In Denmark, Sweden, Italy, and the United States, where the schools are comprehensive at the seventh- and eighth-grade levels, we have classified the population groups according to the socioeconomic level of the neighborhood served by the school (derived roughly from parental occupation, education, and/or quality of housing). In Austria, Germany, and Great Britain the schools are divided into types with differing entrance requirements, curricula, and school-leaving ages, and these types served as basis for our classification. Further details about the populations will be given in the sections reporting the data for the individual countries.

Table 1 presents an enumeration of the student samples. These include the principal research groups of seventh to ninth graders, and additional small pilot groups in other grades that we used to gain comparative information about the reasoning of students of differing ages. We shall not quote the results for these

Table 1. Student Populations and Mean Ages

Country	Population	Principal Group				Pilot Group			
		Girls		Boys		Girls		Boys	
		No.	Age	No.	Age	No.	Age	No.	Age
Denmark	Middle Class	207	14.0	192	14.1				
Sweden	Middle Class	47	14.6	51	14.5				
	Working Class	92	13.9	90	14.1				
Italy	Upper Middle	69	13.2	92	13.1	17	11.4	21	11.8
	Middle Class	58	13.3	52	13.3	(6th grade)			
	Working Class	97	13.6	99	13.9				
United States	Upper Middle	112	13.5	96	13.8	89	13.6	77	13.7
	Middle Class	290	13.6	341	13.6	(8th grade)			
	Urban Low Income	94	13.9	87	14.1				
Austria	Gymnasium	187	14.3	112	14.4	29	12.4	29	12.5
	Hauptschule-A	89	14.3	102	14.4	(6th grade)			
	Hauptschule-B	53	14.4	52	14.7				
Germany	Gymnasium	61	14.6	60	15.0	45	13.3	47	13.6
	Realschule	59	14.9	60	15.5	(7th grade)			
	Hauptschule	54	14.9	53	15.0				
Great Britain	Direct Grant	39	14.3	48	14.2				
	Grammar	56	14.3	61	14.3				
	Comprehensive	87	14.4	85	14.3				

groups, because of their small size, but use them only as they give perspective regarding the principal data.

The American reader may be interested in a few other details of school practice in the European countries visited. First, the children's studies beyond the fifth or sixth grade are departmentalized. Second, the children do not have electives, but participate in all their courses as members of the same group. Third, the groups tend to stay together for several years, and also tend to be taught by the same set of teachers for several years.

METHOD OF PRESENTATION

For all groups studied, the tasks were administered by a team consisting of authors and collaborators, with the tasks being presented by a person (*E*) thoroughly familiar with the language. After being introduced, *E* explained that the class would participate in a research project intended to show future teachers the wide variety of solution methods used by students to approach certain scientific and mathematical tasks designed for that purpose. The two tasks were next presented as described below, first "Proportional Reasoning" and then "Control of Variables." Finally, students were invited to write on the back of one answer page what they thought about the research project and their participation in it.

We made an effort not to press students for time. After about two-thirds of a class had handed in their answers to the first task, *E* called for attention and introduced the second task, allowing students who had not completed the first one to keep their papers and finish them later. Approximately 50 minutes allowed most children to complete both tasks, and this period of time either fitted into the ordinary school schedule, into a double period, or could be arranged with help of the school administration. A few students who worked slowly remained during the recess following the testing session in order to complete their papers. Others remained to ask questions about the research project or to use the equipment. In some classes, the time available after all papers were collected permitted class discussion of the tasks, more explanation of the purposes of the research, and/or demonstration of collisions between spheres, as the students requested.

Occasionally a special assembly, fire drill, or other unscheduled event reduced the time available to less than 40 minutes and may have resulted in incomplete responses. In such cases, one task was dropped from the schedule or the partially used papers were eliminated from scoring. For this reason, and also because some children came late or had to leave early, there are variations in the number of papers scored compared to the number of subjects in each group listed in Table 1.

Since the two tasks were to be used in five different languages, it was necessary to translate the student pages and to formulate the supplementary oral explanation so it could be presented reproducibly. This preparation was carried out in collaboration with science-education research groups in the countries to be visited. We believe that the translations were adequate and did not introduce substantial bias, with one possible exception: the word "target," used in the Control of Variables task to distinguish the stationary sphere from the initially rolling sphere, had corresponding terms in Swedish (*malkula*) and Italian (*bersaglio*) but not in German or Danish. The terminology used to paraphrase "target" may have been more difficult for some students to grasp.

We shall come back to this matter when we compare the performance of populations in differing countries.

PROPORTIONAL REASONING TASK

The task for assessing proportional reasoning was similar to the Paper Clips task (7). After making their own measurement, students could use proportional reasoning to transform certain units of measurement. Subjects were given answer pages and chains of about nine number one "gem" paper clips. They were then told that "Mr. Short" on their papers had a friend, "Mr. Tall," also drawn on a piece of paper, and that the two figures, when measured by a row of large round buttons placed side by side, were found to be four buttons and six buttons respectively, from floor to head (the data was written on the board). Then the items on the answer pages were read aloud and the subjects were urged to be complete in their explanations about the height of Mr. Tall and the width of the car. Individuals with questions were referred to the data on the board and were encouraged to use their own ideas. The height of the picture of Mr. Short was found to be between six and seven paper clips by virtually all the subjects.

When papers were handed in, they were briefly scanned by a team member, who asked students with incomplete or obscure answers to clarify their reasoning orally or in writing.

Response Categories

In this overview of our findings, we shall report scores on the Proportional Reasoning task in terms of four major categories. Four categories are not sufficient to describe all explanations in detail, but they do give a good picture of the students' progress toward proportional reasoning. All papers were separately scored by at least two of the authors, who reached a concensus concerning the interpretation of an unclear statement in the rare cases of disagreement.

Category I (Intuitive): The explanation does not make use of all the data, or uses the data in a haphazard and illogical way. Examples of responses are:

MR. T
10	I just thought about how many paper clips equal one button and tried to figure out how many paper clips there were from there. Answer 10 paper clips for Mr. Tall.
7	Because there are only 7 paper clips in my chain.
12	The way I got that Mr. Tall is 12 inches is I just doubled the 6 buttons. That is how I got the answer.
10	I added 6 and 4 together.
9	Because every button equals 1½ paper clips. That means that actually there are 6 paper clips, but there are 9 actually. (Note: did not justify the button/paper clip relation in terms of the data given even when asked orally.)
	I have no buttons to work with so I cannot do your answers.

CAR {
13   I guessed.

16   I think Mr. Tall is 8 paper clips. So I think his car should be 2 x 8 to get 16, because his car should be big, so he can get into it.

14   Because a paper clip equals a button.

7   Cause you say, what is half of 14.

28   Because it is 2 paper clips which equals 1 button so I add 14 and 14. (Note the reversal of the button/ paper clip relationship.)
}

Category A (Additive): The explanation focuses on a single difference (tall/short or paper clips/buttons) uncoordinated with other differences, and solves the problem by addition. Examples of responses are:

MR. T {
8   If Mr. Tall is 6 buttons and Mr. Short is 4 buttons, that is a difference of two. Now Mr. Short is 6 paper clips tall, so I took the two and added it to 6 and got 8.

8   Because in buttons he was 2 buttons taller, so I thought paper clips are almost the same length. So I put 2 paper clips bigger. Because buttons and paper clips should be about the same.

8   I got 8 because if Mr. Short is 4 buttons, and is 6 paper clips, then that means Mr. Tall should be 8 paper clips, because he is 6 buttons. The main difference is that there is two numbers between the buttons and paper clips.
}

CAR {
12   As in the first example, there are 2 more clips than buttons, so I substracted two and got 12 buttons.

16   In buttons it measures two more than in paper clips, so if it is 14 in paper clips, it will be 16 in buttons. (Note: the reversal of the button/paper clip relationship occurred in about 5 percent of category A responses.)
}

Category Tr (Transitional): The explanation shows only partial proportional reasoning, or makes reference to concrete comparisons or iterations. Examples of responses are:

MR. T {
7-1/2 I divided 4 into 6; 4 is how many buttons Mr. Short is and 6 is the amount of paper clips, and I got 1½. Then I added 6, the amount of buttons of Mr. Tall, to 1½ and got 7½.

9   First I folded Mr. Short in half, then into fourths, then I measured one of these fourths with my paper clips. Then to find Mr. Tall's height in paper clips I multiplied 6 (from his height in buttons) by 1½ and got 9.
}

	9-1/2	I laid out a line of 14 paper clips on the desk and marked off buttons.
	9-1/2	If 6 paper clips = 4 buttons, 4 for each time 6 went into 14. It went 2 times evenly, that was 8 buttons, and 2 clips left. And 2 clips = 1½ buttons. Added together they equal 9½ buttons.
CAR	21	Since 6 paper clips = 4 buttons, then 14 paper clips = N buttons. 6/4 is N/14. Cross multiply, 6 x 14 = 84 ÷ 4 = 21. The answer is 21 buttons. (Note: reversal of the button/paper clip relationship occurred in about 15 percent of responses that would otherwise have been classified in Category *R*.)

Category R (Ratio): The explanation uses a proportion or derives the exact scale ratio from Mr. Tall/Mr. Short dimensions or button/paper clip relationship. No concrete or iterative procedures are employed. Examples of responses are:

	9	1 button = 1½ paper clips, $6/4 - 1½/1$, $1½ \times 6 = 9$.
	9	I got this by putting their height in buttons into a fraction (4/6) and by putting their height in paper clips into a fraction (6/X) and solved it. The result is 9.
MR. T	9	Since 6 paper clips = 4 buttons, then *R* paper clips = 6 buttons, 6/4 as R/6. Cross multiply $6 \times 6 = 36$, $36 ÷ 4 = 9$. The answer is 9 buttons.
	9	Six clips is equal to 4 buttons, 3 clips is equal to 2 buttons, take half and add: $6c = 4b$, $3c = 2b$, $9c = 6b$.
CAR	9-1/3	I put his height in paper clips over the width of his car in paper clips (9/14), then (9/14), then put his height in buttons over his car's width in buttons (6/X), $6/X = 9/14$, $84 = 9X$, $9⅓ = X$.

CONSISTENCY OF RESULTS AND COMPOSITE SCORE ON PROPORTIONAL REASONING TASK

One question we shall answer before proceeding further relates to the consistency of the two parts of the Proportional Reasoning task. Table 2 shows the correlation of scores on the two parts for the more than 3300 boys and girls in the principal groups participating in this study. Each entry is the average percentage of students (calculated from the percentages of 14 subpopulations of seven countries and two sexes) who responded with the *Mr. Tall* explanation identified at the beginning of the row and the *car* explanation identified at the head of the column. For example, the "14" in location *R-Tr* of the table means that an average of 14 percent gave a Category *R* explanation for *Mr. Tall* and a Category *Tr* explanation for the *car*.

Four of the six substantial entries in Table 2 are along the diagonal, representing the same category of explanation on the two parts. A total of 63 percent of all papers was of this kind, 13 percent *intuitive,* 14 percent *additive,* 11 percent *transitional,* and 25 percent *ratio.* The other two substantial entries are in the *A-I* combination and the *R-Tr* combination, indicating about a third of the students that answered in Categories *A* or *R* for Mr. Tall's height fell back to the next lower category when applying their reasoning to the car's width. Thus, the car question was more difficult than the *Mr. Tall* question. Kendall's "tau" rank order correlation coefficient (17) of the two items was 0.48, implying a very good reliability for the task consisting of only two items and offering only four response categories.

Table 2. Correlation of Scores on the Two Questions of the Proportional Reasoning Task (Percent)

		Score on "car" question			
		I	*A*	*Tr*	*R*
Score on Mr. Tall question	*I*	13	0	1	0
	A	11	14	1	0
	Tr	3	0	11	2
	R	5	0	14	25

The only appreciable differences between boys and girls were in two entries: *A-A,* which includes 17 percent for the girls but only 10 percent for the boys; and *A-I,* which includes 14 percent for the girls and eight percent for the boys. The differences are statistically significant, $p < 0.01$ by the chi-squared test which we used for all comparisons in this report. That is, girls tended to respond in the *additive* category more than boys, especially on the Mr. Tall question. The compensating excess of boys was distributed over all the other table entries.

It is gratifying to note that the remaining 10 entries in Table 2 are much smaller than the six we have discussed, adding up to only 12 percent of all the responses. For us, this was evidence that Categories *I* and *A* do not involve proportional reasoning, while Categories *Tr* and *R* do. At the same time, we were reassured that the interaction among students and copying of answers, which certainly took place in many of the classrooms where several students sat at the same table, did not so randomize the responses as to invalidate our method of data gathering.

To simplify the data further, we took advantage of the clustering of responses in Table 2 to create a composite score of *I, A, Tr,* or *R* for each student. The composite score assigned to each of the Table 2 locations is indicated in Table 3. If the score on both explanations was the same (diagonal entries), then this was also used as the composite score. If the two were different, then Category *R* combined with any other category was scored as *Tr, Tr* combined with *A* was scored as *A,* and either *Tr* or *A* combined with *I* was scored as *I.* Since the student's performance on the two parts was well correlated, the composite score does not mislead by concealing great disparities in the levels of the two explanations.

When the entries in Table 2 are combined as just described, one obtains Table 4, which displays the percent of students having a composite score in each of the four categories. Later in this article we shall present the distribution of composite scores for each of the principal populations listed in Table 1.

CONTROL OF VARIABLES TASK

The task for assessing students' reasoning concerning control of variables made use of the collision of two spheres rolling on a track. After identifying the important variables, students could apply the principle that only one variable at a time may be changed if an experiment is to be conclusive.

Table 3. Composite Score on Proportional Reasoning Task

		Score on car question			
		I	*A*	*Tr*	*R*
Score on Mr. Tall question	*I*	*I*	*I*	*I*	*Tr*
	A	*I*	*A*	*A*	*Tr*
	Tr	*I*	*A*	*Tr*	*Tr*
	R	*Tr*	*Tr*	*Tr*	*R*

Table 4. Distribution of Composite Scores on the Proportional Reasoning Task (Percent)

Category	*I*	*A*	*Tr*	*R*
Percent	28	15	32	25

While subjects were previewing the answer pages (see Appendix *A*), *E* showed them a track with raised ends (Figure 1), similar to the diagrams on the answer page. One sphere was allowed to roll down and oscillate back and forth. Then a second sphere was placed at the lowest point as "target," and the experiment of allowing a sphere to roll from a "high," "medium," or "low" position to hit the target was described but not performed. *E* had four equal-sized spheres (two light glass, two heavy steel) which he showed to the class before reading Question 1 aloud and inviting the students to write their answers in the form of a list. After about a minute, several volunteers were invited to read items from their lists. Then *E* called attention to the weight of the sphere—an item that was always mentioned—to direct attention to Questions 2 and 3. After these had been read aloud, the students were asked to answer them completely. Questions about the test were invited and answered individually.

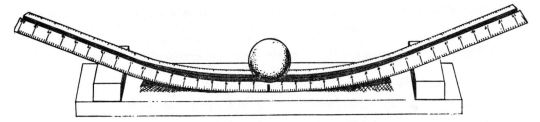

Figure 1. Apparatus for Control of Variables Task

Response Categories

Here also we shall report our findings in simplified form, and group together responses that appeared to us logically similar though they are qualitatively different (18). Since Questions 2A and 2B are similar, we adopted the same three-level classification system for both. We assigned zero, one, or two points to a response according to its level, as described below. For Question 3 we distinguished acceptable from unacceptable answers, assigning one point to the former and zero to the latter, as is also described in detail with illustrative examples below. All papers were read by two or more authors, with disagreements regarding interpretation being resolved through discussion.

Questions 2A and 2B:

0 points (any starting position lacking an explanation or accompanied by an "explanation" that describes the experimental set-up, what the subject wishes to find out, or the phenomenon that will occur) Examples are:

	High, high	Well it seems that if you start it high they'll both get a lot of speed which would be better.
	Low, high	If, when the target was hit by the heavy sphere, it went as far as when hit with the light sphere, you can tell that the spheres can have the same effect even though they are of different weights, as long as they are started at places that will give them different rates of acceleration and speed.
QUESTION 2A	*Medium, high*	If we start the light sphere at a high spot then it will probably have as much speed and force as the heavy sphere starting at the medium spot.
	Medium, high	I would start the heavy sphere at medium to see if, even though the sphere is heavy, will it make a difference? I started the light one at high so it would pick up speed and knock the ball far.

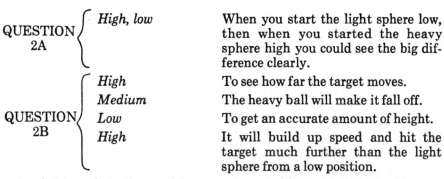

QUESTION 2A	*High, low*	When you start the light sphere low, then when you started the heavy sphere high you could see the big difference clearly.
QUESTION 2B	*High*	To see how far the target moves.
	Medium	The heavy ball will make it fall off.
	Low	To get an accurate amount of height.
	High	It will build up speed and hit the target much further than the light sphere from a low position.

1 point (equal starting positions accompanied by an explanation that ascribes importance, value, or usefulness to this equality)—Examples are:

QUESTION 2A	*High, high*	Start them at the same place and give them the same speed, then measure how far the target goes up the other side.
	High, high	If you want to know the difference the weight of the sphere makes, you should start them in the same place.
	High, high	I would start them at the same place, because this would tell me which one is lighter and which is heavier. The one that hits with a greater force is heavier.
QUESTION 2B	*Low*	It would be better because we could tell how much of a difference it makes.
	Low	So they'll be equal.

2 points (equal starting positions accompanied by an explanation stating the necessity of choosing this equality and possibly mentioning the principle that only one condition should be varied at one time)—Examples are:

QUESTION 2A	*Any equal*	The main reason of this experiment is the weight difference, so you would have to keep all other factors the same.
	High, high	I'd start them both high because that would provide more of a margin of difference to study. They'd have to start at the same point to make the experiment constant because you should have only one variable.
	Medium, medium	You'd have to start both spheres at the same position, otherwise, one would be at a disadvantage. One might gain more speed.

QUESTION
2B
{
Low   Low because a good experiment has only two variables, in this case, weight of sphere and how far target goes.

Low   Because you have to keep all things other than mass constant, as explained above.
}

Question 3:

0 points ("yes" accompanied by any explanation, or "no" accompanied by an explanation that does not refer to either the difference in starting positions or target weights)—Examples are:

Yes   I figure if the target went farther than the metal *A* ball must be better.

Yes   Because it is farther up and its target is lighter than *B*'s.

Yes   Because metal *A* isn't going to hit in the same place every time.

Yes   Because *A* is more powerful and heavier.

No   Because both of them weigh the same.

No   It wasn't a controlled experiment because you really couldn't tell from the information given.

1 point ("no" accompanied by an explanation referring to the starting positions and/or the target weights)—Examples are:

No   This is not a good proof because one target is light and one is heavy. Metals *A* and *B* are set at different points, which would make a difference. The target should be the same weight and the metals *A* and *B* should be let go at the same point.

No   Metal *B* was released in the medium position which doesn't allow it as much room to speed up as metal *A*, so Metal *B* doesn't have as much force on the ball so it doesn't go as far.

No   It doesn't prove it because the targets are different.

No   Metal *A* was at a higher position and hit a lighter target.

CONSISTENCY OF RESULTS AND COMPOSITE SCORES ON THE CONTROL OF VARIABLES TASK

We have represented the correlation of the scores on the three questions in Table 5, which consists of two parts. To the left is the correlation of the response levels to Questions 2A and 2B for subjects who scored zero on Item three, and to the right is the same correlation for subjects who scored one point on Item three. The entries are average percentages like those in Table 2. On the average, one-third of a population scored zero on Item three, and almost all of these, 26 of 33 percent, also scored zero on Items 2A and 2B.

Two-thirds of the subjects scored one point on Item three. One-third of these, or 22 percent of the entire sample, scored zero on both parts of Item two, which was clearly much more difficult than Item three. Of those students who did not score zero on Item two, about half scored differently—usually better on 2B than on 2A. Boys and girls performed substantially the same.

Even though there is a good correlation on the two parts of Item two (73 percent of the students have the same score), this is brought about by the great difficulty of the item, which has the result that almost half of all the students score zero on both parts. We conclude that the correlation is deceptive, and that control of variables on Item 2A represents a greater accomplishment than control of variables on Item 2B. Kendall's coefficient of concordance (19) for the three questions has the value 0.49, an adequate indication of reliability in view of the fact that the questions had only two or three response categories.

To create a composite score for the Control of Variables task, we have added the number of points each student received on his three explanations. This procedure implies that a more advanced answer on one part can compensate for a less advanced response on another. Even though that is not strictly true since the parts differ in difficulty, the total number of points does give a good indication of relative performance, since each total score is achieved primarily through one particular point combination on the parts. The distribution of the total number of points is given in Table 6, where we have also indicated the most frequent point combinations on Items 2A, 2B, and 3 for achieving each score. Thus, most students whose composite score was three had one point on each of the three items. Note the small percentage of students having a composite score of four points. We believe it includes primarily subjects who were capable of scoring five but failed to do so out of carelessness or misunderstanding.

Table 5. Correlation of Scores on the Three Items of the Control of Variables Task (Percent)

		Item 2A		
		0	1	2
Item 2B	0	26	0	0
	1	3	1	0
	2	2	0	1

Score 0 on Item 3

		Item 2A		
		0	1	2
Item 2B	0	22	1	0
	1	10	9	2
	2	5	3	14

Score 1 on Item 3

Table 6. Distribution of Composite Scores on the Control of Variables Task (Percent)

Composite score	0	1	2	3	4	5
Percent	26	25	14	14	6	14
Most frequent pattern of scores on #2A,2B,3	0,0,0	0,0,1	0,1,1	1,1,1	—	2,2,1

RESULTS FOR DENMARK

All classes in Danish schools are coeducational, and there is no ability grouping in the first seven grades. Beginning with the eighth grade, however, there is a division into two tracks, one that will finish school after the tenth grade and another that enters gymnasium and ultimately completes 12 years of school. Our subjects came from Frederiksberg, a city contiguous with Copenhagen, and three suburban communities within commuting distance of Copenhagen. We met with 17 classes of seventh graders and two classes of eighth graders, one from each of the two tracks, in early May 1974. The teachers in these classes

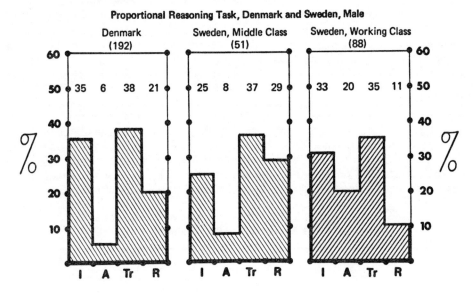

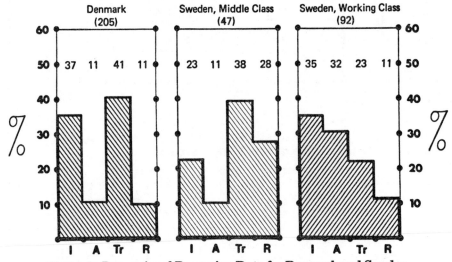

Figure 2. Proportional Reasoning Data for Denmark and Sweden.

specialize in subject areas, and the set of teachers usually remains with one class for several years. Even though there were substantial fluctuations in student performance on the two tasks from class to class, the results for any community or for any one grade were not appreciably different from those for any other. We are therefore reporting only one set of data for Denmark. The frequencies of the various composite categories of responses to the Proportional Reasoning task are given in Figure 2, and those for the Control of Variables task are given in Figure 3. The significance of the shading will be explained later with the help of Table 7.

The distributions on both tasks are similar for boys and for girls. In the Proportional Reasoning task, somewhat more than one-third were *intuitive* and about 40 percent were *transitional.* In the *ratio* category, boys were substantially more numerous, with the difference statistically significant (p < 0.15). In the Control of Variables task, the differences between boys and girls were small, statistically insignificant. About 15 percent scored in the two top categories (four or five points), and more than half scored in the two lowest categories (one or no points).

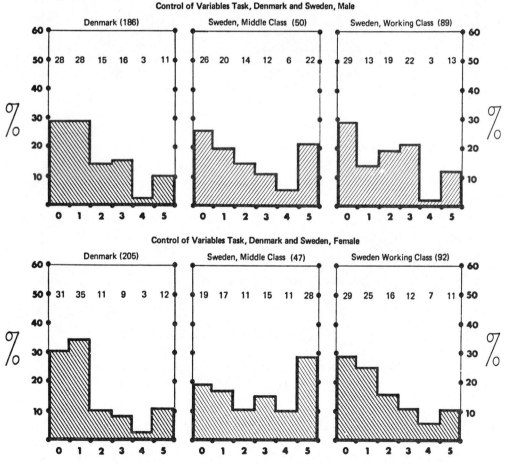

Figure 3. Control of Variables Data for Denmark and Sweden

To display the variation of student reasoning from class to class, we present in Figure 4 a scattergram showing all 19 Danish classes we visited. The abscissa on the graph indicates the percentage of students who answered the Proportional Reasoning task in categories *Tr* or *R*, while the ordinate indicates the percentage of students who scored three or more points on the Control of Variables task. In our judgment, these percentages represent the fraction of students who are at least transitional to formal thought.

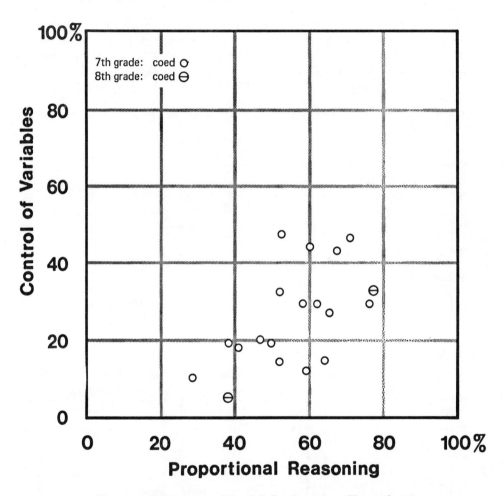

Figure 4. Percentage of Danish Students in a Class who Showed Transitional or Formal Reasoning on Each Task

Grouping students together as a class shows the actual teaching situation faced by the science or mathematics instructor of that class. There is quite a difference between the poorest-performing classes near the lower left on the graph, and the best-performing ones near the center. Actually, the low classes were in schools that also contributed classes with very high averages. The two eighth-grade classes clearly show the effect of tracking, though their average is like that of the seventh graders.

RESULTS FOR SWEDEN

In the coeducational Swedish school, instruction in the seventh and eighth grades is carried out by specialists. The students have several hours of very traditional science instruction in physics, chemistry, and biology each week. Our school visits in Sweden were carried out in Gothenburg and its surroundings during late May, 1974. Gothenburg is an industrial city, with large ship-building, port, and automobile-manufacturing facilities. The labor shortage in Sweden has led to a substantial immigration of Finnish workers whose children, speaking a different language, have some difficulties adjusting to the local schools.

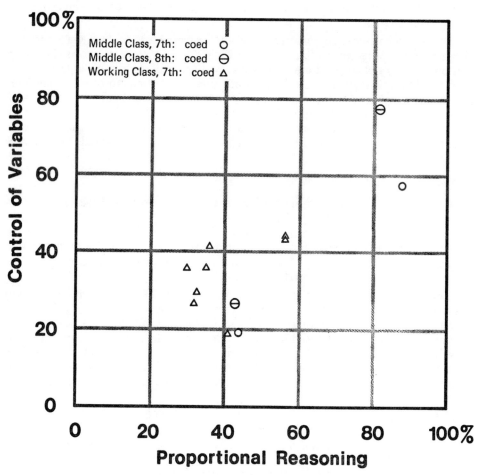

Figure 5. Percentage of Swedish Students in a Class who Showed Transitional or Formal Reasoning on Each Task

Our student sample was taken from *1.* two working-class neighborhoods consisting of fairly large modern apartment houses and *2.* a suburban community with large and small apartment houses. We have designated these as working-class and middle-class populations, respectively. The working-class children were all seventh graders, while the middle-class children were half seventh graders and half eighth graders. Because of the small size of the middle-class sample and the virtually identical scores for the two grades, we combined the data for the two grades and are reporting on a population of greater average age (Table 1).

The frequencies of the various categories of responses on the Proportional Reasoning and Control of Variables tasks are given in Figures 2 and 3 beside the Danish results. In proportional reasoning, almost 30 percent of middle-class students scored in the *ratio* category, but only 10 percent of the working-class students did so, a statistically significant difference ($p < 0.01$). On the Control of Variables task, the differences also favored the middle-class group, but were smaller. The differences between boys and girls of the same socioeconomic group were slight.

To display the variation of student reasoning from class to class, we present in Figure 5 a scattergram similar to that in Figure 4. This diagram shows the group differences very clearly. There were two very successful middle-class groups, one seventh and one eighth grade, which had been pointed out to us as unusual on the basis of test scores and teacher opinion. The other middle-class seventh- and eighth-grade groups, and all eight seventh-grade working-class groups, cluster near one another in the lower portion of the graph. The effect of the increased age and school experience of the eighth graders is seen to be small, as it was in Denmark.

RESULTS FOR ITALY (20)

Italian eighth graders are in the last year of the *media,* a three-year middle school. The *media* is a comprehensive school that may be coeducational or not. The various subjects are taught by academic specialists who usually remain with a group of students for the entire three years. After the *media,* some children discontinue school while others attend secondary schools for technical or vocational studies or university preparation.

The schools participating in our study in February 1974 were located in Rome and were chosen to sample children from various socioeconomic levels. Boys and girls of the working class attended the same schools, but studied in classroom groups that were segregated by sex. Some middle-class children attended coeducational classes while others were segregated. The upper-middle-class children came from a coeducational school and from two specialized schools, a private university preparatory school for boys and a school for girls featuring dance and dramatic arts.

The frequencies of the various response categories on the Proportional Reasoning and Control of Variables tasks are given in Figures 6 and 7. There are many differences among the groups, and we shall point out the most noteworthy of these.

On the first task, the boys at all socioeconomic levels showed more proportional reasoning than the girls. The fractions of *transitional* and *ratio* responses of upper-middle-class, middle-class, and working-class boys were 82 percent, 62 percent, and 55 percent, respectively, while the same fractions for the girls were 62 percent, 37 percent, and 36 percent. It should be added, however, that the differences were much smaller than this in the coeducational classes that are part of our sample. The same figures reveal a substantial effect of socioeconomic level on performance. Girls responded much more frequently than boys with an

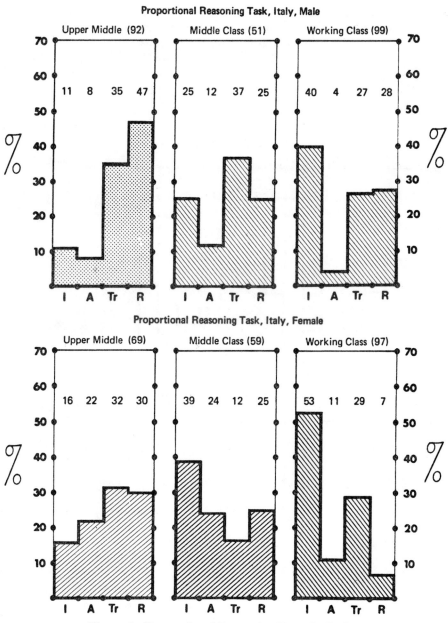

Figure 6. Proportional Reasoning Data for Italy

additive response; for instance, 17 percent of all Italian girls gave *additive* answers to both questions on the Proportional Reasoning task, and another 26 percent gave at least one *additive* answer. Of the boys, only five percent gave both *additive* answers and six percent gave one. Boys were slightly higher than girls on *intuitive* responses (20 percent to 14 percent, statistically significant only at the five-percent level).

On the Control of Variables task, the differences between boys and girls were much smaller and are not significant statistically. The differences between the middle- and upper-middle-class groups were also quite small, but the working-class children found the task very difficult and more than 70 percent scored one point or less. The other groups found the task difficult also, but not quite so difficult, with 40 to 55 percent scoring in the two lowest categories.

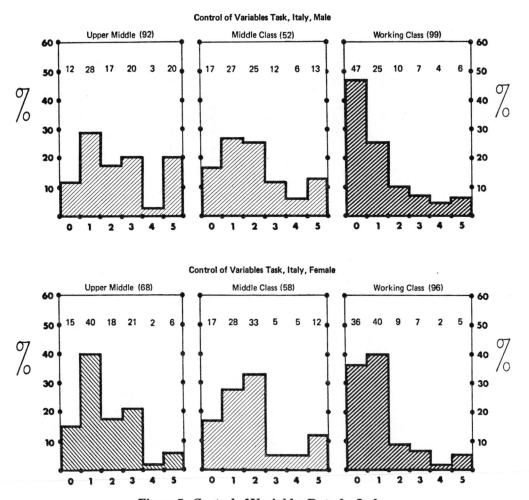

Figure 7. Control of Variables Data for Italy

A scattergram of the class-to-class variation of the Italian students is presented in Figure 8. The overall pattern of points is below the diagonal, again indicating that the Control of Variables task was more difficult than Proportional Reasoning task for the Italian children. The socioeconomic influence is visible, but there is a good deal of overlap among the three levels. One cluster of six points to the upper right of the diagram includes the three coeducational upper-middle-class groups, the two groups of upper-middle-class boys, and one group of working-class boys taught by an outstanding mathematics teacher in a school known for its innovative practices. At the other extreme is a cluster of three points, two classes of girls and one of boys from working-class neighborhoods. Near the middle of the diagram you can find all the other classes, some from each socioeconomic level. Here also are two points representing two upper-middle-class coeducational pilot groups of sixth graders (first *media*) attending the same school as the eighth graders. There is a substantial difference between the two ages.

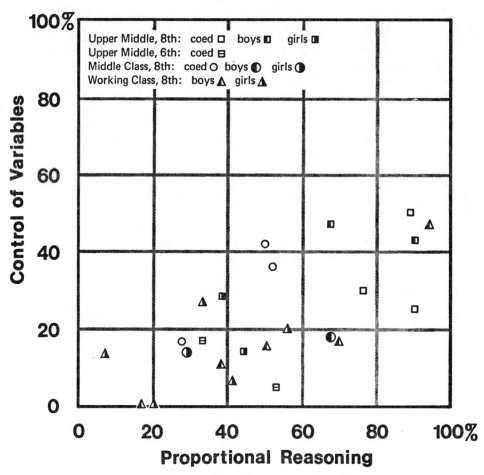

Figure 8. Percentage of Italian Students in a Class who
Showed Transitional or Formal Reasoning on Each Task

RESULTS FOR THE UNITED STATES

The subjects in the United States came from nine different cities and towns in the northeastern and north-central part of the country. Their socioeconomic levels were determined from the character of the community in which they lived. We identified three levels; upper-middle-class suburban, middle-class urban and suburban, and low-income urban neighborhoods. All schools were coeducational, some covered grades seven through eight, some seven through nine, some one through eight, and some seven through twelve. Many practiced a form of tracking by academic achievement or foreign language preference. By selecting classes from several ability levels, we hoped to reduce the sample biases associated with tracking. The teachers were specialists who taught only one subject, but there was no continuity of class groupings or teachers faced by a particular student from year to year. Our investigation occurred between October 1973 and January 1974.

The frequencies of the various response categories on the Proportional Reasoning and Control of Variables tasks are given in Figures 9 and 10. As in Sweden and Italy, formal thought as indicated by the two tasks is statistically related to socioeconomic status. About 40 percent of middle- and upper-middle-class students responded in the *transitional* or *ratio* categories of the Proportional Reasoning task, while fewer than 10 percent of the urban low-income youngsters did so. Boys and girls scored very similarly, with the girls giving more answers in the *additive* category at all three socioeconomic levels, while the boys responded more frequently in the *intuitive* and *transitional* categories. These small differences were statistically significant only for the very large middle-class group.

On the Control of Variables task, the urban low-income boys scored significantly better than the girls (p <.01): 46 percent of the boys earned one or more point, but only 19 percent of the girls did so. Differences between boys and girls in the other socioeconomic levels were very slight. Somewhat more than 50 percent of the upper-middle-class children scored three or above, compared to about 35 percent for the middle-class children.

The variation in explanations among classroom groups is shown in Figure 11, a scattergram similar to Figure 4. One can see that the points are distributed generally above the diagonal, indicating that the Proportional Reasoning task was the more difficult one for the students in the United States. The clustering of urban low-income classes in the lower left corner of the diagram reveals once again that these students had great difficulty in formulating successful explanations; many left the answer spaces on their papers blank and did not give oral statements of their ideas to the team of investigators. Located in the diagram are also two middle-class groups of children with diagnosed learning difficulties.

Toward the upper part of the diagram are groups from the upper-middle-class suburban communities, but they do not fall further to the right (proportional reasoning) than many middle-class groups. In only two classes, both accelerated, did substantially more than half the students show proportional reasoning in the *transitional* or *ratio* categories.

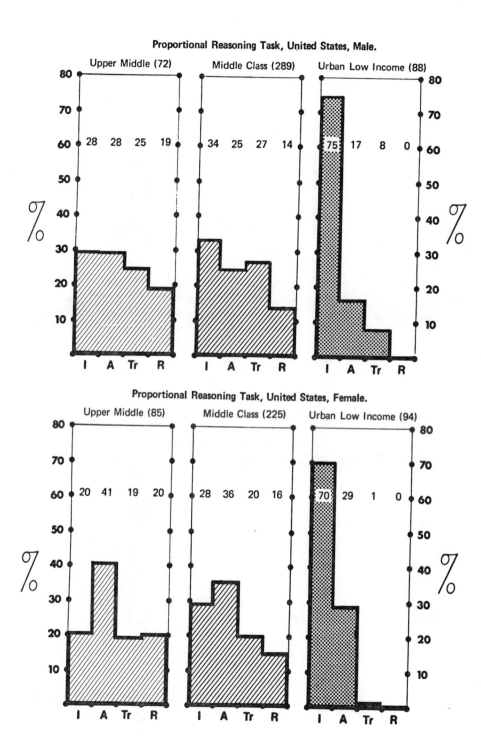

Figure 9. Proportional Reasoning Data for the United States

R. Karplus, et al.

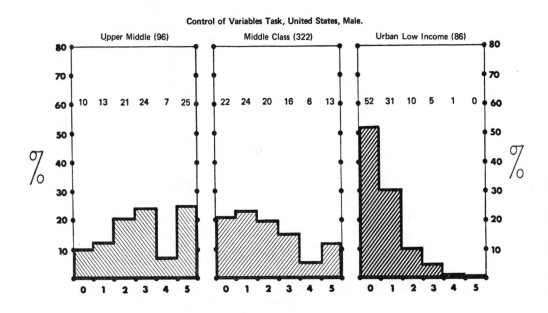

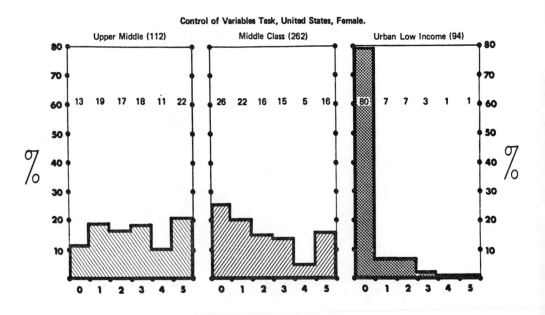

Figure 10. Control of Variables Data for the United States

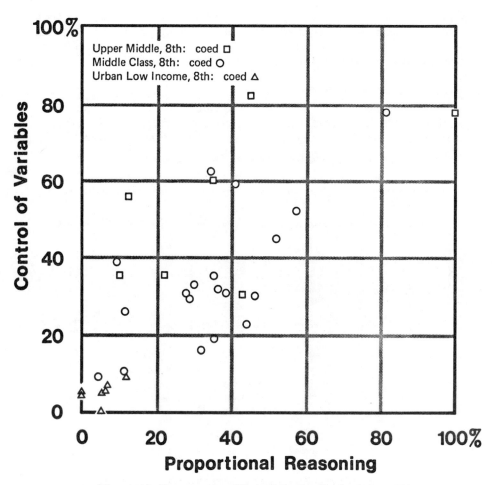

Figure 11. Percentage of United States Students in a Class who Showed Transitional or Formal Reasoning on Each Task

RESULTS FOR AUSTRIA

Austrian students attend secondary school beginning with the fifth grade. Two school types are available, the *Gymnasium* offering an eight-year program leading to university entrance, and the *Hauptschule* offering a four-year program that may lead to a vocational school, a technical school, or apprenticeship and employment. The *Hauptschule* itself is divided into two academic tracks designated as Track *A* (more able students) and Track *B* (less able students). Many *Gymnasiums* are presently for boys or girls only, but there is a trend toward coeducation. The *Hauptschule* is basically a coeducational, departmentalized middle school.

In the city of Vienna both types of schools are easily accessible to all students, with the result that parents can send their fifth graders who meet the academic requirements to any *Gymnasium* in the city. About half the students in any age group in Vienna attend *Gymnasium,* about 30 percent are in Track *A,* and about

20 percent are in Track *B*. In rural areas there are fewer *Gymnasiums;* therefore a larger percentage of students attends *Hauptschule* outside Vienna. All teachers are academic specialists. Even though the two school types are administratively separated, there is an effort to coordinate the curricula so that students from Track *A* of the *Hauptschule* can transfer to the *Gymnasium* after a year or two of special effort.

The frequencies of the various response categories on the Proportional Reasoning and Control of Variables tasks are given in Figures 12 and 13. On the Proportional Reasoning task the *Gymnasium* students performed outstandingly,

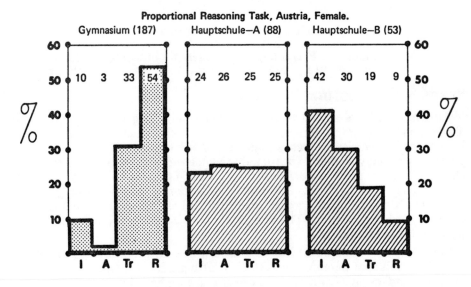

Figure 12. Proportional Reasoning Data for Austria

with more than 80 percent of boys and girls showing *transitional* or *proportional* reasoning. Sex made very little difference here, even though most students tested were in segregated schools. A frequent form of the explanation for Mr. Tall's height and the car's width appealed to the "rule of three," a procedure for solving problems of proportions that is taught in the Austrian sixth grade. Most of the students who used this method applied it correctly. Another frequently used procedure took advantage of the rulers which virtually all students brought with them. After measuring Mr. Short's height and the paper clip length in

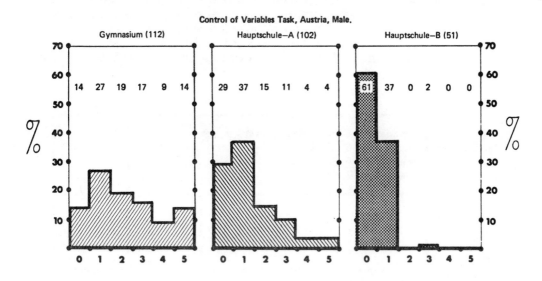

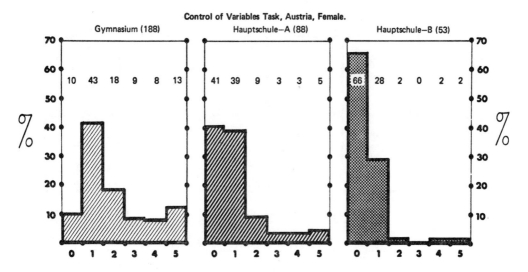

Figure 13. Control of Variables Data for Austria

centimeters, these students calculated the size of a button and the height of Mr. Tall in centimeters, and finally his height in paper clips. We scored such an answer in the *transitional* category since it seemed to us similar to the other concrete procedures for dividing up Mr. Short.

It is interesting to see from the data that the Track *A* (*Hauptschule*) boys scored almost as well as the *Gymnasium* students, but that the Track *A* girls, with only 50 percent *transitional* or *ratio,* scored very substantially lower (p < .001). The Track *B* boys' responses very closely followed the pattern of responses of the Track *A* girls, while fewer than 30 percent of the Track *B* girls scored in Categories *Tr* or *R.* These substantial differences between boys and girls in the same tracks and in the same classrooms are quite surprising. They suggest that the criteria according to which the more gifted students attend the *Gymnasium* may operate very differently for boys and girls, at least insofar as proportional reasoning is concerned.

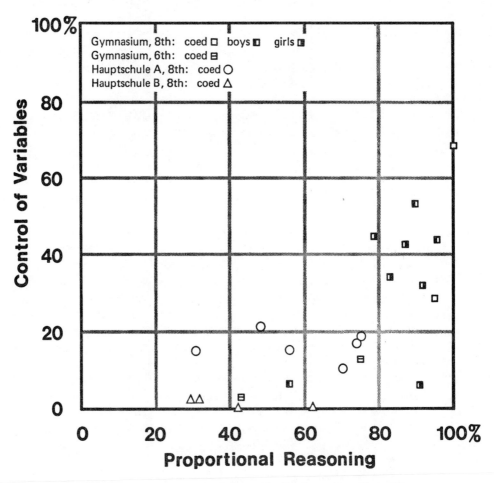

Figure 14. Percentage of Austrian Students in a Class Who
Showed Transitional or Formal Reasoning on Each Task

On the Control of Variables task, the performance of boys and girls in the same courses was similar, with the largest difference in favor of the boys occurring for the *Gymnasium* students ($p < 0.05$) and the smallest for the Track *B* students. The Track *B* students had great difficulty—fewer than five percent scored more than one point. The differences among academic levels were significant ($p < 0.005$) except that between the Track *A* and Track *B* girls. The students' questions during the administration of this task indicated that the wording was difficult for many to understand. Especially Item 2B, which had elsewhere been answered more easily than 2A, was left to the end by many students who asked for an additional oral explanation and encouragement before proceeding.

The scattergram in Figure 14 shows the variation in student reasoning from class to class. The clustering of all points below the diagonal of the graph reflects the children's difficulty with the Control of Variables task. The many *Gymnasium* points along the right side lend emphasis to the children's outstanding success on the Proportional Reasoning task. The differences between the three tracks can again be seen clearly in the grouping of the points in the diagram. The boy/girl differences are concealed, however, and tend to reduce the contrast between Tracks *A* and *B* of *Hauptschule*.

Note that there is one *Gymnasium* class that falls in the group of Track *A* classes, near the lower middle of the graph. We were not able to identify any special circumstance that could explain this observation. Note also the two points of sixth grade *Gymnasium* classes (black circles) that were tested as pilot groups. Their performance was substantially weaker than that of their older schoolmates, but close to that of the two-year-older *Hauptschule* students. They clearly have not yet mastered the Proportional Reasoning task.

RESULTS FOR GERMANY

The German school system presently offers one coeducational, comprehensive neighborhood elementary school for all children in the first four grades. The secondary schooling beginning with the fifth grade consists of three alternative types, with selection for the more challenging courses being based on grades and teacher recommendations. (An optional entrance examination is offered for children whose parents wish to contest the assignment of their child). Most demanding is the eight-year *Gymnasium,* which prepares students for university entrance and enrolls about 20 percent of an age group. Next is the *Realschule,* which offers a six-year course leading to tertiary education in a technical school or specialized academy, and which enrolls about 25 percent of an age group. Least challenging is the *Hauptschule,* which offers a five-year course to satisfy the requirement for compulsory schooling for the remaining 55 percent of an age group; most graduates become apprenticed in a trade or seek unskilled employment.

The schools we visited in April 1974 were located in the university town of Gottingen, in north-central Germany. The *Gymnasium* and *Hauptschule* were coeducational, but the *Realschule* was in transition, with coeducational seventh grades but segregated ninth grades. (As we mentioned earlier, there were no regular eighth-grade classes in 1973-74). Our sample populations were taken from one of the two *Gymnasiums* in the city, from two *Realschulen* (one for boys and one for girls from the same attendance area), and two *Hauptschulen* (one near the city center, one in a recently built suburb occupied by workers in the expanding industrial section of the city).

The nonexistence of regular eighth-grade classes was of serious concern to us. You can see from Table 1 that the average age of the German students was

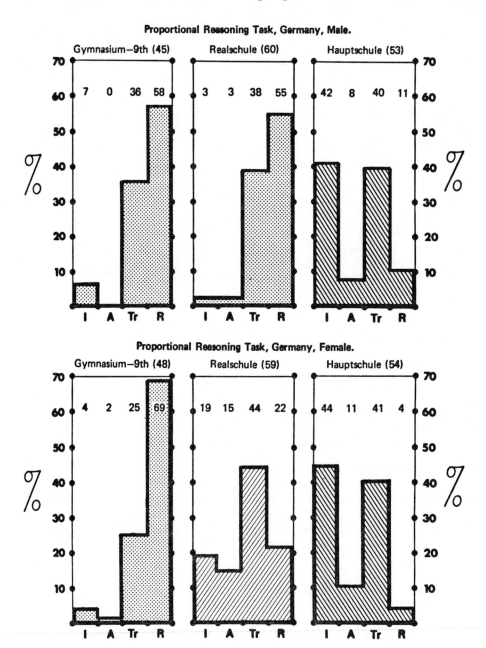

Figure 15. Proportional Reasoning Data for Germany

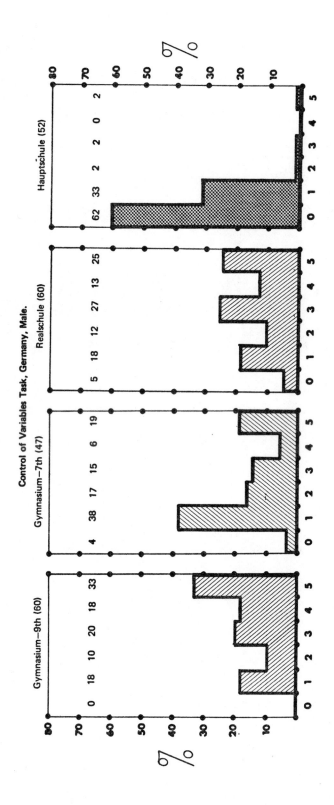

Figure 16. Control of Variables Data for Germany

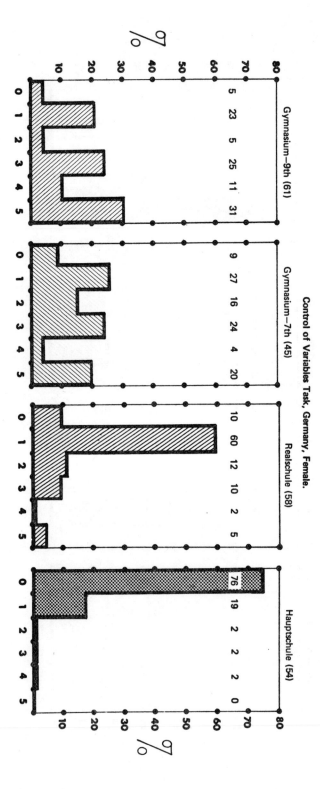

Figure 16. Control of Variables Data for Germany

about a year greater than that of the Scandinavian, Italian, and American students, and about a half year greater than that of the British and Austrian students. To check the effect of age, we compared the scores of German ninth graders under 15 years of age with those of their classmates over 15, and found virtually no difference between them. Since the ninth graders have only had three more months of school than have eighth graders in other countries, we believe that our German principal group is adequate for the purposes of our study. Nevertheless, we sought to test a substantial pilot group of seventh graders to have some results for younger students. In the time available, we were able to administer both tasks to one coeducational seventh-grade *Realschule* class and the Control of Variables task to three additional *Gymnasium* classes.

The frequencies of the various categories of responses on the Proportional Reasoning and Control of Variables tasks are given in Figures 15 and 16. On the Proportional Reasoning task the *Gymnasium* students scored very well, with

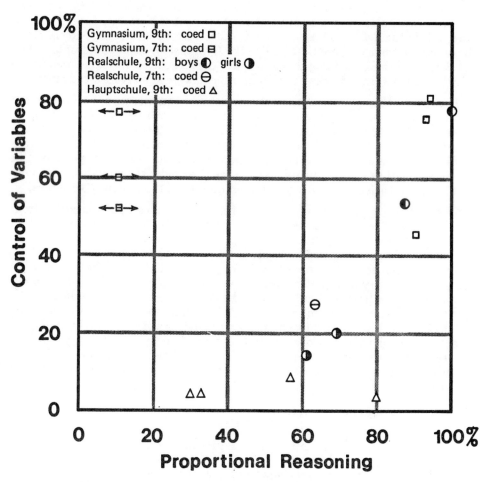

Figure 17. Percentage of German Students in Each Class who
Showed Transitional or Formal Reasoning on Each Task

more than 90 percent giving *transitional* or *ratio* explanations. Boys and girls performed indistinguishably.

The *Hauptschule* boys and girls, who attended coeducational classes as well, also performed similarly, though fewer than half used transitional or proportional reasoning. In the *Realschule,* however, we found very substantial and significant differences between the replies of boys and girls. The boys' performance was very close to that of *Gymnasium* students, while the girls' transitional and proportional answers were only 66 rather than 90 percent of the total, or at a level intermediate between that of the *Gymnasium* and *Hauptschule* students. The "rule-of-three" and rulers were much used here as in Austria.

On the Control of Variables task the results generally followed the same pattern. *Gymnasium* and *Hauptschule* students showed no significant sex differences, with about 70 percent of the *Gymnasium* boys and girls receiving a score of three points or more, while only four percent of the *Hauptschule* boys and girls gave explanations at this level. In other words, the *Hauptschule* students were almost at a total loss on the Control of Variables task. The boys and girls in the *Realschule* again scored rather differently, with 65 percent of the boys scoring three or more (almost like the *Gymnasium* students), but less than 20 percent of the girls doing so. Here again the girls, from the same area in the city as the boys but attending a different school, were at a severe disadvantage.

Because of the age discrepance of our German sample, we are reporting the score distribution of the seventh-grade students in the *Gymnasium* whom we were able to test on the Control of Variables task. These boys and girls, approximately one-and-a-half years younger than the ninth graders, scored very similarly to one another and about 45 percent earned three points or more. In other words, they did not do quite as well as their older schoolmates, but many of them nevertheless showed a good understanding of the control of variables.

The scattergram in Figure 17 presents the variation in student reasoning from class to class. The three classes for which we have only Control of Variables data are marked near the left axis; from the performance of similar students on the Proportional Reasoning task, we would expect them to have scored above 70 percent on that test. The trend here is for points to lie substantially below the diagonal. The very high performance of both seventh- and ninth-grade *Gymnasium* classes can be seen clearly, as can the very low performance of the *Hauptschule* classes on the Control of Variables task. The two classes of girls in the *Realschule* are represented by the two lower circles, while the two boys' classes are represented by the two high circles, very close to the three *Gymnasium* classes represented by squares. A seventh-grade coeducational *Realschule* class is also marked, and showed a somewhat better result than the ninth-grade girls' classes.

RESULTS FOR GREAT BRITAIN

The British secondary school organization is quite complicated, depends somewhat on the community, and is gradually changing. We therefore proceeded somewhat arbitrarily, limiting our study to schools in London, and selecting sample populations from comprehensive schools, grammar schools (publicly administered), and direct grant schools (privately administered). The two latter types are highly selective; students in the top 10 or 15 percent of their

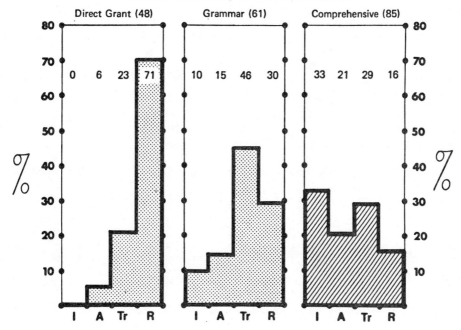

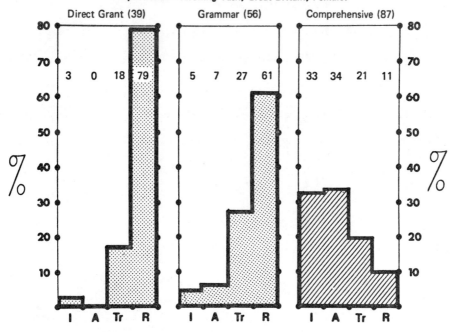

Figure 18. Proportional Reasoning Data for Great Britain

age group are determined by an entrance examination at age 11. The students we tested were completing their third year of secondary school, and averaged a little over 14 years of age. The schools were located in various sections of the city, with a substantial minority of children of recent immigrants to London in the comprehensive schools. Very few students had difficulty with the language. There is academic tracking called "streaming" of students in the grammar and comprehensive schools, but we do not know just how that affected our sample.

The frequencies of the various response categories on the Proportional Reasoning and Control or Variables tasks are presented in Figures 18 and 19. The effect of student selection on both tasks is immediately evident, with the direct grant school students exhibiting almost completely transitional or formal thought on both tasks. About 80 percent of grammar school students gave *transitional* or *ratio* explanations on the Proportional Reasoning task, and somewhat

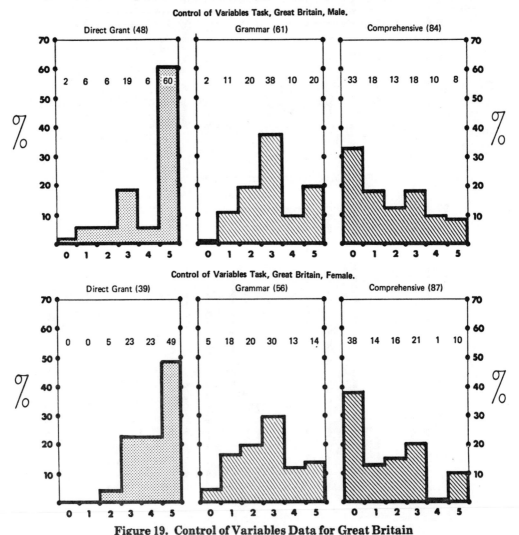

Figure 19. Control of Variables Data for Great Britain

over 60 percent scored three or more on Control of Variables. The comprehensive school students gave explanations that were substantially less advanced, with about 40 percent *transitional* or *ratio* on proportional reasoning, and about 35 percent scoring three or more on the Control of Variables task. The differences between the three groups are statistically significant (p < 0.01). Within each group, however, the only significant boy/girl difference was observed for proportional reasoning in the grammar school, with the girls providing more *ratio* explanations and the boys more *transitional* (p < 0.01). Since we sampled only one boys' and one girls' grammar school, this result must not be generalized.

The enormous range of reasoning from class to class is illustrated in the scattergram in Figure 20 where the points lie near the diagonal but are spread from one end to the other. Thus, the two tasks were of comparable difficulty. The three school types are neatly separated, except for one outstanding comprehensive school class that ranks with the direct grant schools in this respect. We suspect that tracking is responsible, since the other classes in the same school did not give such a performance.

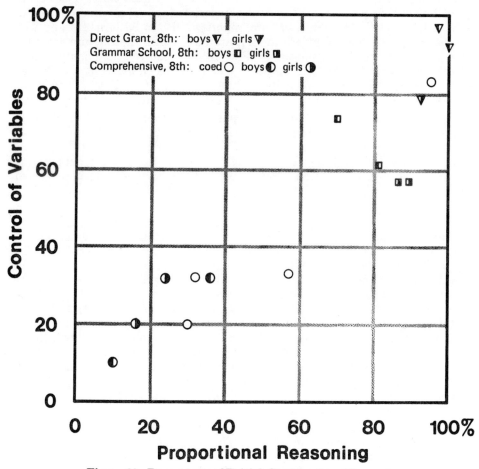

Figure 20. Percentage of British Students in a Class who Showed Transitional or Formal Reasoning on Each Task

COMPARISONS OF POPULATIONS

Because of the complexity of the data, it is difficult to get a comparative overview of the 38 populations (19 male and 19 female) for which we have reported results in the preceding sections. To help in this effort, we have looked for patterns in the frequency distributions in Figures 2 to 20 and have grouped the populations into six sets with respect to each of the two tasks. The patterns identifying each set are listed in Table 7. We have numbered the patterns in order of decreasing evidence of formal thought, or increasing evidence of intuitive thought. Thus Pattern I in each case includes the highest percentages of formal thought, Pattern II somewhat less, and so on. You will notice that certain patterns for the Proportional Reasoning task have been given the same number because they represent closely equal average levels of reasoning and differ primarily in the frequency of occurrence of *additive* responses. Table 7 also indicates how we have distinguished among the various patterns by using differing designs in the graphs of the earlier figures.

We should like to point out a few features of the patterns we have selected. You will note that responses in the *ratio* category to the Proportional Reasoning task are the mode in Pattern I, but are important also in Patterns IIA and IIB. *Intuitive* responses are prominent in all patterns except Pattern I, and make up more than half of the responses in Pattern IV. *Additive* explanations are substantial only in Patterns IIB and IIIB. The patterns on the Control of Variables task show a more gradual change from formal thought in Pattern I to *intuitive* thought in Pattern VI, as the modal score decreases from five to zero points.

The assignment of populations to the patterns is given in Table 8. We have entered the letter *M* in a column to represent the performance of the male population listed at the left, and have entered *F* for the female population. The populations themselves are grouped according to whether the educational systems are comprehensive or selective, and then by socioeconomic criteria or school type.

The overview afforded by Table 8 reveals several interesting results. Most striking is the uniformly high degree of proportional reasoning manifested by the top academic groups in the selective school systems. This achievement extends to the boys but not to the girls in the academic middle groups. The populations in comprehensive school systems, as might be expected, tend to fall into the intermediate patterns with respect to proportional reasoning. Only the Italian upper-middle-class prep school boys—actually a selected group—excel in formal thought on this task, while the urban low-income groups in the United States appear to lack formal thought completely. The United States populations, whose best achievement is Pattern IIB, do not compare favorably with the approximately equivalent European groups, especially European boys. Many girls' preference for *additive* responses, already mentioned in connection with the results for several countries, can be seen again here, in that the distributions of nine girls' populations fell into the high *additive* Patterns IIB and IIIB, while only five boys' populations showed these distributions.

On the Control of Variables task, the differences in selectivity in the participating countries become more evident. The most selective group, the British direct grant population, is clearly far superior to any other population. The relatively less selective Austrian *Gymnasium* is at a disadvantage compared to the German *Gymnasium* and the British grammar schools, with the eighth-grade students achieving at the same level as German seventh graders who were about a year younger. Still, the Austrian third quartile (*Hauptschule A*) equalled the British comprehensive school students who make up 80 percent of the population; the Austrian fourth quartile (*Hauptschule B*), which could virtually not cope with the Control of Variables task at all, shared this difficulty with the German *Hauptschule* (third and fourth quartiles).

The students in comprehensive school systems succeeded rather better on this task than they had on proportional reasoning. The upper-middle-class children in the United States achieved Pattern II, equal to or better than the selected top track students with the exception of the direct grant group. The Italian and Scandinavian students equalled the top or middle track students in Germany,

Table 7. Frequency Distribution Patterns

Patterns on Proportional Reasoning Task	
I	Predominantly Transitional and Ratio
IIA	Predominantly Intuitive, Transitional, and Ratio, very few Additive [a]
IIB	Even distribution in all categories [a]
IIIA	Predominantly Intuitive and Transitional, few Additive and Ratio [b]
IIIB	Predominantly Intuitive, Additive, and Transitional, few Ratio [b]
IV	Predominantly Intuitive
Patterns on Control of Variables Task	
I	Almost exclusively 3-5 points
II	Fairly even distribution with an excess of scores 3-5
III	Fairly even distribution with an excess of scores 0-2
IV	Modal score 0 or 1, with some scores of 2-5 points
V	Predominantly 0 and 1 points
VI	Predominantly 0 points

a: Categories IIA and IIB are closely equal in their "average" level of reasoning, but the distributions are strikingly different.
b: Categories IIIA and IIIB are closely equal in their "average" level of reasoning, but the distributions are strikingly different.

Great Britain, and Austria, rather than equalling the lowest track students as they had on proportional reasoning. Even the boys in the urban low-income group in the United States achieved in Pattern V on Control of Variables, whereas they had been able to do virtually nothing with the Proportional Reasoning task.

A comparison of the students' relative performance on the two tasks is made most easily with the help of the seven scattergrams in Figures 4, 5, 8, 11, 14, 17, and 20. One can see that the United States is the only country for which the points fall generally above the diagonal, indicating that the Proportional Reasoning task was more difficult than the Control of Variables task as evaluated by our scoring procedure. For England and Sweden the points lie relatively close to the diagonal and give evidence that the tasks were of comparable difficulty both for higher and for lower achieving groups. For Denmark and Italy the points are distributed broadly below the diagonal, which means that the Control of Variables task was somewhat more difficult than the Proportional Reasoning task, but that the difference was not very great. In Germany and Austria, finally, the points tend to lie near the bottom and right edges of the diagrams; the implication is that controlling variables was very hard to understand for all but some of the most able students.

This last observation raises questions about the adequacy of the translation of the Control of Variables task into German. We have already explained that the language on the answer pages gave many children difficulty, so that they asked for supplementary oral clarification. The best way to resolve this concern would be to use slightly differing forms of phrasing the task and then to determine how greatly these differences affect the students' performance. As a matter of fact, trials of this kind did precede our designing of the questions in their original English forms. Unfortunately, our travel schedule did not permit us to carry out similar pilot studies in each of the countries visited.

In the absence of such an investigation, we have looked for clues within our present data. Consider first the German boys and girls attending the *Realschule*. Whatever language problems there might be, both groups were certainly affected in the same way. The substantial score difference observed with respect to both tasks (see Figures 15-17) can only be explained in terms of student selection by the schools, the teaching program, and sex-related cultural biases. Next, we ask whether any German-speaking groups performed without apparent language handicap on the Control of Variables task. The answer is affirmative, the ninth-grade German *Gymnasium* students and the ninth-grade *Realschule* boys scored according to Pattern II, as did the approximately comparable upper-middle-class students in the United States and the British grammar school students. Furthermore, the Austrian *Gymnasium* students and the German seventh graders scored in Pattern III like the middle-class students in the United States and in Italy, and somewhat better than the comprehensive school students in London. In other words, these German-speaking subjects performed quite adequately, though not up to their exceptional achievement on the Proportional Reasoning task. We therefore conclude that the translation did not introduce any substantial bias into the data, and that the very great difficulty experienced by the *Hauptschule* students in Austria and Germany is due to student selection, emphases of the teaching program, and possibly cultural biases.

DISCUSSION AND CONCLUSION

For the sake of clarity, we shall divide this section into four parts dealing, respectively, with proportional reasoning, control of variables, implications of both tasks insofar as formal thought is concerned, and the comments and opinions expressed by the subjects.

Proportional Reasoning

The great diversity of methods used by our subjects to find the height of Mr. Tall is similar to what we had observed previously in schools of the San Francisco Bay Area (7,8). The second question concerning the width of the car had not been used before. It is interesting to note, therefore, that about two-thirds of the students used the same method on both questions (Table 2). Most of the remainder combined methods in one of two ways: *1.* the *additive* procedure for Mr. Tall, followed by an *intuitive* method—often scaling, as in Examples I-9 and I-10—for the car; *2. ratio* for Mr. Tall, followed by a *transitional* approach for the car. An additional five percent of subjects used *ratio* for Mr. Tall followed by *intuitive* for the car. We believe that most students in this small though not negligible group had learned an algorithm for proportions that they could apply to the direct first question but could not invert to answer the second question. Lack of time, loss of interest, and possibly copying of the first answer may also have contributed to the *R-I* response pairs.

As we have pointed out before, the car question was more difficult than the question concerning Mr. Tall and had the intended effect of challenging the students to reveal how thoroughly they could apply proportional reasoning. Our procedure for constructing the composite score used in most of the figures and tables in this article, therefore, has made this assessment of proportional reasoning more stringent than it was in our earlier studies.

We were repeatedly struck by the similarity of individual responses, which occurred again and again and could be recognized clearly regardless of the language used. Even though the frequency distributions varied rather widely, we believe that all the population groups we sampled share a basically similar development of their patterns of reasoning. As a matter of fact, this conclusion is not really surprising, since all the countries we visited are culturally and technologically similar.

Most of the explanations in categories *A* and *R* were simple and straightforward, and could be identified easily. A few of the *additive* replies explicitly mentioned that paper clips and buttons were about the same size to justify the procedure (see Example *A*-2), but of course did not take into account the size difference indicated by the two measurements of Mr. Short. Thus they did not coordinate all the known data.

The explanations in categories *I* and *Tr* showed a great deal of variety, and we chose the illustrative examples to communicate this variety. Because of the contradictions they contain, the misconceptions in many replies could be interpreted only with difficulty (see Example *Tr*-1: Did the student add six to one-and-one-half instead of multiplying, or did he mean to add two times one-and-one-half for the two additional buttons?). In Category *I*, for instance, there are some guesses (Examples *I*-1 and *I*-7), operations with numbers regardless of their significance (Example *I*-4), introduction of extraneous data (Examples *I*-2 and *I*-8), and unexplained scale relationships (Examples *I*-3, *I*-5, *I*-9, *I*-10, *I*-11).

Finally, there were a few subjects who believed that the problems could not be solved without seeing Mr. Tall, the car, and/or the buttons (Example *I*-6).

One recurring pair of responses classified *A-I* concluded that the car was seven buttons wide because that was half the paper clip measurement of 14, very possibly using a division by two as the inverse operation to the addition of two. There were even some explanations in Categories *I-I* that stated "two times as big in paper clips" for Mr. Tall, but then added $2 + 6 = 8$ rather than multiplying $2 \times 6 = 12$.

Included in the *transitional* category is a substantial number of responses that determined the conversion from buttons to paper clips by using a ruler to measure Mr. Short and the paper clips, then expressing all dimensions in inches or centimeters. This procedure occurred primarily in Austria and Germany, where each pupil carried at all times a small supply kit containing pens, pencils, rulers, and erasers. Even though rulers being used were easily visible to neighboring students, this approach was not copied very often. Another group of responses we included in the *transitional* category started out with 7½ for the height of Mr. Short in paper clips (probably a counting error), and concluded that Mr. Tall was about 12 paper clips high because each button measured about two paper clips. Since we had previously found that use of a two-to-one ratio is not indicative of proportional reasoning at the level of formal thought (3), we felt that it would not be appropriate to include these responses in category *R*.

The rather large percentage of students replying in the *additive* category was originally one of the surprising results of the Proportional Reasoning task (7,8). Wollman suggested that this effect might be induced partially by the two-to-three ratio of heights, which acts to discourage comparison by dividing and draws particular attention to the excess of one dimension over the other (21). In Appendix *B* we describe a small study using the car question before the Mr. Tall question, and show that the latter question did appear to influence about 20 percent of the subjects. It was therefore interesting to observe that the composite *additive* responses occurred at a frequency of more than 20 percent in 14 of the 36 populations we investigated, in five of the seven countries. Only Germany and Denmark lacked such populations entirely, with the percentages there ranging from zero (German *Gymnasium*, boys) to 15 (German *Realschule*, girls).

All six United States populations had large frequencies of *additive* responses, ranging from 17 to 41 percent. The Austrian groups divided rather curiously: three populations had between three and five percent *additive* responses, while the other three gave between 26 and 30 percent. In Italy the upper-middle-class and middle-class girls gave more than 20 percent *additive* replies, while the corresponding boys' groups and the working-class students contributed low percentages. In Sweden and Great Britain, two groups of lower socioeconomic or ability levels gave more than 20 percent *additive* answers.

To what extent are the *additive* explanations developmentally intermediate between *intuitive* and more advanced procedures? Clearly the frequency distributions in Pattern I (see Table 7) reflect such advanced proportional reasoning that few *additive* answers would be expected. Conversely, frequency distributions in Pattern IV indicate so little development that one would also expect relatively few *additive* responses. Patterns II and III are therefore the im-

Table 8. Comparison of Populations on the Two Tasks

Population	Frequency Distribution Pattern											
	Proportional Reasoning						Control of Variables					
	I	IIA	IIB	IIIA	IIIB	IV	I	II	III	IV	V	VI
Comprehensive School Systems												
Upper Middle Class												
Italy	M		F				F		M			
United States			M,F				M			F		
Middle Class												
Sweden		M,F						M,F				
Italy		M		F				F	M			
United States		M			F				M,F			
Denmark					M,F				M,F	F		
Working Class & Low Income												
Sweden					M,F					M,F		
Italy				F							M	
United States						M,F					M,F	F

Table 8. Comparison of Populations on the Two Tasks (cont'd)

Population	Proportional Reasoning						Control of Variables					
	I	IIA	IIB	IIIA	IIIB	IV	I	II	III	IV	V	VI
Selective School Systems												
Top Groups												
Direct grant (Great Britain)	M,F						M,F					
Grammar (Great Britain)	M,F							M,F				
Gymnasium (9th, Germany)	M,F							M,F				
Gymnasium (7th, Germany)	M,F								M,F			
Gymnasium (Austria)	M,F								M,F			
Middle Groups												
Realschule (Germany)	M		F					M		F		
Hauptschule A (Austria)	M		F							M	F	
Comprehensive (Great Britain)					M,F					M,F		
Low Groups												
Hauptschule (Germany)				M,F								M,F
Hauptschule B (Austria)			M		F							M,F

portant ones to consider.

The present data strongly suggest that additive reasoning does not lie on an invariant developmental sequence but is strongly influenced by instruction and represents an effort by students to deal with a task in an ad hoc rather than systematic way. Thus, the German *Hauptschule* students, the Danish students, the Swedish middle-class students, and the Italian middle- and working-class boys all show frequency distribution Patterns IIA or IIIA (see Table 8), which have substantial *intuitive* and *transitional* contributions but lack *additive* ones. In fact, all of these groups have a larger percentage of individuals responding *additive-intuitive* to the two questions on the Proportional Reasoning task than respond *additive-additive*.

At the same time, the Swedish working-class students, the American upper-middle- and middle-class students, the Italian upper-middle- and middle-class girls, the Austrian *Hauptschule-B* students and *Hauptschule-A* girls, and the British comprehensive school students all had frequency distribution patterns IIB or IIIB, which included more than 15 percent of *additive* responses and a comparable percentage of *additive-intuitive* answer pairs. Most of these students attend different schools and therefore may experience different courses of study than the student groups in the same countries that gave *additive* responses. It appears that the educational programs in the United States, Austria, and Great Britain tend to encourage *additive* procedures while those in Germany and Denmark tend to discourage it on the part of the students who have not yet mastered proportional reasoning. In Italy and Sweden the results are mixed, depending on sex or socioeconomic level.

Control of Variables

Our scoring procedure for the three questions on this task was simple to apply but partially obscured the qualitative nature of the answers given by an individual. The application of the concept of a controlled experiment occurred most frequently on Question 3, somewhat less frequently on Question 2B, and least frequently on 2A.

Many responses to Question 2A receiving zero score dealt with the positions and actions of the two spheres independently (Examples 1 and 4), while others used the light sphere's height to compensate for the heavy sphere's weight (Examples 2 and 3). Zero score explanations to Question 2B tended to relate to the heavy sphere only (Examples 6 and 7) or communicated the need to maximize the difference in target distances (Example 9). All of these answers—many of them quite lengthy and carefully stated—suggest that the respondent was not conceptualizing the experimental comparisons that were described in the questions.

In other words, rather than considering the effect of the *weight* on target *displacement,* many subjects focused on the effect of the *sphere* on the target *position.* The former point of view addresses a relationship between two abstract variables (weight and displacement) which requires a formal operation; the latter concerns a relationship between two tangible variables (sphere and position) which requires only a concrete operation. In the concrete approach there is no need for two spheres, hence the two trials in Question 2A are seen as unrelated.

Scores of one point on Questions 2A and 2B were awarded for explanations that identified the equality of starting positions as important but not as necessary. Here also there may have been a semantic problem in that the subjects were asked to explain their ideas but were not urged, as would have happened during an interview, to give the strongest possible justification. These intermediate replies, therefore, reflect the subject's spontaneous view of how well-grounded an explanation was required.

The maximum score of two points was given for explanations that indicated the necessity of equal starting points for the two spheres. As indicated in Table 6, only about 20 percent of the students earned a total score of four or five by doing this consistently on all questions. And most of these merely asserted that equality was necessary (Examples 1 to 5), but did not refer to the ambiguities that would arise if the starting positions were unequal (Example 6). Some of them did, however, refer to the more general principle that only one variable at a time could be changed (see Examples 1, 2, 4, 5). Again, an interview would have allowed for probing questions after the first answer, while the group administration of the task was limited to the three standard questions printed on the answer page.

Question 3, though identified as an inadequate comparative experiment by about two-thirds of all students, was not well understood by others who were given zero points. Some explanations, for instance, seemed to concern the reason why metal A hit the target farther (Example 2). Others brought in extraneous information (Examples 4, 5, and 6), much like some of the Category I answers to the Proportional Reasoning task, while still others referred to the concrete happening described in the question (Example 1).

Of the explanations that earned one point, only a minority referred to both starting position and target weight as uncontrolled variables (Examples 1 and 4). The others either mentioned only the position (Example 2) or the target weight (Example 3). In addition, many supplied comments regarding the mechanism by which the position or target weight would have its effect (Example 2).

It would seem that successful design and analysis of experiments depends on an individual's ability to recognize multiple causes. Question 1 of the Control of Variables task asked the students to enumerate the variables that might affect how far the target goes. Virtually all students named some variables, and many participated in the brief discussion that preceded the remainder of the task. The quality and number of the written items in answer to Question 1 varied greatly from student to student; very few appeared to add to their list during the discussion.

Some of the variables frequently mentioned could be classified as experimenter manipulated, such as the rolling sphere's weight and starting height, the target weight, and the angle of the ramp. Other variables, such as the sphere's striking speed and force of impact, could not be directly manipulated by the experimenter. Some students listed objects (the sphere, the track) and others made descriptive statements ("the heavy sphere will hit it hard"). The items bore little relationship to the answers to Questions 2 and 3. We therefore conclude that designing a controlled experiment requires substantial logical insight beyond an awareness of variables, and that many eighth graders lack this insight.

Formal Thought

Our findings support and add further detail to the findings of others that the development of formal thought among adolescents is at best a gradual and incomplete process (8,12,22-26). As we have pointed out, the nature of our tasks, with their lack of feedback or encouragement for the subjects, provides a rather stringent assessment. Piaget's clinical interview characteristically proceeds as a dialogue in which the subject gradually gains information from the investigator or by using a piece of apparatus, and one can observe how he responds to the new data. In such a situation, many subjects may ultimately show reasoning patterns more advanced than we were able to collect. Yet, in an earlier study of proportional reasoning that compared interviews and group paper-and-pencil tasks, we found that the differences were moderate (8). Hence we believe that the main features of our findings have general significance. Furthermore, our tasks are rather similar to what is expected of secondary school students in science and mathematics classes (27,28). Literary and historical analysis also, though qualitative rather than quantitative, depend on students' abilities to evaluate evidence, identify causes, and marshall a conclusive argument, and therefore require reasoning patterns like those evaluated by our tasks.

Formal thought in our study manifests itself as fluent use of ratio on the Proportional Reasoning task (composite score R) and a score of five points on the Control of Variables task. We actually found 251 such students, or about seven percent of our sample. At the other extreme, there were 422 students, or about 14 percent, who used intuitive reasoning on the Proportional Reasoning task and scored zero points on the Control of Variables task. Both groups included about equal numbers of boys and girls.

To show how reasoning on the two tasks was correlated, we present the contingency table in Table 9. To give equal weight to all countries in spite of differences in sample sizes, we have again used the averages of the percentages for the 14 populations (seven countries, two sexes) as we did in Tables 2, 4, 5, and 6. A close examination of Table 9 shows that most of the subjects who scored three, four, or five points on the Control of Variables task responded in Categories *Tr* or *R* on proportional reasoning, an average of 28 percent of a population. This is the level of performance that we presented in the scattergrams for the individual countries and identified as being at least transitional to formal thought.

It is also clear from the table that very few of the students who scored zero points on Control of Variables task answered in the *ratio* category on the Proportional Reasoning task. Subjects with one or two points, however, were approximately equally likely to respond in any one of the four categories *I, A, Tr,* or *R*. The failure of the two scores to correlate more closely reflects once again the fact that students in several countries found the Control of Variables task more difficult and scored only one or two points even though they gave evidence of proportional reasoning.

The features we have just described are summarized in Table 10 where we have grouped categories *I* with *A, Tr* with *R*, scores 1 with 2, and scores 3 with 4 and 5. Each of the resulting six entries for the average percentage is followed, in parentheses, by the expected figure calculated on the assumption that the two tasks are completely uncorrelated. The lowest and the highest groups on control of variables are substantially correlated with the corresponding performance on proportional reasoning, but the middle group (scores of 1 or 2) is not. It is this lack of correlation for the middle group and the great differences in correspondence of the two scores from country to country that casts doubt on the existence of a single operational structure underlying all manifestations of formal thought. Rather, it would seem that proportional reasoning can be substantially advanced through the teaching programs used in some of the countries we visited, while control of variables can be enhanced through others (29-32).

Our findings also have implications for teachers. According to Tables 9 and 10, about 37 percent of eighth-grade students lack formal mental structures in both proportional reasoning and control of variables, and 13 percent appear to be nonoperational in these areas (Category I and score 0). Another 36 percent have formal structures in only one of the areas, either proportional reasoning (30 percent) or control of variables (six percent). It would seem, therefore, that teaching goals and teaching practice might well recognize these differences and

Table 9. Contingency Table for Composite Scores
On the Proportional Reasoning and Control of Variables Tasks (Percent)

		Control of Variables Score						
		0	1	2	3	4	5	Total
Proportional Reasoning Score	I	13	7	3	2	0	1	27
	A	6	4	2	2	0	1	15
	Tr	6	8	5	5	2	4	30
	R	2	6	4	5	3	8	27
	Total	26	26	14	14	5	14	100

Table 10. Condensed Contingency Table for Scores on the Two Tasks (Percent)[a]

		Control of Variables Score			
		0	1,2	3,4,5	Total
Proportional Reasoning Score	I,A	19 (11)	18 (17)	7 (15)	43 (43)
	Tr,R	8 (15)	22 (23)	27 (19)	57 (57)
	Total	26 (26)	40 (40)	34 (34)	100 (100)

a: Figures in parentheses are percentages expected in the absence of correlation.

address themselves more consciously to the task of building formal operational structures (33-35). During our school visits we became aware of some efforts by individual teachers in this direction, but did not observe any major, concerned activity. The syllabi and most of the textbooks in general use seemed poorly suited to this task.

A very interesting question raised by our study concerns the effects of selective versus comprehensive school systems as regards formal thought. The scattergrams in Figures 4, 5, 8, 11, 14, 17, and 20 summarized the approximate level of formal thought encountered by a science or mathematics teacher in various class groups. We have already pointed out the extent to which selectivity or socioeconomic level influenced these graphs. In the absence of large-scale longitudinal studies we cannot distinguish the impact of specific teaching programs from the effects of original selection. We are not convinced of the superiority of either approach and believe that improvements could be made within both. In our opinion, the most urgent problems derive from the difficulties of the low-achieving groups and their needs for learning activities strongly rooted in concrete operations.

Student Opinions

At the conclusion of the two tasks, the subjects were invited to use the backs of their answer pages for describing their opinions of the research project and our visit to their class. About two-thirds of them did so. The vast majority expressed positive attitudes, but a handful out of about 3000 students were dissatisfied because they disliked tests, had not been consulted in advance, or felt they were being exploited. Students liked the visit because it allowed them to avoid their regular assignments, challenged them in unusual ways, enabled them to participate in an international activity, and/or acquainted them with new scientific ideas. Many wished us a good trip home (they had been impressed by our coming from California to work with them) and hoped we would come again.

Some students indicated that their teachers needed to be more tolerant of a diversity of approaches to a problem; many German *Gymnasium* students, however, expressed surprise at our project because they considered it obvious that different persons would use different methods for solving the tasks. A few subjects expressed curiosity about our findings and asked to see our report, and some even pointed out the difficulty we faced of adequately sampling students in their country. Quite a few students confessed that they did not really understand the purpose of our project. Here are a few comments quoted from English language answer pages:

> I think that it was an unusual experience, but it helped me to think before I try something. I think it should be done more often to help students understand things like math, and so forth, better.

> I like science and the teacher, but sometimes science is difficult, so keep on coming.

> This visit was very considerate of you and very nice. I am sorry you all could not see my best work. My head isn't together today. One day I wish to be a third grade teacher and I am going to work on up until I make it. I really enjoyed your visit. Goodbye for now.

> Since I understand why you came here I cooperated. It is a survey to help teachers understand their students better.

This experiment of having us write our ideas is good because the teachers need to know how their students' ideas vary and that not everyone thinks the same.

I don't know why you made these tests, but I think they were good. Because they got me thinking on how you can figure out questions and problems with your own head and not the teacher. I liked the sphere problem. I would have liked to use the metal right in front of me, but on the paper was alright.

To Dr. Karplus: I, . . . myself have really enjoyed you here today. And I have really learned how to say and write what I and only I think, with no one else to try to tell me. So I hope you come back again.

I would like to see the results of the experiment.

I really appreciate your coming. I might be at college some day.

It was nice to have Mr. Karplus here. The papers he gave us was hard and not understanding for me, but I think I will understand sometime when I go to college.

I didn't mind your testing. It was different from anything else I've done.

I think it was good because the teachers will have some interesting problems to teach and they'll know that there isn't only one answer to a problem.

My opinion of this test is that I don't mind it. I found it reasonably easy. Anyway, it got us off biology.

Even though the experiments were quite simple, it was hard to write down your thoughts on paper. In general it was quite hard.

It was a good change from our usual science lesson. It was interesting and made you think.

Very good idea.

It makes an interesting change to have tests like these once in a while, tests which are interesting with a hint of problems. It also makes a change for visitors to take a lesson. This gives the lesson a mysterious hint about it (you never know what's going to come next). Very interesting!

I thought the first paper was quite hard because the man in the picture was standing with his legs apart which made him much shorter.

I hope your research helps teachers in teaching people.

I don't like tests, quizzes, etc. . . . I usually don't do too good on them.

I liked it. Most of it has to do with logical reasoning.

I liked these two puzzles, I thought they were very interesting. I'm glad we could have change from our normal science course. It is good that you take the time to do this research and I wish you luck. (P.S. Maybe my kids will be taking your course and I can tell them I helped in the research of it!)

I think these tests should have been given during a study period instead of during science class.

ACKNOWLEDGEMENTS

We are very grateful to the many students and teachers who cooperated with our investigation, and to the science educators and school administrators who helped us make the necessary arrangements with the participating schools. In making these arrangements, we benefited from the assistance of Mr. Edmund Mahoney (Arlington, MA), Miss Mary O'Connell (Arlington, MA), Mr. Charles Crisafulli (Belmont, MA), Mr. Douglas James (Boston, MA), Prof. Nils Svantesson (Gothenburg), Dr. Helmut Mikelskis (Göttingen), Dr. John Spice (London), Mr. Sylvester Webb (Philadelphia), Mr. Richard Bliss (Racine, WI),

Mr. Frank Ruberto (Revere, MA), Prof. Matilde Vicentini-Missoni (Rome), Prof. Clotilde Pontecorvo (Rome), Prof. Aldo Visalberghi (Rome), Präsident Hermann Schnell (Vienna), Frau Landesschulinspektor Dr. Erna Janisch (Vienna), Herr Bezirksschulinspektor Alfred Prosl (Vienna), Dr. Reuben Pierce (Washington, D.C.), and Mrs. Maxine Broderick (Westwood, MA). We should also like to thank Dr. Marilyn Appel, Mr. Thomas Baker, Prof. Robert Fuller, Miss Barbara Karplus, Mr. Paul Andrew Karplus, Mrs. Marilyn Krupsaw, and Mrs. Kerstin Waern for helping us adminster the tasks. Discussions with Dr. Warren Wollman contributed to our formulation of the tasks and Mr. Joseph Walker furnished some of the equipment. Drs. Björn Andersson, H. Bleichroth, Anton Lawson, and Warren Wollman made enlightening comments on our manuscript. In addition to ackknowledging the professional assistance to our project, we express our deep appreciation for the hospitality, friendliness, and interest with which we were received everywhere.

We are grateful for the hospitality of the Division for Study and Research in Education at the Massachusetts Institute of Technology, where we began this investigation. A fellowship from the John Simon Guggenheim Foundation helped defray the costs of the research.

REFERENCES

1. Inhelder, B. and Piaget, J. *The Growth of Logical Thinking from Childhood to Adolescence.* New York: Basic Books, Inc., 1958.

2. Lunzer, E. A. and Pumfrey, P. D. "Understanding Proportionality." *Mathematics Teaching,* 34, pp 7-12, 1966.

3. Lovell, K. and Butterworth, I. B. "Abilities Underlying the Understanding of Proportionality." *Mathematics Teaching,* 37, pp 5-9, 1966.

4. Lovell, K. "Proportionality and Probability." In *Piagetian Cognitive Development Research and Education,* edited by M. F. Rosskopf, L. P. Steffe, and S. Taback, Reston, VA: National Council of Teachers of Mathematics, 1971.

5. Lovell, K. "A Follow-Up Study of Inhelder and Piaget's 'The Growth of Logical Thinking'." *British Journal of Psychology* 52(2), pp 143-153, 1961.

6. Kohlberg, L. and Gilligan, C. "The Adolescent as a Philosopher: The Discovery of the Self in a Postconventional World." *Daedalus,* 100, Fall 1971, pp 1051-1084.

7. Karplus, R., Karplus, E., and Wollman, W. "Intellectual Development Beyond Elementary School IV: Ratio, The Influence of Cognitive Style." *School Science and Mathematics,* 74, Oct 1974, pp 476-482.

8. Wollman, W. "Intellectual Development Beyond Elementary School V: Using Ratio in Differing Tasks." *School Science and Mathematics,* 74, Nov 1974, pp 593-611.

9. Wollman, W. "Controlling Variables: Assessing Levels of Understanding."

Science Education, 61(3), pp 371-383, 1977.

10. Suarez, A. "Die Entwicklung der Denkoperationen beim Verständnis Funktionaler Zusammenhänge." Unpublished doctoral dissertation, Eidgenössische Technische Hochschule, Zurich, 1974.

11. Andersson, B. "Some Simple Experiments Illustrating School Children's Entry into the Stage of Formal Thought." *Kognitive Impuls,* Gothenburg: School of Education, 1974.

12. Lawson, A. E. and Benner, J. W. "A Quantitative Analysis of Responses to Piagetian Tasks and Its Implications for Education." *Science Education,* 58(4), p 545, 1974.

13. Kuhn, D., Langer, J., Kohlberg, L., and Hahn, N. "The Development of Formal Operations in Logical and Moral Judgment." In *Genetic Psychology Monograph,* Provincetown, MA: Journal Press, pp 97-188, 1977.

14. Case, R. "Structures and Strictures: Some Functional Limitations on the Course of Cognitive Growth." *Cognitive Psychology,* 6, pp 544-573, 1974.

15. Comber, L. C. and Keeves, J. P. *Science Education in Nineteen Countries.* Stockhom: Almqvist & Wiksell, 1973.

16. See reference 15, Appendix IX; the items are on Test 1A, #8, 9, 14; on Test 1B, #10, 18; on Test 10A, #3, 4, 10, 11; on Test 11B, #4, 6, 12.

17. Downie, M. M. and Heath, R. W. *Basic Statistical Methods,* third edition. New York: Harper & Row Publishers, Inc., 1970, pp 123-124.

18. Our scoring procedure differs somewhat from that used in reference 9.

19. See reference 17, pp 124-125.

20. Pascucci (Formisano), M. "Sviluppo Cognitivo e Persiero Formale, Primi Risutati di una Ricerca Sperimentale Interculturale." Tesi di laurea in Pedagogia, Universita degli Studi di Roma, 1973/74, contains a more extended description of the effects of sex, mixed versus segregated classes, and sociocultural variables.

21. Wollman, W. Personal communication.

22. Elkind, D. "Quantity Conceptions in Junior and Senior High School Students." *Child Development,* 32, p 551, 1961.

23. Lovell, K. "Some Problems Associated with Formal Thought and Its Assessment." In *Measurement and Piaget,* edited by D. R. Green, M. P. Ford, and G. B. Flamer, New York: McGraw-Hill Book Co., 1971.

24. Piaget, J. "Intellectual Evolution from Adolescence to Adulthood." *Human Development,* 15, pp 1-12, 1972.

25. Lunzer, E. A. "Formal Reasoning: A Re-Appraisal." In *Topics in Cognitive Development,* Vol. II. Edited by B. Z. Tresseisen, D. Goldstein, and M. H. Appel, New York: Plenum Press, 1978.

26. Karplus, R. "Opportunities for Concrete and Formal Thinking on Science Tasks." In *Topics in Cognitive Development,* Vol. II. Edited by B. Z. Tresseisen, D. Goldstein, and M. H. Appel, New York: Plenum Press, 1978.

27. Shayer, M. "Some Aspects of the Strengths and Limitations of the Application of Piaget's Developmental Psychology to the Planning of Secondary School Science Courses." Master's thesis, University of Leicester, 1972.

28. Ingle, R. B. and Shayer, M. "Conceptual Demands in Nuffield O-Level Chemistry." *Education in Chemistry,* 8(5), Sept 1971.

29. Lawson, A. E. and Wollman, W. "Encouraging the Transition from Concrete to Formal Cognitive Functioning—An Experiment." *Journal of Research in Science Teaching,* 13(5), pp 413-430, 1976.

30. Wollman, W. and Lawson, A. E. "Teaching the Procedures of Controlled Experimentation: A Piagetian Approach." *Science Education,* 61(1), pp 51-70, 1977.

31. Wollman, W. "Controlling Variables: a Neo-Piagetian Sequence." *Science Education,* 61(3), pp 385-391, 1977.

32. Suarez, A. and Rhonheimer, M. *Lineare Funktion.* Zurich: Limmat Stiftung, 1974.

33. Karplus, R. and Lawson, C. A. *SCIS Teacher's Handbook.* Berkeley: University of California, 1974, pp 48-50.

34. Karplus, et al. *Workshop on Science Teaching and the Development of Reasoning.* University of California, Berkeley, 1977.

35. Lawson, A. E. and Wollman, W. "Physics Problems and the Process of Self-Regulation." *The Physics Teacher,* 13(8), pp 470-475, 1975.

APPENDIX AI. Answer Page for Proportional Reasoning Table

Name _____ Grade _____ Age _____ Sex _____

What is the height of Mr. Tall, measured in paper clips? _____
Please explain carefully how you found this answer.

Mr. Tall's car is 14 paper clips wide.
How wide is Mr. Tall's car, measured in buttons? _____
Please explain carefully how you found this answer.

Mr. Short

What is the height of Mr. Short, measured in paper clips?

APPENDIX AII. Answer Page for Control of Variables Task

1. A sphere rolls down a ramp and hits a target. What might affect how far the target goes?

   _____

   _____

   _____

   high x
   medium x
   low x
   heavy target

2. Suppose you want to know how much difference the weight of the sphere makes in how far the target goes. You are going to use the two spheres on the heavy target.

   (a) Where would you start the heavy sphere?_____
   Where would you start the light sphere?_____
   Please explain your answers carefully.

   _____

   _____

   (b) Suppose someone starts the light sphere from the low position. Where would you start the heavy sphere to find out how much difference its weight makes?_____

   Please explain your answer carefully.

   _____

   _____

3. Suppose you have spheres of two different kinds of metal, metal A and metal B. You want to find out which metal is better for hitting a target far. You do the experiments in the two drawings.

   high (A)
   medium (B)
   ← This target went farther.
   light target
   heavy target

   What you see is that the metal A sphere hits its target much farther than the metal B sphere hits its target. Does this prove that metal A is better for hitting the target far?   Yes___ No___

   Please explain your answer in detail. _____

   _____

   _____

APPENDIX B

In one of our earlier studies of proportional reasoning we found a substantial effect on the students' responses of the order of two questions, one of which was harder than the other.[8] In the present investigation, one may wonder to what extent a subject's first encounter with the Mr. Tall question may influence his approach to this and to the car question. We have already pointed out that students whose answers to the two questions differed usually gave a better answer to the Mr. Tall question. Yet Wollman pointed out to us that the larger numerical differences on the car question, if solved first, might well lead students to use a proportional rather than an additive approach.

To determine what influence the question order might have, we prepared an answer page on which the two questions were exchanged, and adapted the presentation of the task to the new order. This modified task was then administered to 77 boys and 90 girls in the pilot group of eighth graders in the United States listed in Table 1. They were in two classes in an upper-middle-class community and in five classes in middle-class communities in the same schools as students in the principal United States group.

The composite scores resulting from this investigation are presented in Table 11 for the pilot group as well as for a group of middle-class and upper-middle-class students matched as closely as possible to the pilot group in ability levels and community. There is a statistically significant difference for boys ($p < 0.05$) and for girls ($p < 0.001$) between the two forms of the task. The reversed order leads to a substantial reduction in the *additive* responses for both sexes, and a substantial increase in *ratio* answers for boys, in *transitional* answers for girls.

When the individual explanations to each of the two questions are examined, one finds that the largest decrease is in *additive* responses to both questions and that there is an increase in papers with a *transitional* answer to the car question and *ratio* answer to the Mr. Tall question (composite Category Tr). For boys there is a further substantial increase in papers with ratio answers to both questions and a decrease in papers with *transitional* answers to both questions.

Table 11. Composite Scores on the Proportional Reasoning Task for Reversed and Standard Presentation (Percent)

	(N)	I	A	Tr	R
Boys, Reversed Order	(77)	34	9	30	27
Boys, Standard Order	(77)	29	27	27	27
Girls, Reversed Order	(90)	21	22	38	19
Girls, Standard Order	(90)	24	47	14	15

The attitudes of most classes differed when the reversed order was presented. After the question about the width of the car was posed, many students gasped at the task and were initially at a loss as to how to proceed. Some of them then answered the Mr. Tall question first, even though it was placed second on the answer page. The students perceived it as a very difficult assignment, yet many more ultimately applied transitional or proportional reasoning.

It is clear from this investigation that the Mr. Tall question elicits a particularly high number of *additive* responses from students in the United States. The standard order of questions on the Proportional Reasoning task, with the Mr. Tall item first, therefore appears to be more challenging than the same task with the reversed order, and we used it for that reason. The fact that many student populations in our principal study nevertheless applied relatively little *additive* reasoning is evidence of their good understanding of the concept of proportion. Some students in the early stages toward achieving proportional reasoning, however, were undoubtedly misled and thus their number was underestimated by our approach.

Proportional Reasoning in The People's Republic of China: A Pilot Study[1]

Robert Karplus

Proportional reasoning, an essential part of learning and understanding science and mathematics, has been investigated in numerous studies in many countries (Karplus, Karplus, Formisano, & Paulsen, 1977 and 1978; Karplus, 1978). As part of an effort to exchange information regarding the development of reasoning with scholars in the People's Republic of China, the author administered the Paper Clips task (Karplus, 1978). In this task students had to deduce a scale factor from given and measured data to predict the height of a hypothetical figure (Comrade Tall). The task included two similar items.

THE SUBJECTS

Forty-nine fourth- and fifth-grade students from the laboratory school of Shanghai Teachers University participated in the investigation. They included 22 girls and 27 boys, who were studying the fifth-grade mathematics syllabus that included concern with ratio and proportions. Most were the children of professional workers such as teachers at the University itself. The 14 fourth graders were gifted students who had been moved one year ahead in their mathematics studies. The 35 fifth graders were selected by their teachers from the school's total fifth-grade enrollment of about twice that number.

METHOD

The Paper Clips task was administered in the usual fashion, using Chinese paper clips and a suitably scaled illustration of "Comrade Short" so his height would be about six and one-half paper clips. An instructor from the University's foreign language department translated the author's oral explanations and instructions. The papers were handed in by the students after five to 20 minutes of effort, whenever they had completed their work.

RESULTS

Forty-six of the 49 students used proportional reasoning by applying ratios successfully on at least one of the two items of the task, and 37 used this approach on both items. No student used the incorrect additive reasoning that had been observed in many student samples in the United States and Europe (Karplus, et al., 1977; Karplus, 1978). For comparative purposes, therefore, the percentages of students using ratios on the first of the two items (which was the

---

1. First published in *The Genetic Epistemologist* 7 no. 3 July, 1978.

easier) and on both items are reported in Table I, for all the student samples for which these data are available.

Table I. Responses Using Ratio on the Paper Clips Task

Population Sample	Ratio Approach		Population Sample	Ratio Approach	
	Item 1	Items 1 and 2		Item 1	Items 1 and 2
China[a] Age 11-12 4th-5th Grade	92	76	Denmark[a] Age 14 7th Grade	37	16
United States[a] Age 11-12 6th Grade Upper Middle	21	—	Sweden[a] Age 14-15 7th-8th Grades Middle Class Working Class	54 25	29 11
United States[a] Age 13-14 8th Grade Upper Middle Middle Class Urban Low Income	34 25 2	20 15 0	Austria[b] Age 14-15 8th Grade Top Group Middle Group Low Group	74 52 35	49 33 16
United States[a] Age 16-18 11th-12 Grades Upper Middle Middle Class Urban Low Income	75 48 16	53 31 11	Germany[b] Age 15 9th Grade Top Group Middle Group Low Group	89 66 41	64 39 7
Italy[a] Age 13-14 8th Grade Upper Middle Middle Class Working Class	70 43 31	40 25 18	Great Britain[b] Age 14 8th Grade Top Group Middle Group Low Group	90 70 29	70 45 13

a: Comprehensive School System
b: Selective School System

In addition to responding to the task items, three students wrote comments as follows:

fifth grade boy: "Say hello to my little friends in America."

fifth grade boy: "American teacher, may I ask two questions? I also want to give my opinion on these problems. One: in what grade do American pupils solve these problems? Two: how long does it take them to solve these problems? My opinion: these problems are good. They are detailed and clear and easy to understand. They help us develop our intelligence. They are very interesting."

fourth grade girl: "These two problems are related. These two problems are not difficult. I can work them out if I set my brain to work."

These comments are similar to the remarks reported by Karplus and Karplus, (1978).

DISCUSSION

The high fraction of Chinese students using ratios is remarkable, especially in view of their age, even though the sample is not necessarily representative. The percentages are close to those of the English eighth-grade students in Direct Grant schools, who were selected from the top five percent of the population. Unfortunately, the short time available for the investigation in Shanghai did not permit a detailed examination of the students' teaching materials. Visits to secondary schools in other communities, however, revealed that seventh graders (about 14 years of age) in an agricultural area were studying simple trigonometry, a subject that requires a good understanding of ratios. In their physics classes, seventh and eighth graders in a city school worked on the mathematical treatment of hydraulic presses and other machines, whose mechanical advantage also illustrates proportionality.

ACKNOWLEDGEMENTS

I am indebted to Robert Cremer for translating the task into Chinese. Chin Han-feng of Shanghai Teachers University made the necessary arrangements to secure the cooperation of the Chinese students and their teachers. I am very grateful for their cordial friendship and hospitality, which made this pilot study possible. The travel in China was arranged under the auspices of the California State Board of Education, to which I am indebted for the opportunity to make the visit to the People's Republic.

REFERENCES

Karplus, R. "Formal Thought and Education—A Modest Proposal." Presented at the Eighth Annual Symposium of the Jean Piaget Society, Philadelphia, May 1978.

Karplus, R. and Karplus, B. "Comment on Karplus, Karplus, Formisano, and Paulsen, 'A Survey of Proportional Reasoning and Control of Variables in Seven Countries.' " *Journal of Research in Science Teaching,* 15(5), 1978.

Karplus, R., Karplus, E., Formisano, M., and Paulsen, A.C. "A Survey of Proportional Reasoning and Control of Variables in Seven Countries." *Journal of Research in Science Teaching,* 14(5), pp 411-417, 1977.

Karplus, R., Karplus, E., Formisano, M., and Paulsen, A.C. "Proportional Reasoning and Control of Variables in Seven Countries." This volume.

Information Processing Models and Science Instruction[1]

Jill H. Larkin

In this paper I want to show how models from psychology, in particular information-processing models, can be used to discuss the following intriguing question: "What makes an expert? What makes an expert in any field?" But because my original training is in physics, and because this book emphasizes science education, I shall discuss in particular, "What makes an expert in physics?" To make clear how I want to address this question, let me point out that there are two ways of asking, "What makes an expert?"

First one can ask directly applied questions about the learning and teaching processes through which an unskilled beginner can acquire the skills of an expert. Work in this area typically uses very global models of effective instruction, [for example, those of Ausubel, (1968) and Gagné, (1970)] to guide the design of practical instruction. The effectiveness of this instruction is then tested, usually by means of an hypothesis-testing experimental paradigm, for example, a paradigm in which a group receiving experimental instruction is compared statistically with a control group who did not. The result of such work is ultimately some instructional materials of demonstrated effectiveness; and, consequently, some indirect support for the model of instruction which guided the development of these materials.

There is a second way to ask: "What makes an expert?" One can ask basic questions about the mechanisms of expertise and how it is developed, for example: "What are the skills and capabilities an expert possesses, and how do they differ from the skills and capabilities of a beginner?" "Through what processes does a beginner acquire the skills and capabilities of an expert?" Work relevant to such basic questions involves the construction and testing of detailed models of how human beings process information. Evaluating information-processing models often requires data more detailed than that obtained through statistical comparison of groups. Thus, at least early stages of research in this area often proceed in the following way:

1. Detailed observations are made of how a few individuals perform the tasks of interest. Commonly this is done by asking the person to "think aloud" while performing the task, and recording his comments for future analysis by means of audio or video tape.

2. On the basis of these detailed data, a model of the mental processes involved in the task is constructed. This model is written as a "program" which

1. This paper was presented at the AAAS Symposium on Models of Learning and Their Implications for Science Instruction, February 1978.

can be implemented by a computer (or by a human being following its directions), and which actually produces samples of behavior relevant to the task of interest. For example, such a model for solving physics problems acts on a physics problem to produce a series of steps in a solution.

3. The observable output produced by the model is compared with the work of human beings performing the same task.

If, after successive revisions, the model's work is in good agreement with that of human beings, one has some confidence that the model captures at least some features of the problem-solving mechanisms used by people. Clearly the degree of confidence depends in part on the level of detail at which the model accounts for human problem-solving behavior. This level of detail varies greatly, depending on the nature of the process being modeled and the purpose of the research. But in my work, for example, I work at a level of detail with which I can account for the order in which individual physics principles are applied in solving a problem.

Such detailed models explicating mechanisms for expertise can have substantial implications for science education for the following reasons: As Michael Polanyi (1967) points out, much of human knowledge is "tacit" in that, under ordinary conditions, it can not be explicated in words. We can recognize the face of a friend without being able to describe how we do it. Similarly, much of an expert's skill (e.g., in solving physics problems) is tacit, in that the processes he uses are *not* obvious either to himself or to a casual observer. If these tacit processes remain unexplicated, then, to help a beginner learn, there is little one can do beyond providing examples and practice, and hoping that the beginner will somehow "pick up" these unspecified skills. But if one can begin to build explicit models for formerly tacit processes, then it becomes possible to teach these processes, either directly or through appropriately *selected* practice and example. In addition, explicit models for tacit processes can aid in identifying and remedying errors in the developing skills of learners.

Indeed, some of these instructional payoffs have been exploited in systematically designed instructional instruction (Reif, Larkin, and Brackett, 1976; Larkin and Reif, in press), and more recently in a few experimental computer-based programs. For example, Brown (1977) at Bolt, Beranek, and Newman, and Goldstein (1976) at MIT have developed instructional programs which are based on an explicit model of skilled performance in games requiring use of logical inference. The instructional programs then act as "coaches" diagnosing learners' difficulties and suggesting to them useful modifications and their strategies.

I shall not, however, discuss these instructional payoffs here, but instead will review a small sampling of recent research on information processing mechanisms relevant to expertise in science. In particular, I shall focus on work which is relevant to problem solving, partly because that is the area in which I myself have done research, but primarily because science is fundamentally concerned with solving problems. Thus the difficult task of teaching students to solve problems effectively is a central concern in science education.

I will discuss work from two increasingly overlapping fields: cognitive psychology, which is concerned with modeling complex human behavior; and artificial intelligence, which is concerned with programming computers to perform tasks "intelligently." I shall discuss three ideas, which are reflected in work from both cognitive psychology and artificial intelligence, and which I think have substantial implications for science education. These ideas are the following: *1*. condition-action units (often called "productions") which specify an action, together with the condition under which this action is executed; *2*. functional units in memory (variously called "frames," "schemas" and "scripts") which allow related bits of information to be stored, accessed and used coherently; and *3*. low-detail representations and procedures, which are used for comprehensive planning of detailed or complex work.

CONDITION-ACTION UNITS

In modeling psychological processes, it is essential to specify not only the actions which can be taken, but also the conditions under which each action can be taken. This importance of conditions is made explicit in many computer-implemented models which are written in terms of "productions," units consisting of an action together with a condition specifying when the action is to be taken (Newell, 1973; McDermott, 1978). According to such models, the human mind has stored an immense number of these condition-action units. Whenever the current situation satisfies one of the conditions stored in memory, the corresponding action is implemented. To give a very simple example, when the condition of seeing a red light is satisfied, the action of stopping a car is taken.

More generally, the condition-action unit is a useful way of describing complex psychological processes at a variety of levels of detail. For example, Collins (1976) is working to develop a global theory of Socratic tutoring by specifying actions (questions the Socratic tutor can ask) each with an associated condition specifying when that kind of question is asked.

The following examples illustrate the use of condition-action units in explicating tacit processes which I think are relevant to problem solving in science.

Early studies of chess (de Groot, 1966) found no differences in the grossly observable action taken by master chess players and by weaker players. Both types of players considered about the same number of alternatives, and looked ahead about the same number of moves. However, master chess players were strikingly superior to weaker players in their ability to remember and recreate piece positions on a chessboard they have briefly seen—if this position is part of a real game. The work of Chase and Simon (1973) suggests that this superiority is due to a master player's large memory store of piece positions commonly seen in real games. (This hypothesis is consistent with the fact that master players are not superior to weaker players in recalling position of randomly placed pieces.) Chase and Simon then suggest that these stored piece positions are actually parts of condition units. Thus, when a master player considers alternative moves or strategies, he can start by considering those possibilities suggested by his memory store. It is then not surprising that the master considers a "better" set of alternatives than does a beginner. In short, it seems possible that the master's superior ability in chess lies not in general strategies (e.g., considering more alternatives, planning farther ahead), but in a large store of condition-action units, allowing him to identify a good set of alternatives for further consideration.

Klahr and Siegler (1977) have used condition-action units to describe the developing abilities of children to reason about the balance scale. Their work suggests that young children have very simple sets of condition-action units. For example, "if there is more weight on the right, then predict it goes down on the right." with increasing experience and age, these condition-action units become more sophisticated and complex. For example, the preceding unit might be modified to: "if the distances from the fulcrum are the same, *and* the weight is larger on the right, then predict it will go down on the right." Older children, who can reliably predict how the balance beam will tilt, have the most extensive set of condition-action units. For simple conditions (e.g., distance same, weight different), they have simple actions (e.g., predict the heavy side goes down). For more complex conditions, they have more complex actions (e.g., applying physicist's "torque law"). Interestingly, however, these proficient children do *not* routinely use the general torque law, even though it would always give correct predictions. I suspect that when an action is complex (like applying the torque law), then it is advantageous to remember also some simpler actions (together with their special simple conditions). The reason is that these simpler actions are less demanding and less prone to error.

These two examples, expertise in chess and developing abilities to predict behavior of a balance beam, have illustrated the importance of condition-action units in complex skills. Skill involves not only being able to implement potentially useful actions, but also being able to recognize conditions under which a given action will be useful.

This idea of condition-action units central to skilled performance certainly suggests the following implications for science education:

1. When we teach students useful actions (e.g., how to apply Newton's laws), we should be comparably conscientious about teaching the conditions under which these actions are useful. Perhaps students' most common complaint about their problem-solving ability is that they "don't know how to get started," or "don't know how to decide what to do." In order words, these students may have learned some actions; but failed to learn appropriate conditions for applying them. When students don't learn *appropriate* conditions, they too often learn *inappropriate* ones. For example, not infrequently, physics students think that principles describing circular motion are applicable to motion along any curved path; and so they try to use these principles to describe the flight of baseballs!

2. Because many scientific principles are general, science teaching often encourages students to apply a single general principle under a wide variety of conditions. However, using general principles in simple situations can result in work which is clumsy and error prone. Research of the kind I've briefly mentioned here suggests that skilled persons avoid needlessly cumbersome applications of general actions by recognizing special conditions under which simpler actions do as well. By judiciously exposing learners to selected, special cases, one might spare them needless drudgery, and help them to acquire the skilled person's economy in applying actions appropriate to the complexity of the situation.

In concluding this section on the implications for instruction of condition-action units, let me briefly mention some impressive work on practical instruction done by Landa (1974, 1976). His instructional design in a variety of subject-matter areas (Russian grammar, mathematics, problem solving in science) assumes that learners can quite readily learn to execute actions. He then focuses on teaching conditions to guide the use of these actions.

LARGE-SCALE FUNCTIONAL UNITS

I now turn to a second global idea which is prevalent in models of human information processing as well as in "artificially" intelligent computer programs. This idea is the use of large-scale functional units stored in memory. These units make available coherently bits of information which are often used together. Further, such units provide a pre-assigned "default" structure. Individual items or "slots" in the structure can be changed as the structure is used, but many parts of the large unit simply remain as they were stored in memory.

In the following paragraphs I discuss some recent work (including some of my own) which illustrates the utility of large-scale functional units in the formation processing required by complex tasks.

Shank and Abelson (1977) model how people understand stories by using a variety of large-scale units, one of the simplest being a "script." A restaurant script, for example, specifies much of the common knowledge about restaurants—that a person entering a restaurant is hungry, has money to pay, sits at a table, reads a menu, etc. If the reader of a story has such a script stored in memory, then a story involving a restaurant is comprehensible even if only a small number of items from the script are mentioned explicitly. After noting a few items which indicate the existence of a restaurant, the reader accesses his restaurant script and assumes that it is part of the story.

In solving problems, it also seems likely that people work with similarly large-scale, preset units. For example, Hinsley, Hayes, and Simon (1976) read to people brief sections of algebra word problems. When read the brief phrase, "A river steamer," people were able to respond with statements like: "It's going to be one of those river things, with upstream, downstream, and still water." or "It is going to be a linear algebra problem of the current type" and later, "or else it's a trig problem—the boat may go across and get swept downstream." Thus from a tiny element of the problem, "A river steamer," the person responding was able to activate an entire unit describing a particular kind of problem (upstream, downstream and still water) and probably even a large part of its solution (trigonometry problem).

In an experiment of mine, I wanted to assess whether experienced physicists had more large-scale functional units for solving physics problems than did beginning students (even when these students did know the individual physics principles). I asked several "experts" and several "novices" (beginning physics students) to think aloud while solving a set of problems. My prediction was the following: If experts have more large-scale units specifying how to solve pieces of a problem, then the equations the expert generates should be clustered in time, each cluster corresponding to the application of one large-scale method. In contrast, if the novices have fewer such large-scale methods, the equations they generate should be more randomly spaced in time, corresponding to access of principles individually rather than as part of larger-scale methods. Figure 1 shows graphs made by analyzing the tape-recorded verbal comment of an expert and novice subject, each solving five problems from mechanics. The graphs show the percent of sequential pairs of equations as a function of the time interval Δt separating the two equations in the pair. The dotted curves are the graphs which *would* occur if the subject had generated equations randomly in time. The error bars indicate one standard error in the observed frequency.

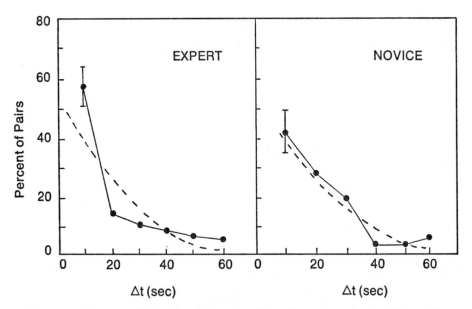

Figure 1. Percent of Equation Pairs, as a Function of Intervening Time ΔT, in Problem Solutions of an Expert and a Novice Subject

In Figure 1, the novice's graph reflects the predicted random distribution of his equations in time, supporting my suggestion that, for him, principles are accessed individually. In contrast the expert's graph indicates a large number of equation pairs with very small intervening times reflecting a significant clustering in time of the equations he generated. This clustering thus supports the idea that the expert does recall and use principles as part of large-scale coherent units.

Let me consider a final example of a study suggesting the importance of large-scale units in skilled performance. Simon and Simon (1978) constructed two models, specified as computer programs, which simulated the order in which an experienced individual and a beginner applied principles in solving some simple problems about linear motion. A central difference in these models is the following: the model of the novice works backwards algebraically, applying principles individually to write equations involving the quantities to be found. In contrast, the model of the expert applies principles without evident regard for the ultimate goal. I suspect that this expert, like those I've observed, does not apply principles individually but as part of a large-scale unit. For the simple problems Simon and Simon considered, the processes used to select this method were not apparent. But having selected the method, he applied principles from within it with confidence that they would be helpful.

In the examples I've mentioned, understanding stories, solving algebra word problems, and solving physics problems, there are consistent suggestions that skilled performance on these tasks involves large-scale functional units, which are stored in memory and which allow bits of information to be accessed and used coherently.

This idea of large-scale functional units suggests the following implications for science education:

In one class session, or in one section of a book, a student learns a rather small bit of information. Thus it is not surprising that, at least initially, students work with small units of information. For example, most diligent physics students quickly learn to apply individual principles. However, they often do not even know what is meant by a "force method" or an "energy method." (These "methods" are large-scale units, each involving many principles, which experts mention and use explicitly in solving problems.) Thus it is probably worth some explicit instruction to aid students in consolidating the bits of information learned day by day into functional units.

What kind of instruction might help students to acquire large-scale units? Almost certainly it would be useful to allow time for review and consolidation, even at the expense of "covering new material." Further, functional units (e.g., force and energy methods) could be identified explicitly and illustrated with examples and exercises. Currently, in most physics textbooks, it is difficult to recognize that methods exist, except by inferring them from the structure of chapters.

PLANNING AND LOW-DETAIL REASONING

A third general idea which seems central to many models of skill in complex tasks is the idea of planning complex actions at a low level of detail before working out all details.

For example, Greeno (1976) has developed a program to simulate the processes of good high school students in solving geometry. The program routinely substitutes for the definite goal of the original problem, an "indefinite" subgoal. For example, when the problem requires proving that two triangles are congruent, the program sets as an indefinite subgoal proving that any corresponding parts of the two triangles are congruent. Thus the students Greeno observed seemed to begin their solutions with somewhat vaguely defined goals.

The use of low detail or qualitative reasoning is also quite striking in several artificial intelligence computer programs designed to solve problems of varying kinds. For example, in the Carnegie-Mellon General Problem Solver, GPS (Newell and Simon, 1972), the problem-solving program could work by addressing first the problems simplified by suppressing many details. The resulting simplified solution could then be used as a guide for more detailed work.

More recently, two artificial intelligence programs which solve physics problems (Novak, 1977; de Kleer, 1977) both work by translating the problem into some visual or pictorial representation, and then use the picture to guide the generation of the detailed equation needed to solve the problem quantitatively.

In an experiment of mine, I tried to assess the importance of low-detail quantitative reasoning by asking a group of skilled problem solvers (professors and advanced graduate students) to solve some quite difficult problems. In these problems (Larkin, 1977) these problem solvers invariably worked in the following way:

1. They began their solution by making a qualitative representation of the problem situation (usually a labeled sketch).

2. They then tentatively selected a method and applied it to the problem to make qualitative statements about the situation. For example, in applying a force method, subjects would discuss how forces balanced or failed to balance so

as to account for the motion of various objects.

3. They then checked these qualitative statements to assess whether any intractable difficulties were likely to arise in applying the method.

4. Only then did these skilled problem solvers begin to apply the method quantitatively to produce mathematical equations.

In all cases, once a problem solver had selected a method and begun writing any equations, he was then successful in applying that selected method to solve the problem.

Thus in several diverse settings it seems that low-detail, qualitative, often vague reasoning is crucial to effective problem solving, particularly at the beginning of the problem-solving process. What are the implications for science instruction?

Students often think that qualitative reasoning (e.g., with diagrams or sketches) is illegitimate in the context of science. The reasons for this bias are clear. For want of space and time, textbook authors and lecturers commonly present only the final precise, mathematical form of an argument. They omit the initial stages in which qualitative low-detail reasoning is used to plan more detailed work. Thus students infer that all reasoning in science should be quantitative, and they are often puzzled about how such reasoning is done. For example, students often say things like, "I can follow all your reasoning, but I can't see how you decided what to do."

The examples of research I've mentioned here suggest that the qualitative reasoning so often omitted is crucial to the process of "deciding what to do." What is needed is instruction aiding in using qualitative reasoning effectively to plan more detailed work.

SUMMARY

I've talked about three ideas which seem to be central to information processing models in several areas. These ideas are:

1. The importance of specifying not just actions, but also the conditions under which those actions can usefully be executed.

2. The importance of large-scale functional units—"scripts" or "methods" which a problem solver can use in place of piecemeal assemblying individual items of information.

3. The importance of low-detail qualitative reasoning to plan problem solutions before execution of details.

All three of these ideas could, I think contribute substantively to good science instruction, which too often focuses on what to do (and not on when to do it); on individual principles (rather than coherent methods); and on precise mathematical techniques (rather than more global qualitative reasoning).

REFERENCES

Ausubel, D.P. *Educational Psychology: A Cognitive View.* New York: Holt, Rinehart, and Winston, Inc, 1968.

Brown, J.S. and Burton, R.R. "A Paradigmatic Example of an Artificially Intelligent Instructional System." *Proceedings of the First International Conference on Applied General Systems Research: Recent Developments and Trends.* Binghamton, NY, Aug 1977.

Collins, A. "Processes in Acquiring Knowledge." In *Schooling and the Acquisition of Knowledge,* edited by R.C. Anderson, R.J. Spiro, and W.E. Montague, Hillsdale, NJ: Lawrence Erlbaum Associates, 1976.

Gagné, R. *The Conditions of Learning,* second edition. New York: Holt, Rinehart, and Winston, 1970.

Goldstein, I. "An Athletic Paradigm for Intellectual Education." Logo Memo 37, AI Memo 389, Artificial Intelligence Laboratory, Massachusetts Institute of Technology, 1976.

Greeno, J.G. "Indefinite Goals in Well-Structured Problems." *Psychological Review,* 83, pp 479-491, 1976.

de Groot, A.D. "Perception and Memory Versus Thought: Some Old Ideas and Recent Findings." In *Problem Solving: Research, Method, and Theory,* edited by B. Kleinmuntz, New York: John Wiley and Sons, 1966.

Hinsley, D.A., Hayes, J.R., and Simon, H.A. "From Words to Equations: Meaning and Representation in Algebra Word Problems." CIP Working Paper #331, Carnegie-Mellon University, 1976.

Klahr, D. and Siegler, R.S. "The Representation of Children's Knowledge." In *Advances in Child Development,* Vol 12, edited by H. Reese and L.P. Lipsitt, New York: Academic Press, 1977.

de Kleer, J. "Multiple Representations of Knowledge in a Mechanics Problem Solver." *International Joint Conference on Artificial Intelligence,* 5, pp 299-304, 1977.

Landa, L.N. *Algorithmization in Learning and Instruction.* Englewood Cliffs, NJ: Educational Technology Publications, 1974.

Landa, L.N. *Instructional Regulation and Control.* Englewood Cliffs, NJ: Educational Technology Publications, 1976.

Larkin, J.H. "Problem Solving in Physics." Working Paper, University of California, Berkeley, 1977.

Larkin, J.H. and Reif, F. "Understanding and Teaching Problem Solving in Physics." *European Journal on Science Education,* in press.

McDermott, J. "Some Strengths of Production System Architectures." In NATO A.S.I. Proceedings: Structural/Process Theories of Complex Human Behavior, The Netherlands: A.W. Sijthoff, International Publishing Company, in press, 1978.

Newell, A. "Production Systems: Models of Control Structures." In *Visual Information Processing,* edited by W.G. Chase, New York: Academic Press, 1973.

Newell, A. and Simon, H.A. *Human Problem Solving.* Englewood Cliffs, NJ: Prentice-Hall, 1972.

Novak, G. "Representations of Knowledge in a Program for Solving Physics Problems." International Joint Conference on Artificial Intelligence, 5, pp 286-291, 1977.

Polanyi, M. *The Tacit Dimension.* Garden City, NY: Anchor Books, 1967.

Reif, F., Larkin, J.H., and Brackett, G.C. "Teaching General Learning and Problem Solving Skills." *American Journal of Physics,* 44, pp 212-217, 1976.

Simon, D.P. and Simon, H. "Individual Differences in Solving Physics Problems." In *Children's Thinking: What Develops?* edited by R. Siegler, Hillsdale, NJ: Lawrence Erlbaum Associates, 1978.

Shank, R. and Abelson, R. *Scripts, Plans, Goals, and Understanding,* Hillsdale, NJ: Lawrence Erlbaum Associates, 1977.

A Tale Of Two Protocols[1]

Dorothea P. Simon and Herbert A. Simon

TEXTBOOK VERSUS "REAL-WORLD" PROBLEMS

In giving students practice in solving textbook problems, our objective is to help them acquire skills they will be able to apply when they encounter problems in the real world. Transfer of skills to real-world problems requires both that the skills be relevant to the problem and that they be *recognized* as relevant. But textbook problems are highly structured—all or almost all irrelevancies are stripped from them, and their solutions usually involve two or three equations or concepts just studied by the class. Real-world problems, on the other hand, come to the problem solver poorly defined—surrounded by a vast mass of information that is possibly relevant but often irrelevant.

Awareness of the cognitive processes involved in solving both kinds of problems may help us plan instruction that can help students bridge the gap between "book larnin' " and real-world problem solving. One possible technique for bridging this gap, which has been suggested by our recent work in this area, is to give students practice with problems that have some of the ambiguity and complexity of structure of the latter.

In order to gain some understanding of how people approach relatively complex, ill-structured problems, we constructed such a problem and asked a number of subjects to solve it while thinking aloud. We discuss here primarily the tape-recorded thinking-aloud protocols of two subjects who were experts in the domain of the problem—one a professional physicist, the other a chemical engineer—whose protocols, as we shall see, take strikingly different paths. Our interest lies in understanding why this was so and what lessons can be drawn for the skills required to solve ill-structured problems that are meant to reflect real-world complexity.

The problem, which we entitled "A Desperate Plight," is as follows:

THE TOM SWIFT PROBLEM: *A DESPERATE PLIGHT*

When Tom Swift and his crew were shipwrecked on the moon, they were able to salvage oxygen supplies ample for a lifetime, several miles' length of aluminum pipe, a solar-powered water pump with variable capacity, and a sack of seed corn. They immediately set about to see what could be found by way of water and food supplies.

1. This research was supported by Research Grant MH-07722 from the National Institute of Mental Health.

They found a large, and apparently fathomless, spring only three miles from their landing place. Two miles from the spring, and at the same altitude, were three acres of potentially fertile soil, and five miles in the opposite direction from the spring, and also at the same altitude, were eight acres more. One or the other of these plots, adequately irrigated, might raise just about enough food for them. The aluminum piping would stretch from the spring to either plot, but not to both.

The agronomist of the party estimated that water, not land, was the limiting factor in producing food, with expected output proportional to the total amount of irrigation water. The smaller plot would yield about six bushels of corn for every acre-foot of water applied, and the larger plot about ten bushels. (One acre-foot = 43,560 cu ft) The pump was able to perform about 20 foot-pounds of work per minute over a wide range of speeds and delivery rates.

Which plot should they irrigate in order to grow as much corn as possible? What other parameters would you need to know in order to make a decision? How much corn could be produced at most? Carry out your analysis as far as you can, using literals for any unknown parameters.

PROBLEM DIFFICULTY

Before analyzing the protocols of special interest, it may be instructive to present briefly the chief difficulties the other subjects had with this problem. All worked at least 45 minutes on it and some much longer. Throughout, we consider primarily work on the first question, "Which field should they irrigate?" since several people did not go beyond that point.

Our subjects were four graduate students in psychology and four Carnegie-Mellon science or engineering faculty members. All but one of them approached the problem as they would a standard textbook problem, as we had expected. Two of the students, who had had only a year of undergraduate physics, either could not remember or did not see how to apply the equations for work and force, and failed to arrive at a solution. One of the students, with an undergraduate degree in physics and electrical engineering, and one science faculty member neglected friction entirely and spent a good deal of time calculating— translating feet into miles, pounds of water into acre–feet, etc. A member of the engineering faculty answered the first question by asserting flatly, "That's easy. The amount of water is inversely proportional to the distance, so irrigate the closer field." The fourth of the students, who has a solid background in physics, chemistry and math, produced a very interesting protocol and a novel twist on the solution, which he worked out by analogy to the laws of electricity, but further discussion of that we will leave for another time.

Of the two expert subjects whose solution strategies we consider in some detail, one did a skillful job of solving the textbook problem embedded in the cover story; the other pulled out what he saw as the "real problem" presented by the story situation and, guided by an explicit and general problem-solving method, invented interesting possible solutions to the subproblems that presented themselves in the course of his analysis. Because of the length of the protocols we cannot reproduce them in full here but summaries and selections are presented to illustrate the discussion.

THE ABSTRACT SOLUTION

We consider first the protocol of PH3 working out the solution to the first question of the problem statement—"Which plot should they irrigate in order to grow as much corn as possible?"

PH3 is a physicist engaged in teaching and planning instruction for undergraduate physics courses. Before we gave him the problem, we spent an hour with him discussing the general lines of our research; we explained that we had found the problems in an introductory physics text too easy for an expert (Simon and Simon, 1978) and in order to try to tap more details of expert performance had constructed a much harder problem.

From the very beginning PH3 extracts from the "real world" cover story an abstract problem in hydraulics. He quickly sets his first goal as finding the relationship between the amount of water that can be pushed through the pipe per unit time and the energy expended. He recovers from memory a law (Poiseuille's Law) that expresses the relation between the velocity of laminar flow through a pipe and the force required to maintain the flow,[2] and relates this to the power of the pump. The resulting relation allows him to calculate the rate of flow as a function of the length of the pipe, which in turn permits him to determine the relative amounts of water that can be pumped to the two alternative sites and to calculate from the givens of the problem which site would produce the larger crop of corn.

A surprising (to him) but correct result emerged from his equations, namely that the flow rate is inversely proportional to the *square root* of the length, not to the length itself as the engineer had stated. PH3, who at first found this result counterintuitive, himself explains it:

> Somehow that doesn't seem intuitively right. And I guess the point is that if the pressure were constant, independent of flow rate, then the volume flow rate would be proportional to one over L. But the point is the pressure isn't constant. The pressure is determined by the flow rate and the fact that the rate of doing work is constant.

The problem as given (pump of constant power) was an unusual one. Normally engineers and physicists concerned with water supplies are faced with situations where the pressure at the pump is given, hence constant, while the variables to be solved are the pressure drop per mile of pipe as a function of the diameter of the pipe. The surprise of the physicist at his "counterintuitive" result, and the error of one engineer on the identical point, demonstrate that the problem representations that experts generate may not be completely general, but may incorporate implicit assumptions about "normal" conditions for values of variables. A system that does not satisfy these implicit conditions may be represented and analyzed incorrectly or, until reanalyzed, may create a feeling that the system behavior is counterintuitive. PH3's straightforward solution to the problem is outlined below. His very minor difficulties along the way have to do with the precise formulation of some of the variables[3] and later in not realizing immediately that the given area of the plots is irrelevant.

---

2. He does not consider the possibility that the flow might be turbulent, in which case another, empirical, law would apply.

3. One of these he soon corrects; the others become elements of the constant term and do not affect the result.

MAIN EPISODES OF PH3'S PROTOCOL

1. Read problem, note givens, draw diagram

2. Set Goal I: Find relationship between volume flow of water and other parameters
   A. Set up equation for amount of water (volume flow rate—dv/dt) in relation to pressure drop, length, and an expression that includes diameter of pipe and effect of viscosity
   B. Set up equation for work pump has to do (dw/dt) in terms of volume flow rate and pressure drop
   C. Find relationship between volume flow rate and other parameters to find how dv/dt depends on length
   D. Announce result: "Important thing to recognize is that dv/dt is proportional to one over square root of length."

3. Set Goal II: Find relative amount of corn for each plot
   A. Find amount of corn as a function of area, yield, and amount of water, evaluate crop for each plot
   B. Note error in specifying subgoal II-A: Area irrelevant
   C. Find crop as a function only of yield and amount of water
   D. Announce result: Use second plot—larger, farther plot
   E. Check and confirm result

4. Check and confirm result of GOAL II: "Water you can pump is inversely proportional to square root of distance."

5. Explain non-intuitive result of GOAL II: "Pressure isn't constant; it is determined by the volume flow rate and the fact that the rate of doing work is constant."

The goals PH3 sets relate directly to the question asked in the problem statement. He has stored in long-term memory the relevant physical concepts from mechanics and hydraulics, manipulates the variables and relations with consummate skill to arrive at a direct answer, checks his reasoning along the way, and finally examines and explicates the result. In short PH3 behaves like an ideal student solving an abstract physics problem dressed up in a fanciful cover story.

A "REALISTIC" SOLUTION

The approach to the problem taken by CE8 is entirely different from that of any of the other subjects. CE8 is a chemical engineer engaged not only in teaching university courses in the subject, but also in research on automation of chemical—process design. He undertook to work on the problem with no preliminary information as to our specific research aims in this project. Following is an outline of his protocol.

MAIN EPISODES OF CE8'S PROTOCOL

1. Read problem, list given resources, draw diagram

2. Name goal, FEED COLONY, note existence of a difference and operators

3. Generate and elaborate operators (subgoals)
   A. PRODUCE CORN
      1. List "constraints" (requirements)—given and evoked from memory
      2. Generate alternate problem solving methods—steady vs unsteady state
      3. Make assumptions about moving resources

B. MOVE WATER. List constraints, generate and dismiss alternate methods to move water

C. MOVE OXYGEN. List constraints, note problem of oxygen diffusion. Add constraint—insure contact of corn with O_2 and water

D. CONTAIN O_2 IN CONTACT WITH SOIL AND CORN
  1. Generate methods to maintain contact
  2. Note change in problem from moving water to providing oxygen
  3. Generate methods for maintaining contact

E. PRIME MOVER FOR OXYGEN AND WATER, generate methods

4. Consider question "How much corn?" Explore use of objective function to relate costs to benefits, considering constraints and resources

5. Consider question "Which plot to irrigate?" Generate methods to irrigate both plots

6. Find water flow
  A. Generate alternate methods to MOVE WATER—siphon, canal
  B. Note how to calculate flow in pipe using reference tables

7. Note how to calculate yield from givens and flow rate

8. Summary: Suggest using airlift pump, siphoning and digging canals

9. Comments on problem-solving method

CE8 is not satisfied with the question as stated. He considers the *real* problem (the "NASA problem"?)[4] to be: How can the colony be fed, or more specifically how can corn be grown on the surface of the moon, with the stock of things that Tom Swift and his friends have available to them? Starting with this broader goal of producing corn, rather than the narrower goal of delivering water to a specified locality, CE8 begins to elaborate, by means-ends analysis, the functional requisites for solving the problem. To grow corn requires not only water but also oxygen,[5] nutrients, sun, time, and temperature. These requirements are obtained not from the written problem statement but from qualitative knowledge of agriculture and botany stored in his long-term memory and evoked by the goal statement as he has formulated it.

Fixing upon the subproblem of supplying oxygen, CE8 again draws upon real-world knowledge (this time of astronomy) to observe that there is no oxygen on the moon, and therefore some means must be found to bring and keep in contact with the corn plants the oxygen the party has brought with them[6] as well as the water they have found. A considerable part of CE8's protocol is devoted to devising schemes for solving this problem. In the course of working on it, he reformulates the original problem to combine the tasks of water and oxygen transport, and drawing on his knowledge of physics and engineering he devises several

---

4. Unbeknownst to us, it appears that CE8 had at one time actually been a member of a NASA research team that investigated life-support problems on the moon. This may account for the way he reformulated the goal—not for the way he attacked the problem.

5. CE8 confuses a plant's need for carbon dioxide with his own need for oxygen. Only one other subject even mentioned the problem of gas diffusion to the plants and then considered it only briefly and dismissed it.

6. The problem statement describes the oxygen as "ample for a lifetime," a statement whose plausibility CE8 does not question. To this extent, he reacts in "science fiction" rather than "real life" mode—but in any event quite differently from "textbook" mode.

possible solutions to this joint problem that do not depend at all on the capacity of the pump. (See Appendix, lines 50-57).

CE8 goes beyond the written problem statement in other ways, for example, suggesting methods not even hinted at in the problem statement for lifting and transporting fluids. The goal of answering the question of which field to irrigate becomes only a small part of the entire problem, and is, indeed, reformulated instead of being solved in its original form. (See Appendix, lines 24-40.)

CE8 did not solve the textbook type of problem that we had thought we were presenting to our subjects. Rather, he analyzed the functional requirements of the situation. In other words, he was behaving more nearly as a real Tom Swift would if he were faced with the prospect, or the reality, of surviving on the moon under the conditions given.

COMPARISON OF SOLUTION STRATEGIES

It is not our object to evaluate the relative merits of these two approaches to the Tom Swift problem—real life, and especially professional life in math, science, and engineering constantly presents situations where one or the other or both approaches are required. Our interest here is to identify the set of problem-solving skills each of the subjects was employing and thereby to discover some of the similarities and some of the differences between these two modes of problem solving.

Both subjects use means-ends analysis in their solution attempts. They create goals and in the course of searching for ways to attain the goals create subgoals; they then repeat the process until all requirements for achievement of particular subgoals are met.

The major differences between their performances are *1.* in the initial representation of the problem to be solved, which resulted in striking differences in both the specific knowledge and the strategies called on for solution, and *2.* in the degree to which each showed awareness of his problem-solving processes.

The representation that this problem evoked in PH3, under the experimental conditions described, was that of an abstract hydraulics problem. The knowledge he called on, from the givens and from long-term memory, was that required to determine mathematically the relation of distance to the other parameters, and his worksheets display a series of equations.

PH3 at no point mentioned the problem-solving method he was using, and from the direct way he went about the task, we may assume that he was unconscious of his method—that this is just the way he goes about solving a physics problem of the textbook sort. His solution strategies, as inferred from the solution path, may be outlined as follows:

1. Note question given in problem statement
2. Abstract problem and identify problem domain
3. Identify given facts
4. Retrieve relevant physical laws
5. Retrieve and supply missing physical constants
6. Instantiate laws
7. Solve equations

These are the strategies that science teachers commonly aim to instill in students. To the extent that such strategies are so thoroughly overlearned that they become unconscious, as in our PH3, they constitute the "skills" of solving science problems—skills that are essential for professional scientists or technicians.

The principal worksheet of CE8, on the other hand, contains only a single equation. He begins by organizing the knowledge given in the problem statement, first (as almost all subjects did), by drawing a diagram; second, by listing resources, showing for each species (e.g. water), amount available and location; and finally, writing down the goal: "To adequately feed the Swift Colony." His subsequent solution attempts consist in progressively expanding the network of associations evoked by this initial representation of the problem.

The technique he used to elaborate his representation is evident throughout the protocol shown in the Appendix and is summarized in his own words below. Briefly, he *1.* lists "operators" (subgoals) and "constraints" (requirements), *2.* adds to the list as he goes along, *3.* relates the items on the list to each other and to the givens, *4.* generates solution mechanisms, *5.* evaluates them, and finally *6.* summarizes the considerations most likely to yield a workable solution.

His strategies may be outlined thus:

1. Define problem
2. Apply means-ends analysis recursively to elaborate
   requirements for solution
   a. Retrieve functional requirements from memory
      and problem statement
   b. Retrieve possible solution mechanisms
      (operators and methods) from memory

In applying his strategies CE8 abstracts no problem variables and retrieves only one formal physical law; instead he considers almost exclusively the qualitative *functional* requirements for dealing with the *situation* described by the cover story. He introduces many relations of the form: "Corn grown is a function of temperature, water, and oxygen," but he never specifies the exact (mathematical) functional form of these relations, much less the values of parameters.

This kind of problem-solving method would seem to be very productive in dealing with real-life problems—problems which, as presented, contain information that may or may not be relevant and fail to contain important relevant information. These strategies of formulating goals and searching for functional requirements to achieve them are rarely explicitly taught to undergraduate students in math and science courses.

IMPLICATIONS OF THE PROTOCOLS

The art of designing instruction in general skills for solving problems is still in the experimental stage. Much remains to be learned about the processes themselves, and this area provides the focus of our own research plans. Cognitive psychology has reached a point, however, where we may venture some hypotheses as to what these processes are, how they develop, and how they may be incorporated into instructional designs.

Technical Skills

We know from our own research and that of others that technical skill in science rests on a solid base of knowledge specific to the problem domain. In the strategy outline for our PH3 we find several crucial steps that demand retrieval of task-specific information: "Retrieve relevant physical laws," "retrieve missing physical constants." Information must be stored in memory and stored in such a way that it is accessible when sought.

We also know something about the strategies required to solve textbook problems. The solver must be able to "understand the relations" (in the sense of Reif, Brackett and Larkin, 1975, p 309), to represent the problem in abstract form, and to perform the mathematical manipulations that permit solving for an unknown.

These are the skills that science courses aim primarily to teach. The usual textbook problem is normally presented in a cover story that presumably will lead to transfer of skills to a wider context, but we should not delude ourselves that such transfer will automatically be made when the student is dealing with a real-world type of problem of the sort discussed above.

Knowledge Requirements for Real-World Problems

Knowledge relevant to the domain of the problem is as crucial to real-world problem solving as it is to textbook problem solving. CE8's solution strategies, as outlined above, include retrieval of functional requirements and of possible solution mechanisms. Here, however, "relevant to the domain" implies a greatly expanded problem space—a space that, for this problem, encompasses knowledge not only of agriculture but of astronomy, gas laws, engineering concepts, etc.

But more knowledge of these broader aspects of the domain is not enough to explain the different approaches to the problem by our two subjects. Each at some point in his protocol refers to himself as "a farm boy," and surely PH3 also has knowledge of the gas laws and at least most of the other facts considered by CE8, although probably not of all the engineering considerations. How can we explain the radically different solutions that were generated for the same problem? There are at least two possible explanations. One is problem "set" and the other familiarity with and consciousness of one or more problem-solving methods that may lead to a solution.

Psychological Set

It has long been known that performance is greatly affected by the "set" with which a subject begins a task—his attitudes and his knowledge as to expectations, difficulty, and so on. As stated earlier PH3 began work on the task expecting a textbook type of problem of some difficulty. In fact he says at one point, "I feel as though I'm a freshman doing a final examination paper." It is not surprising, then, that he proceeded in the manner he expected of his students.

CE8, however, to whom we had given the problem with no explanation as to the purpose of this experiment or what we expected to learn from his protocol, did not feel obligated to perform the task as if he were an expert student. In addition, CE8 was aware of our interest in general problem-solving processes, was thoroughly familiar with the literature in that field, and, indeed, was professionally engaged in research in which he used those methods. Hence, he very likely came to the task with the set: "Here is an opportunity for applying a general schema for problem solving."

Consciousness of Method

Although both subjects used means-ends analysis in their solutions, PH3 nowhere indicated that he was aware of the method he was using. CE8, on the other hand, made several references to his strategy, and his tape ends with comments on his approach. With his permission, we quote in full the last minute or so of his protocol for what it reveals about the method he used and its conscious application to a "really real" (as contrasted with a contrived "textbook") situation.

> Just some additional comments about this approach . . .

> You could tell I initially set this up as a means-ends analysis problem. I wrote down the resources, wrote down very carefully my operators. I'm of course a very strong advocate of this approach, because you see all the things examining the constraints on the operators reveal. They reveal all those assumptions that one jumps to when one gets a problem like this. Hidden down in a very careful listing of the constraints on a strictly functional level are all those other hidden little assumptions that get revealed with a careful examination of the constraint set. So I'm obviously a strong advocate of means-ends analysis.

> By the way, our program which does chemical process design would have fun with this, because it would ask for exactly the data I have listed here. It would ask for the resources; it already has a large set of operators, the operator PRODUCE CORN being a reaction, MOVE WATER? and so forth. [End of tape]

A SUGGESTED APPLICATION

Thus our two expert subjects, facing the same problem, exercised two quite different sets of problem-solving skills and tapped two different domains of knowledge in long-term memory. Standard methods of instruction in science give priority to one of these sets of skills and associated knowledge and tend to ignore the other. Without conscious attention and practice, the ability to transfer narrower, technical, professional skills to the discovery of the functional requirements and practicable solutions of real-life problem situations is unlikely to develop rapidly or to reach a high level. Hence, our analysis of the two protocols reveals a possibly serious gap in science education for professional practice.

What would we do if we thought it important for our students to be able to behave sometimes like the second subject, CE8? We would provide training in explicit means-ends analysis processes—given a goal, what are the requirements, how can we reach the desired goal with the resources given. Both subjects, of course, used means-ends analysis, but what did CE8 do that was different? For one thing, he formed his initial representation by starting higher in the goal hierarchy. From this representation emerged a broader goal than the stated question of the problem.

This broader goal was a functional one, which led him away from trying to solve the problem with the use of algebraic and differential equations; it led instead to a consideration of qualitative information he had stored in memory about concepts, relations, methods and devices not mentioned in the problem statement.

We would teach a problem-solving style that makes explicit the procedure of defining the broadest possible goal and using qualitative analysis to arrive at a solution. This would involve *1.* moving from goals to lists of requirements (recursively) without specifying initially the form of the mathematical functions that connect them, *2.* underscoring the importance of knowing, and *3.* reasoning about what depends on what without necessarily knowing the *form* of the dependence.

As a means of inducing "transfer," we would occasionally provide students with the appropriate psychological "set" to question the goal specified in the problem statement by giving them a problem embedded in a sufficiently complex and ambiguous story situation. For example, an otherwise routine exercise in electric circuits might be presented in a setting where students could bring to bear economic considerations—could we achieve the same result at less cost, or a better result at the same cost, by doing thus and so. Or they might be able to suggest more efficient devices that would still be practicable to achieve the higher goal, or note the need to acquire further resources before the goal could be reached.

CONCLUSION

In this paper we have demonstrated the usefulness of thinking-aloud protocols for discovering the different ways in which experts can approach the same simulated real-life physics problem and how the representation selected influences problem-solving strategy.

From the protocols themselves, from our inferences about problem-solving strategies, and from the conditions under which the protocols were produced, we have drawn some implications for science instruction that has as its goal transfer of skills to real-life situations. The typical story problems used in science textbooks in no way reflect real-world situations because they lack the ambiguity and complexity of most real professional problems. As a result, there is no assurance that the skills students exercise in solving textbook problems will transfer readily to the quite different situations they will encounter in professional life.

We have suggested that "problem set"—the expectations as to the desired outcome—with which a student approaches a problem is a large determining factor in the representation chosen and hence in the kinds of knowledge and strategies used.

And finally, we have demonstrated the usefulness in an actual professional setting of the conscious employment of a particular problem-solving method.

Other papers in this volume discuss some experimental methods that currently exist or are being developed for training, on the one hand, in the technical skills of science problem-solving, and for improving, on the other hand, generalized problem-solving techniques. Our protocol analysis underlines the need for convergence of these two lines of cognitive research and development, and leads to a suggestion for hastening this convergence.

APPENDIX. Selected Statements From CE8's Protocol

(Reads from problem statement)

1. O.K. Let's list down resources.
2. And the resources are water.
3. And so the attributes of this resource are going to be species—that's water.
4. Another attribute of this is amount available. And the amount available from this spring is infinite, according to this statement.
5. And another attribute of the spring is its location.

(Reads from problem statement)

6. So another resource is three acres . . . Well, the resource is land; let me call it Land *A*.
7. Its amount is three acres; its location two miles from spring.

(Reads from problem statement and draws diagram)

8. So the resource is Land *B*.

(Rereads problem)

9. So, it looks like the goal here—let's write down a goal—is to adequately feed the Swift population, Swift colony.

. . . .

10. O.K., now operators, we need operators. We have a difference here.
11. O.K., we have some operators. And one is called PRODUCE CORN.
12. The constraints on PRODUCE CORN are: We need land, we need water, we need sun, and we need seed corn, and we need oxygen. O.K. And various kinds of nutrients, which I will put in the land.
13. Go back to my resource list and add oxygen. Amount available infinite, in essence, here for this thing, I'll say—"for a lifetime."

. . . .

14. And at the landing site I have the oxygen supply, and the pipe.
15. O.K. Another resource—aluminum pipe.

(Considers possible uses of aluminum pipe; recaps resources and operators)

16. O.K., so we have one called PRODUCE CORN, and one called MOVE WATER.
17. And another operator called MOVE OXYGEN.
18. Because that's not really mentioned in great detail how we're going to get the oxygen over there.
19. Besides, we're going to have problems here. We have an oxygen supply, but if we let the oxygen out into the air, we've got problems, because it'll just diffuse out away from the plants, and we won't be able to grow things.
20. The oxygen's going to have to be held in some container or some device for insuring adequate contacting of the oxygen with the corn.
21. So another constraint relative to the corn is that the oxygen and water must contact for a given period of time in order for the chemical reactions to take place.

22. And so, associated with time we're going to have to develop an operator associated with containment.

23. So let me add another operator here, and that is CONTAIN OXYGEN IN CONTACT WITH SOIL AND CORN, with the constraint added in that the sun can reach them during this.

(Considers mechanisms for doing this.)

24. So we're going to have problems, and I don't think we're going to be able to pull this off unless we somehow cover this land.

25. Hence, all this moving of water and all of that is quite irrelevant.

26. If we can't generate a mechanism by which we can satisfy the constraint OXYGEN IN CONTACT WITH THE CORN for adequate time periods, the transport of water will be nonminimal. It won't be a tight constraint.

27. So this problem has now changed here, to where we were worrying initially, apparently, about moving the water, to one really where we have to develop a container for the oxygen.

28. Now here, I believe, that would be the real problem, and if we were stuck there, that would be the case.

        . . . .

29. That indeed would, it seems to me, to be the problem. And I see no way, given the amounts of material shown here, that we could carry that out.

30. In fact, if one were to do that, I would probably then move to quite a separate kind of device.

31. I'd probably move the soil to a location where I could do very high production corn producing, rather than this relatively low-yield corn producing on the land.

32. I'd go to a hydroponic kind of mechanism, where I have the oxygen . . . I'd bring the water to it in fact.

33. I'd locate next to the spring to minimize the energy I use there.

34. I'd build a relatively small—since I wouldn't have much material for building this—solar house to contain the oxygen.

35. I'd run the concentration up very high to accelerate growth, perhaps even extract nutrients, fertilizers, and so forth, from the existing soil, and perhaps even carry out a hydroponic kind of soil-free growth.

36. So I would say that's the way to solve this problem. And the other solutions associated with pumping out to the Land *A* or *B* are really not relevant, given the fact that they're on the moon.

        . . . .

37. Let me go back now to this other problem which is addressed here—"Which plot should they irrigate in order to grow as much corn as possible?" And I'll assume we know how to contain oxygen in contact with the soil and corn for adequate time, and I'll just write down as far as the constraints, they're known and solved—by some unknown means.

38. The operator MOVE OXYGEN requires oxygen at source, and a pipe, and transport mechanism.

39. In addition—let me go back now. I can see I didn't put the prime mover in the MOVE OXYGEN-MOVE WATER.

40. You have to have water at source, aluminum pipe from source to sink, and then some prime mover.

41. That could be the pump, but there are many other kinds of prime movers.

42. We might be able to get a little thermal syphon thing going here. Even though the altitude is the same, we could generate density gradients perhaps, lower gravity field, density gradients—

43. Well, probably not going to be able to move it. We don't have the size of the aluminum pipe, so it's going to be hard to do pressure drop. We can do a friction-free calculation, and see eventually if that's an adequate prime mover.

44. O.K. Back to the MOVE OXYGEN operator. Oxygen at the source, pipe, prime mover.

45. We could also, since it's not spelled out in great detail in the problem, move the oxygen at the source, physically move it to the land site.

46. Otherwise we may have to do double duty on the pipe—that is—during a certain fraction of the time we would move water down the pipe to the land. We would be irrigating during this time.

47. During another fraction of the time we could be moving with oxygen.

48. In fact we could use, if the oxygen's under pressure . . . We could take the oxygen supply and allow it to be heated by the radiation, the solar radiation.

49. That would raise the pressure of the oxygen either by, if it's liquid oxygen, we could get the vapor pressure which would be very high, or if not, we could simply use pv=nrt, gaseous expansion, and use that to drive the water down the pipe, either by using an airlift kind of foam phenomenon, or simply pure displacement with alternate flow of liquid and gas.

50. Probably the airlift would be better, then we could supply both water and oxygen simultaneously, and we wouldn't need the pump at all.

51. We could use the pump to do useful work in making our cover that's going to go over all this land.

52. O.K. So we don't even need the pump, because we could use airlift driven by the oxygen supply.

(Considers "How much corn" using cost-benefit analysis but abandons, seemingly for lack of adequate data)

53. "Which plot should they irrigate?" I think it's actually possible to irrigate both plots.

54. There's nothing that says you have to irrigate continuously. Since this is aluminum pipe, I would do the following: I'm going to water both plots . . . We're going to water Land *A* for some period of time, and then we're going to take the plumbing and drag it over . . . toward plot Land *B*.

. . .

55. So I'm going to be able to irrigate both sides, and now I'm going to be pinched by the water flow.

56. So let's just do a little bit on the water flow here. One doses an energy/

balance in order to find the flow. We can write Bernoulli's equation, which is, in essence we're going to equate the work done by the pump, is going to be equal to the work to overcome delta Z. That's any change in altitude, which you're implying there is none here.

57. Unless we can set up some kind of siphon and have it siphon it over.

58. We could actually do that. Maybe we wouldn't need the pump again.

59. We could get a siphon going. We just dig a hole at the other end.

60. Well, again, you're not going to get much head, and hence not much flow rate. So we probably will need the pump.

61. I'll leave the delta Z in there just in case we can work out—dig a hole at the other end and try to get a little siphon going.

62. In fact, maybe we just dig a canal over there . . . It wouldn't take much work to dig a little trench two miles long, adequate certainly to carry out the irrigation.

63. So let me add another subset to the MOVE WATER—dig a canal.

64. That may take too much work. Initially, at least, we'd use the pipe. In the long run, we'd probably dig the canal.

. . .

65. The main thing we're going to have to overcome is the friction due to flow. So I'm going to use my Cameron hydraulic data book, and no self-respecting Tom Swift would be caught without this, loaded with all kinds of friction-loss data in copper and aluminum tubing.

(Assumes one-inch smooth-wall tubing, 10-gal/min flow rate; indicates how to make calculation from tables, but does not carry out calculation for given yield and distance.)

66. Oh, I forgot about the oxygen . . . Let me go back here a second. This is for pure water flow. It's a different calculation if we could augment it with an airlift pump.

(Considers how to use a handbook with correlations to arrive at efficiency of airlift to get the right ratio of oxygen to water.)

67. So, actually this work from the pump is a bit of a lower bound on what we can do, because we can do better than this.

68. We can probably do better than this with airlifting, siphoning, and digging canals.

REFERENCES

Reif, F., Brackett, G. C., and Larkin, J. H. *Principles of Physics*, preliminary edition. New York: Wiley, 1975.

Simon, D. P. and Simon, H.A. "Individual Differences in Solving Physics Problems. In *Children's Thinking: What Develops?* Edited by R. S. Siegler, Hillsdale, NJ: Lawrence Erlbaum Associates, 1978.

Mapping a Student's Causal Conceptions From a Problem-Solving Protocol

John Clement

INTRODUCTION

The problem of how to describe the structure of a student's current conceptions in a given area is a fundamental one for researchers who seek to develop a theoretical foundation for cognitively oriented instruction. Recent developments in the physics teaching community, for example, have emphasized cognitively oriented approaches to teaching and the need for understanding the cognitive processes that underlie one's ability to "do" physics. Ideally, we would like to have a picture of the kinds of knowledge structures and reasoning processes that are present in beginning students and that are present in experts. Detailed descriptions of experts' knowledge structures, including those tacit knowledge structures not represented explicitly in the curriculum, would presumably help to define more clearly what is to be learned by the student. Detailed descriptions of beginning students' preconceptions and misconceptions would have value not only as a sophisticated evaluation tool, but would also make it more possible to take common preconceptions and misconceptions into account during instruction.

This paper attempts to show that it is possible to study systematically certain types of beginning students' conceptions in physics; specifically, causal conceptions in mechanics. The paper examines the conceptions a freshman student uses to understand a simple physical system involving the horizontal motion of a cart launched across a table. The task given to the student does not ask him to find a long series of actions which will solve a problem. Rather he is asked for a prediction and explanation of the effects on the system resulting from a single action. Protocols of such explanations are particularly interesting because they tend not to be limited to formal, deductive arguments, but to include informal arguments that reflect the structuring of the subject's physical intuitions.

The methodology used in this study involves two phases: obtaining problem-solving protocols via taped interviews, and analyzing these protocols to produce a model of the conceptions that underlie the student's responses in the interview. Several considerations are important to the success of this technique. An important consideration during the interview process is the attempt to encourage the student to express himself verbally as he thinks through a problem. The interviewer must also search for questions which match the level of the student's conceptions. An important consideration during the analysis phase is the attempt to model the student's conceptions at a level that is neither simpler nor more complex than the level reflected in the student's comments. In addition, the analyst must be ready to encounter conceptions that are qualitatively different from the standard conceptions used by physicists. Even when the student uses a standard term like "friction," only repeated probes by the interviewer can indicate whether the student's meaning for the term is equivalent to the physicist's.

For a discussion of the methodology of protocol analysis, see Witz and Easley (to appear), Easley (this volume), and Newell and Simon (1972).

Figure 1. Launching a Cart

TRANSCRIPT ANALYSIS

The example discussed in this paper concerns a first-semester freshman engineering student named Mark who has had a course in high school physics but who has not yet had a course in college physics. The verbatim transcript sections shown below are from a videotape of Mark working on the following problem: he launches the cart from an elastic band attached to the table as shown in Figure 1 and watches it roll to a stop. When asked, he says that the cart won't go as far if the band isn't stretched as far, and that the car attains its maximum speed near the point where the band goes slack. Both of these predictions are in agreement with physical theory. The interviewer then asks Mark to predict the effect on the motion of putting a metal weight in the cart. (Adding the weight to the cart will in fact reduce the distance traveled; this can be shown theoretically by energy considerations.)

TRANSCRIPT[1]

I = Interviewer (Clement)   S = Student

Section 1

> (S rolls cart back and forth several inches with right hand)

I: "What would happen if we put the weight in, (puts 500 g   1
        weight in cart) do you think—and used the same stretch for
        the rubber band? Do you think it would affect the maximum
        speed?"

S: (12 seconds pause) "Um, yes I do (lifts cart up with weight in   2
        it)—'Cause it seems like an awful, like a pretty heavy
        weight." (Begins to set up cart for another launch.)

I: "How would it affect it, before you try it out? . . ."   3

S: "It would affect it, it would slow it down, I would think."   4

---

1. Dashes (—) indicate pauses.

I: "But why, why would that happen?" 5

S: "Well, you got a large, a larger mass that you have to pull.— 6
(Rolls cart back and forth several inches with band stretched). It's just not gonna have as much zing to it, you know, that, you don't get—the strength of the elastic is gonna not really be strong enough to pull it as fast as it did before."

We can represent the major conception behind Mark's statements here using the diagram shown in Figure 2. This conception could be expressed verbally in the statement: "If one increases the mass of the cart, one expects the maximum speed of the cart to decrease." (We assume that the same conception embodies the expectation that decreasing the mass would mean an increase in speed.) Thus the subject's conception is thought of as an expectation of the results of a contemplated action. The numbers on the arrow in the diagram indicate specific lines in the transcript from which the presence of the conception is inferred. We call this a semi-quantitative conception because it anticipates the *direction of change* in a dependent variable (cart speed) that will be caused by a change in an independent variable (mass of the cart).

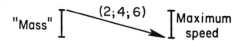

Figure 2. Map of Conceptions for Section 1 of the Protocol

This type of conception is more elaborate than the strictly qualitative conception that the weight will affect the speed in some unspecified way, but it does not go so far as to provide a quantitative mapping from particular values of the independent variable to the dependent variable. Thus the diagram is an attempt to represent the conception as it exists in Mark at this time—without making it more simple or more complex than it really is. The system of notation is adapted from a notation used by Driver (1971). She found that the eighth graders interviewed in her study used semiquantitative conceptions spontaneously to explain various physical phenomena. This type of conception is also related to the "qualitative operations" discussed by Inhelder and Piaget (1958) that precede the use of proportional laws by children as a means of understanding simple physical systems. We have found that engineering students enter college with many physical conceptions at this same semiquantitative level.

Next, the interviewer asks whether putting the weight in the cart will also affect the total distance traveled by the cart.

Section 2

I: "OK. Before you try it out, how do you think it would affect 7
the, uh, distance that it goes?"

S: (Lets go of cart so that band pulls it forward but stops it with 8
left hand after it travels only eight cm.) "It wouldn't go as far, I don't think."

I: "Why would you predict that—"   9

S: "(a) Wait a minute, wait a minute. (17 seconds pause) (b)   10
        Hm—It'd probably go almost as far. (c) (Rolls cart back and
        forth, letting it coast a little in each direction) Just the
        weight bearing down on the surface [of the table] is probably
        why I think it wouldn't go as far. (d) But it, it'd probably
        make it almost quite as far, I would think."

I: "OK. What are the factors that make it go farther, or—as far   11
        with the weight in it?"

S: "(a) One is you have the added, the added mass, (moves cart   12
        back and forth, letting it coast) so it's more weight on the ta-
        ble. (b) But then again you have the added mass—um, that's
        already moving. Once you started moving it, it's gonna help
        it move it, even, (holds hands 10″ apart and moves them si-
        multaneously several inches back and forth above the cart)
        it's gonna help it keep going a little bit (moves cart back and
        forth slightly)."

I: "With more mass?"   13

S: "(a) Yeah. But I don't think that'll overcome all the friction   14
        (brings side of right hand down on back edge of cart wheels
        several times until it moves) that it'll give it. (Starts cart
        rolling with hand, moving other hand slightly ahead of cart
        until it stops rolling on its own) (b) See, once it starts rolling,
        it rolls pretty good."

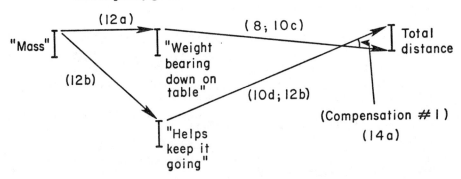

Figure 3. Map of Conceptions for Section 2 of the Protocol

The diagram in Figure 3 gives a model of the conceptions that are active in
Mark as he generates his comments in lines 7-14. Basically, his conceptions take
the form of a *network of causal expectations*. The diagram indicates that Mark
considers two competing factors that will affect the distance traveled by the
cart. First, the increased mass will increase the "weight bearing down on the
surface [of the table]." This will in turn decrease the total distance traveled.
This type of relationship is indicated in the diagram by the notation

A|➔|B|↘|C, translated: "Increasing A causes an increase in B," and "Increasing B causes a decrease in C." Secondly, however, increasing the mass in the cart according to Mark also "helps it keep going more," which has an opposite effect and causes an *increase* in the total distance. Thus Mark expresses his understanding of a relationship of *compensation* between two competing intermediate variables affecting the total distance in opposite ways. These conceptual structures appear to be intuitive in the sense of not being based simply on verbal facts learned in school. Since both of these variables are increased in the experiment, Mark has some difficulty in predicting the result, as evidenced by the way he changes his answer in line 10.

At this point Mark diverts his attention momentarily to observe that the cart seems to roll somewhat more easily when he pushes it to the right than when he pushes it to the left, and he conjectures that the table may not be perfectly level. He then returns to the problem at hand, and in response to a question from the interviewer integrates his conceptions from the first and second sections to produce another compensation relation.

Section 3

S:	"(a) Um, I would say the added, the added weight probably doesn't do all that much except to give it a little, not—you know, a little bit of added friction, (b) and I'm sure it won't go as fast off the start, just 'cause the strength of the elastic band isn't—(stretches band between fingers)"	15
I:	"Would that affect it?"	16
S:	"What?"	17
I:	"How far it goes?"	18
S:	"The strength of the band?"	19
I:	"No, it won't go as fast at the start, you said."	20
S:	"(a) Yeah. Uh, oh—(scratches head) How far it goes?—Yeah, it'll affect how far it goes, (b) but this added mass is gonna tend to, (rolls cart back and forth several inches) I think, keep it rolling maybe a little better even if it doesn't have that maximum speed."	21
I:	"Uh huh.—Why does it keep it rolling, just to have more mass? Does that just seem that way, or is it like—?"	22
S:	(Continues rolling cart back and forth, letting it coast.) "I don't know, you know, like, when a car goes down a hill?"	23
I:	"Yeah."	24
S:	"Or when you go down sledding? if you have two people on a sled, it goes better than if you have one person?"	25

I: "Hm."   26

S: "Just, uh—(moves cart back and forth) it seems the added   27
          mass would just give it more of a, (moves left hand with back
          of hand leading in quick motion over the cart's original path)
          more momentum maybe, once it gets started."

Figure 4 shows how the two preceding diagrams can be combined to account
for many of Mark's comments in this section. He expresses a second relationship
of compensation in line 21 between increased "momentum" and decreased
maximum speed. These two factors have opposite effects on the total distance
traveled. Thus he is able to *coordinate* his conception involving maximum speed
from the first section with the conceptions he expresses in the second section.

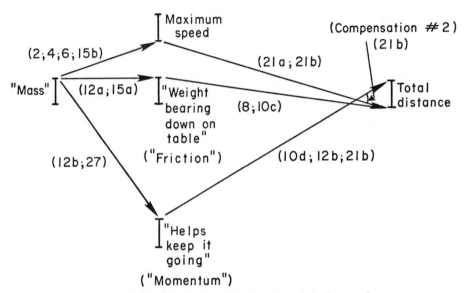

Figure 4. Map of Conceptions for Complete Protocol

This coordination, however, appears to be at least partially provoked by the interviewer when he asks Mark specifically about whether maximum speed will
affect total distance. Had Mark made this connection in a spontaneous manner,
he would have been linking two separate conceptual structures spontaneously
via a variable common to both structures. We suspect that this kind of coordination process is an important one and we are attempting to study it in other protocols. The process may underlie the ability to plan the linkages between theory
domains that are necessary in solving multi-step problems in mechanics. In the
previous section (lines 10-12) where Mark suddenly recognizes the first compensation relationship, he indicates that he can make such coordinations spontaneously under certain circumstances.

Mark was one of a group of 15 freshmen who were given the cart problem, but
only his protocol will be discussed here. All of these students were able to give
semiquantitative predictions and Mark's response was fairly typical although the

number of variables he relates semiquantitatively (five) is somewhat above the average for the group. Many, but not all, of the other students detected compensation relationships and generated analogous cases in a similar way.

It is difficult to score Mark's performance on the question as correct or incorrect, because he never does make a strong prediction about the effect added mass will have on the distance the cart travels. For the purpose of studying his conceptions, however, labeling his answer as "correct" or "incorrect" is actually an irrelevant concern. Indeed, the interesting part of his response is not whether it is correct, but the picture his *explanations* give us of his highly structured conceptual model of the situation.

However, some issues cannot be resolved on the basis of this short interview. First, one can ask: "Are Mark's conceptions a result of his high school physics training, or are they primarily self-constructed?" We suspect that the answer here is, "primarily self-constructed," but we really cannot be sure of this on the basis of the written transcript. A second question is: "Since Mark physically manipulates the cart to some extent, are some of his conceptions actually constructed by him on an empirical basis during the interview, rather than being preconceptions?" It seems unlikely that Mark's conceptions are constructed "from scratch" during the interview, but we cannot be certain about this issue either, on the basis of the transcript. Given that we cannot firmly resolve these issues concerning the origins of Mark's conceptions, the transcript is nevertheless a rich source of information concerning the nature of his current conceptions.

A primary source of confidence in our model of Mark's conceptions (as represented in Figure 4) are the multiple entries of transcript line numbers attached to each semiquantitative relationship. There are at least two points in the transcript supporting each relationship shown in the diagram. This means that the diagram is based on repeated *patterns* in the content of Mark's comments. This gives us a measure of confidence in the validity of our micro-theory of what his conceptual structures are like, because we have exhibited a network of ties between specific aspects of the theory and specific observations from transcript data.

STRENGTHS AND LIMITATIONS OF MARK'S CONCEPTIONS

It is instructive to consider some of the strengths and limitations of Mark's conceptions. He seems to shift back and forth on the question of the effect of adding mass, first leaning toward predicting a decrease in the total distance traveled and then toward an increase. This is understandable, since a conceptual system at the semiquantitative level is powerful enough to *1.* model several of the causal relationships that are operating in the experiment, and *2.* predict the effect of a change in either maximum speed or friction on distance traveled; but his conceptual system is not quite powerful enough in this case to predict a definite answer for the question of how adding mass will affect the distance traveled. Thus his uncertainty in this case is an effect produced by the limitations of the knowledge structures available to him, not an effect produced by careless reasoning.

Several other limitations of Mark's conceptions can be noted. He uses the terms *friction, mass,* and *momentum,* but it is not at all safe to assume that these terms carry the same precise meaning for him that they do for the physicist. Thus, we must not make the mistake of assuming a one-to-one correspondence between the external use of a standard term and the internal use of a standard concept. For example, he uses the word 'friction' in lines 14a and 15a. The physicist here would use this term to refer to a force pushing back on the car. Mark, however, does not give evidence of having this concept. Instead, he says in line 10c: "Just the weight bearing down on the surface is probably why I think it wouldn't go as far." This comment is similar to those of several other freshmen we have interviewed who appear to conceive of friction as purely a downward force which is seen to retard motion solely by reason of being in a different direction from the motion. So one should give Mark credit for recognizing that increasing the mass will cause more of a retarding influence, but one cannot assume that his conception of friction is identical to the physicist's conception, even though Mark uses the same term that the physicist uses.

Similarly, it is not clear that Mark differentiates between the meanings of his terms *weight* and *mass*. This limitation becomes clearer later on when he is asked to predict what would happen if the same experiment were tried in outer space. He predicts that adding mass will have no effect on the maximum speed, contrary to the physicist's point of view. Also he uses the term *momentum* at the end of the protocol, but it is interesting to examine the way in which he separates his comment that the added mass will reduce the maximum speed attained during the first part of the motion from his comment that the added mass "helps it keep going" in the latter part of the motion. The physicist conceives of the inertia of a quantity of mass as the resistance of that mass to any change in velocity, whether it be acceleration from rest or deceleration from a state of motion. However, the protocol does not indicate that Mark has an integrated conception which assimilates these two situations. Rather, it indicates that he has separate conceptions for thinking about each of them.

Finally, Mark generates two analogous situations to explain the way in which added mass "helps it keep going." He refers to the way "a car goes down a hill" and to the way a sled "goes better" with two people rather than one. This type of spontaneous analogy construction on the part of the student in order to make sense of relationships between more abstract concepts like momentum and force is certainly to be encouraged. But it is clear from the transcript that the analogies are not evaluated with the level of precision required in physics, and this is an ability that Mark needs to develop further. See Clement (1977b, 1978) for further discussion of spontaneous analogies.

Given these limitations in Mark's performance, we were still impressed with the complexity of his knowledge structures in this area and with the potential power of his semiquantitative reasoning processes. First, he does not simply make a prediction from his global impression of the cart's behavior. He engages in an *analysis* of the situation by successfully isolating several variables that will affect the cart's motion and by considering compensation relationships between these variables.

Second, we can contrast Mark's analysis of independent factors affecting the cart with the type of symbol manipulation response we would expect from other students who have had a course in mechanics but who have "gotten through" largely by memorizing formulas. Typically, these students have not developed causal conceptions at a deeper level of understanding that provide a foundation for understanding the quantitative relationships in the formulas. Such students have difficulty answering a question about a formula that goes deeper than the superficial level, such as: "Can you give an intuitive justification for the formula, F = ma ?" An example of an answer that would indicate some intuitive understanding of the principle is the following: "I can think of pushing a certain mass to accelerate it. If I want a larger acceleration I'll need a larger push. If I want the same acceleration with a larger mass, I'll also need a larger push. Therefore the formula makes sense." Or consider the following excerpts from a response given by Ron, a more advanced physics student:

I:	"I'm wondering if you can explain a rationale for the formula F = ma, that makes sense to you . . ."	1
S:	"Alright . . . let's choose two objects. If we exert the same force to both these objects, which differ in mass, the result is gonna be the lighter object will be displaced at a greater rate . . ."	2
S:	"And by increasing 'm,' 'a' will have to decrease.	3

These are semiquantitative arguments that indicate that the formula is grounded intuitively at this level for the student and is a principle that he believes in with some conviction. Mark's arguments are of this same form. He is engaging the problem at a deeper conceptual level than a student who can merely juggle formulas algebraically.

Third, Mark's conceptual system is very different from a memorized set of isolated facts. We can describe this system, represented in Figure 4, as a network of causal expectations. Each A | → | B relation in this network represents a dynamic, action→result expectation. These form an interconnected network of action-oriented conceptions rather than a collection of isolated facts.

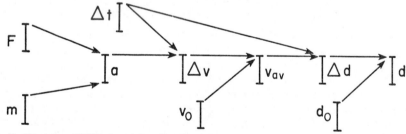

Figure 5. System of Conceptions for Understanding the Movement of a Particle Under a Constant Force

Fourth, although semiquantitative knowledge structures lead to predictions that are less precise than those from structures involving quantitative functions, they can still allow for a significant degree of predictive power. For example, the diagram in Figure 5 models a system of semiquantitative conceptions that can underlie a physicist's understanding of the movement of a particle under a constant force. Such a conceptual system can provide a *causal understanding* of this type of motion in terms of finite intervals at a pre-calculus level. As such, it can underlie and support the more refined relationships represented in the formula:

$$\frac{F}{m} \frac{\Delta t^2}{2} + v_0 \Delta t + d_0 = d$$

It is intriguing to note that such a conceptual system seems well-adapted for the task of selectively assimilating those problem situations where the use of the associated formula is appropriate and thus providing the subject with a semantic basis for deciding *when to use* the formula. Simply memorizing the formula provides no such knowledge.

We can list some characteristics of the inferences that appear to be possible using semiquantitative structures:

1. the predictions made are transformational rather than static in the sense that one predicts that a *change* in A will produce a *change* in B;

2. such A→B anticipations can be linked to inference chains of potentially unlimited length;

3. such A→B anticipations can reflect empirically observed relationships as well as causal theories (these two kinds of relationships may require different cognitive explanations at a deeper level);

4. there is an implicit logic for how such anticipations can be combined; for example A| $\overset{\rightarrow}{}$ |B implies A| $\underset{\rightarrow}{}$ |B and (A| $\overset{\rightarrow}{}$ |B, B| $\overset{\rightarrow}{}$ |C) implies A| $\overset{\rightarrow}{}$ |C. Significantly, the students we have interviewed all seem to be at home with these implications—these appear to be *natural inference patterns* that make sense to them intuitively;

5. several independent variables can be linked to a single dependent variable, and this can be represented by tree-like structures such as A| $\overset{\rightarrow}{\nearrow}$ |C;
   B|

6. these tree-like structures can provide a basis for representing the control of variables in an experiment and for compensation relationships between independent variables;

7. more complex causal conceptions can be represented by the analyst as lattices or feedback networks using the same elements. Thus the potential power of causal networks of this type appears to be far from trivial.

We can now summarize our discussion of the several weaknesses and strengths in Mark's conceptions concerning the cart experiment. His concepts of mass, friction, and momentum do not seem to be as refined as those of the physicist. His use of analogies is not as precise. He almost never refers to quantitative functions. On the other hand, he is able to isolate several of the most important relevant variables, including the effect of increasing weight on the momentum of the cart. He also identifies compensating variables. His conceptions appear to form a network of causal chains which give him a first-order understanding of the system, and this integrated conceptual structure is quite different from a set of memorized formulas. Finally, certain types of natural and potentially powerful reasoning processes seem to be associated with knowledge structures of this type.

IMPLICATIONS FOR INSTRUCTION

Mark's conceptions are *not* equivalent to those that a physicist would use, but our analysis has shown that he is definitely not a "blank slate." This raises the possibility that certain of Mark's conceptions can serve as starting points for learning—that Mark can build on what he already knows by modifying his existing knowledge structures rather than starting from scratch. The way in which students' preconceptions are taken into account in a particular course will depend on the educational goals of the course. In the case of courses where the student's primary goal is to gain an understanding of Newtonian physics, the student needs to become aware of the similarities and differences between his intuitive preconceptions and the Newtonian point of view. By analyzing Mark's conceptions we have already identified three specific points where he should be able to build on his current ideas: discriminating between mass and weight; elaborating his concept of surface friction by establishing a causal relationship between normal forces and retarding forces; and integrating two concepts of inertia as resistance to acceleration and deceleration. It is unlikely that these changes in his beliefs at a causal level will happen automatically simply as a result of memorizing formulas. It is clear that he will need to construct new conceptions at a semiquantitative level in addition to learning new quantitative relationships. These are the conceptions that will provide a semantic underpinning for the equations he learns. Building on his existing conceptions as outlined above would appear to be a more fruitful approach to this task than attempting to build a new conceptual system from the ground up. Thus, we suspect that networks of semiquantitative conceptions of this type represent an important level of knowledge in students that must be taken into account in standard physics courses if superficial formula memorization approaches are to be avoided.

However, the ability to identify students' conceptions should also have particular value in courses which take the development of methods of scientific inquiry as their primary goal. Such courses may encourage students to: 1. articulate their preconceptions as hypotheses; 2. design experiments (or thought experiments) to test them; 3. modify and refine their conceptions; and 4. select the most successful theories from those shared in class discussions. Students who are successful in completing such cycles go through an important learning experience regardless of whether their working hypotheses are correct from the physicist's point of view. Conceptual mapping techniques should make it possible to 1. identify appropriate topics for such courses which tap rich sets of preconceptions in students and 2. document changes in students' knowledge structures in such courses.

Clearly it is not possible for an instructor to analyze each student's preconceptions at this level of detail and respond accordingly to each student individually. But while it would be hard to find another student whose conceptual system matches Mark's exactly, individual elements of his system do appear to be identical to the conceptions of many of the 15 other freshman students we have studied using the cart problem. Thus there appears to be a small set of common preconceptions concerning momentum, friction, force, velocity, etc., and there is a need to compile a catalogue of these common preconceptions as an aid to teachers.

The fact that Mark does not give a decisive and correct answer to the question posed about whether the cart will travel further means that his *performance* on this question would traditionally be interpreted as poor. Yet, as we have seen, Mark does some impressive thinking about the problem in terms of his own conceptions. A wrong answer in this case does not imply that no useful thinking has taken place. This indicates that a special kind of patience and sensitivity is going to be required from teachers who wish to help students develop principles that are anchored in physical intuition.

DIRECTIONS FOR FUTURE RESEARCH

Models of student knowledge in the form of semiquantitative structures like those discussed here capture important aspects of intuitive conceptions of force and motion. As with all models, however, we can hope to eventually expand and refine them as we become more sensitive to the fine structure of the phenomena being observed. I will comment on several limitations of the analysis given here as an indication of directions for future research: first, the question of how the conceptions are accessed or activated; second, the need to give a more detailed description of what the "units of representation" or "units of meaning" are in Mark's knowledge structures; and third, the need to somehow account for Mark's hand movements which seem to parallel his explanations in such an important way.

The question of how Mark activates or accesses his conceptions is a difficult one to answer on the basis of the short transcript analyzed here. Since it seems that subjects are often not conscious of the access process, we will need case studies involving the same subject working on several related problems to make inferences about which aspects of a situation are involved in triggering a particular conception.

With regard to the second question, the static collections of symbols used in these diagrammed cognitive models are necessary for purposes of notation, but, like equations in physics, they can be used to represent elements that are either static or dynamic. Although the semiquantitative diagrams that we have used as a first-order model of the student's conceptions are discrete relational structures, this does not commit us to a position on the issue of whether the internal cognitive structures they represent are best thought of as static symbols in the form of propositions or dynamic (time-varying) patterns of functioning in the form of action-oriented schemes. Cognitive models involving continuous, dynamic structures have been proposed recently by Witz and Easley (1971, and forthcoming), and Shepard (1978).

Consider the case where Mark says:

12(b) "But then again you have the added mass—um, that's already moving. Once you started moving it, it's gonna help it move it, even (holds hands 10″ apart and moves them simultaneously several inches back and forth above the cart) it's gonna help it keep going a little bit (moves cart back and forth slightly)."

In the first order theory of Mark's knowledge structure already given, we represented the conception behind this statement with a structure in the form $A \uparrow \rightarrow \uparrow B$ where A is a conception of "mass" and B is a conception of "tendency to keep going." However, when conceptions like these are eventually modeled at a deeper and more detailed level, it may be in terms of either dynamic action schemes and kinesthetic imagery with characteristic time constants governing their coordination, or it may be in terms of static relational structures with "slots" to be instantiated and inference rules for replacing one symbol with another.

These choices reflect a current area of controversy over the nature of internal representation and the nature of the units of meaning in cognitive psychology. The resolution of this issue will come only after much more empirical and theoretical work has been done, but the transcript analyzed here does suggest that while nondynamic models may account for some features of the behavior, dynamic, action scheme models will be needed to account for other features. That is, we must consider the possibility that going through an action vicariously over a period of time and vicariously experiencing its effect, are activities intrinsic to this kind of causal knowledge.

Several factors lead us to consider this possibility. The semiquantitative relations used to model Mark's conceptions are of the form $A| \rightarrow |B$, where this means that a *change* in A leads to a *change* in B. It is very natural to propose that this $A| \rightarrow |B$ notation represents an action-based scheme for doing action A and anticipating the direction of change in B. The vicarious operation of such a scheme without external actions could involve internalized actions and kinesthetic and visual imagery. The fact that Mark's statements are accompanied by hand motions and hefting-like manipulations of the equipment also suggests that action-based structures involving kinesthetic feedback are central to Mark's thinking in this interview. This is especially true whenever he talks about the way the added weight will affect the cart's motion by being "a large mass that you have to pull" at the start, and by "helping it keep going" once it gets started. During almost all of the time he gives his explanations he is looking directly at the cart. The hand movements take place over periods of seconds and are often repeated several times. He also tends to redescribe and rephrase several of his explanations as he stares at the cart.

All of these observations suggest the presence of intuitive nonverbal conceptions which become active in Mark over periods of two to 10 seconds and which are responsible for his awareness over these time periods of the visual and kinesthetic effects of some imagined action involving the cart. This suggests that the knowledge structures responsible for Mark's physical intuitions, like overt actions, must *function continuously over a period of time* and *involve the motor-kinesthetic and visual systems at some level* in order to be meaningful. If these preliminary indications are confirmed, it will be necessary for theorists to be as creative and open-minded as possible in order to develop more detailed models of physical intuition that account for these aspects of the behavior.

SUMMARY

We have tried to show that it is possible to study systematically students' conceptions in physics, specifically in the area of mechanics, by using protocol analysis techniques. This can be done even when the basic concepts that the student uses are physical intuitions that are not equivalent to standard concepts used by the physicist. Such an analysis provides a much richer source of information about students' knowledge structures than do written tests.

With regard to methodology I believe that we are at a stage in the science of studying complex cognitive processes where the primary need is to develop viable qualitative models of cognitive structures. The analysis of many more protocols at various levels of detail should provide a needed background of rich phenomena as a fertile ground for the development of such models. The inclusion of verbatim sections of transcript in such studies provides an important constraint on the model construction process, since the requirement that a proposed model be consistent with as many aspects of the transcript as possible is a demanding one in the case of extended protocols.

In the analysis given here, the student's conceptions were modeled as a network of causal expectations. It was suggested that causal conceptions of this type represent an important level of knowledge in students that can provide an intuitive foundation for understanding many quantitative laws and that students' preconceptions are natural starting points for building such a foundation.

The fact that this type of conceptual mapping is possible opens up the potential for describing differences in the knowledge structures of an individual at two different points in time. This in turn holds potential for the development of more sophisticated evaluation tools; and the development of new instructional strategies which take typical preconceptions into account and which foster a deeper level of understanding in students.

REFERENCES

Clement, J. "Some Types of Knowledge Used in Physics," Technical Report. University of Massachusetts at Amherst, 1977a.

Clement, J. "The Role of Analogy in Scientific Thinking: Examples from a Problem-Solving Interview," Technical Report. University of Massachusetts at Amherst, 1977b.

Clement, J. "Catalogue of Spontaneous Analogies Produced by Students Solving Physics Problems," Technical Report. University of Massachusetts at Amherst, 1978.

Driver, R.P. "The Representation of Conceptual Frameworks in Young Adolescent Science Students," Ph.D. Dissertation. University of Illinois, Urbana-Champaign, 1973.

Easley, J.A., Jr. "The Structural Paradigm in Protocol Analysis." This volume.

Inhelder, B. and Piaget, J. *The Growth of Logical Thinking from Childhood to Adolescence.* New York: Basic books, 1958.

Newell, A. and Simon, H.A. *Human Problem Solving.* Englewood Cliffs, NJ: Prentice-Hall, Inc., 1972.

Shepard, R.N. "The Mental Image." *American Psychologist,* 2, pp 125-137, 1978.

Witz, K.G. and Easley, J.A. Jr. "Cognitive Deep Structure and Science Teaching." Presented at a conference entitled "Operations et Didactiques," Centre de Recherche en Didactique, University of Quebec, Montreal, 1971. Reproduced in Appendix 11, "Analysis of Cognitive Behavior in Children," Final Report, Project No. 0-0216, Grant No. OEC-0-70-2142(508). United States Department of Health, Education and Welfare, Office of Education, 1972.

Witz, K.G. and Easley, J.A., Jr. "New Approaches to Cognition." In *Neo-Piagetian Perspectives on Cognition and Development,* edited by L.V. den Daele, J. Pascual-Leone, and K. Witz, New York: Academic Press, in press.

On Learning to Balance Perceptions By Conceptions: A Dialogue Between Two Science Students

Jack Lochhead

In this paper I present the complete transcript of a conversation between two college students who are grappling with a formal reasoning task. The problem asks them to construct a mathematical formula which describes how to balance a multiple hook balance beam. The dialogue lasted slightly over an hour; the transcript is essentially unedited, only a few brief and irrelevant passages have been deleted.

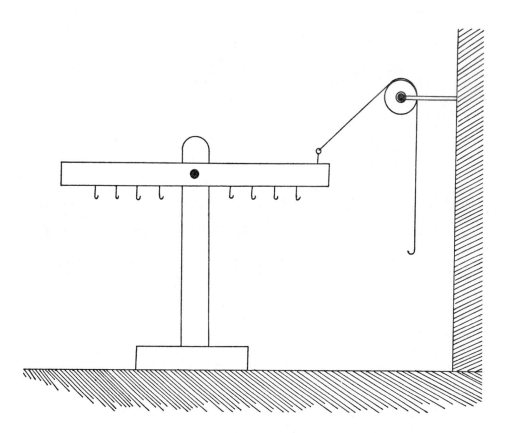

By presenting a complete transcript this paper breaks many traditions and some will dismiss it as unscientific phenomenology. However I am convinced that we cannot hope to make meaningful theories for higher order human learning processes without completing extensive empirical investigations. The bulk of this paper therefore presents raw data—the words and actions of two students engaged in the process of learning. My intention is to offer the reader the opportunity to observe phenomena which are usually difficult to see and to therefore convey a feeling for the learning process which more condensed descriptions lose. I will begin by setting the scene and describing some of the things I have learned from the dialogue.

Inhelder and Piaget (1958) have described a series of experiments through which they investigated characteristics of formal operational thought. From this work, Karplus, Renner and others have devised tests whose purpose is to determine if a particular student regularly employs formal reasoning or depends instead on the less sophisticated precursor—concrete operations. These same investigators have shown that between 25 and 50 percent of the college population is deficient in formal reasoning skills as measured by these tests. Arons (1973, 1977) Fuller (1977) and Renner (1972) describe college-level programs which are designed to encourage the transition from concrete to formal reasoning. Similar programs are now in operation at many colleges and universities across the country.

Although many people have worked on the problem of promoting the transition to formal reasoning, no one has so far described actually observing such a transition take place. This is true in part because the full transition is a gradual process which takes place over a period of months and which cannot be localized to an interval of several minutes' duration. However, the transcript presented in this paper shows that certain aspects of the transition can be observed. The dialogue reveals a relatively brief period of time during which two students made significant changes in the way they thought, and it is possible to identify the precise moment when their discussion first begins to include formal arguments.

Several weeks before the experiment, both students were tested on a variety of standard formal-reasoning tasks including the paper clip man (Karplus, 1978), the length of a building's shadow, and the flexibility of rods. They solved none of these tasks entirely correctly, although each showed partial understanding of one problem. One week after the experiment described in this paper, both students were shown their work on these tests. They found it inconceivable that anyone could fail to solve the problems and were unable to explain why they had. This is the classic transitional amnesia which Piaget has observed so often. Furthermore, during the same week, one of the students reported that he was beginning to understand the function of x in algebra. I worked closely with both students over several months, and it is my subjective opinion that the learning which took place during this hour was one of the most important single events in the development of their reasoning abilities. It is therefore tempting to view the dialogue as being symbolic of the concrete-formal transition.

The problem which the two students (whom I will refer to as George and Joe) are given to solve is to determine a mathematical equation that describes the equilibrium state of a multiple hook balance. The ways in which children understand this type of balance have been studied extensively by Inhelder and Piaget (1958), Klahr (1976), Driver (1973) and Karmiloff—Smith and Inhelder (1975).

The balance beam is now a standard tool both for diagnosing students' levels of cognitive development and for promoting that development. It has been used from elementary school (Elementary Science Study) to University (Lochhead, 1977). George and Joe come to this task with one semester of prephysics, a course that places heavy emphasis on the importance of ratios. Furthermore, George had actually worked with the balance and had determined an incorrect rule involving the addition of length and mass. He had been told that the proper rule involved multiplication but he replied that multiplying did not feel right to him and remained unconvinced.

Probably the most striking single characteristic of the dialogue between George and Joe is repetition. In the short term they constantly repeat and rephrase each other's statements. In the longer run they seem to regularly oscillate between understanding and confusion. Any experienced teacher will recognize that such oscillations are a fundamental characteristic of human learning, but so far learning theories have neglected to explain them. I believe they are associated with the learner's need to construct multiple representations of new knowledge and to test these representations against each other. Thus, as soon as one mode of representation has been mastered, the learner is apt to switch to another in order to broaden his grasp. In the dialogue between George and Joe, many of their apparent reversals are associated with changes in both the form and level of the descriptions they employ. What on first reading seems to be haphazard fluctuations appears on more careful investigation to be a systematic buildup of increasingly powerful models. In fact these models follow a sequence similar to the four stage series described by Klahr (1976).

The series starts with Model I in which the only variable considered is the number of weights on each side and evolves through two partially consistent models to Model IV in which weight and distance are properly coordinated. By representing these models as decision trees Klahr shows how each logically follows from its predecessors and involves increasing complexity as well as superior predictive power.

As the sophistication with which George and Joe view the problem increases, so does the precision of their descriptions. The clarity of their language mirrors the clarity of their thoughts. What is remarkable here is that they are able by themselves without any outside "expert" input to detect and correct many inappropriate uses of terminology. These corrections take place spontaneously once the need for them has been discerned. On the other hand when I, the instructor, attempt to draw distinctions which have not yet become meaningful to the students, they gain little or seriously misinterpret the intent.

Much of Joe's learning seems to be complicated by his use of poorly understood jargon. He is aware that certain words and forms of mathematical representation carry academic prestige and he invests a considerable amount of time in developing such formalisms. It is my opinion that this search for status is inherent in learning and its proper control is one of the most difficult strategies to master. The act of learning inevitably creates insecurity because it can only occur after the learner has recognized failure within his current schemes. Joe's desire to find security in subscript notation is one example of this problem.

The current emphasis on promoting formal thought has tended to denigrate concrete reasoning. However, many important reasoning processes are not formal. During the early portions of the dialogue, George and Joe demonstrate that they already possess many sophisticated problem-solving skills. They begin by breaking the problem into parts. They learn by trial and error, but they do so efficiently and avoid trials which will merely confirm what they already know. They are able to build up semi-quantitative models involving direction of change relationships and they use these effectively in deriving more mathematical concepts. Most important, they learn; and they do so without an instructor. My brief role as an instructor in fact appears to be mostly a hindrance. It is well-known that teachers tend to talk too much and too fast. What is much harder to see is how their monologues tend to interrupt the students' flow of thought. When I return to the room after an hour's absence I assume that more has been accomplished then the evidence suggests. Because I see a few indications that George and Joe have reached a correct formula, I assume that they have. This is probably the most common error that teachers make. One incidence of a correct response is not sufficient evidence of understanding. The cyclical nature of learning and the multi-representational characteristic of understanding mean that a student's grasp of a subject can only be inferred from varied and interrelated responses. In the context of the dialogue that precedes it, my interaction with George and Joe seems ludicrous, yet it is probably typical of nearly all instruction.

THE PROBLEM

George and Joe are presented with the apparatus shown in Figure 1. They are asked:

1. to figure out how the system works;
2. to formalize that knowledge so they can write out instructions for making the apparatus balance;
3. to construct a mathematical formula that describes the behavior of the system.

The balance in this experiment was "homemade" and used bent paper clips as hooks. In the dialogue the term *paper clip* is often used instead of *hook*. To the right of the dialogue a vector-like notation is used to keep track of the current state of the balance. For example (0100 0200) means that one weight is on the third left hook and two weights are on the second right hook.

TRANSCRIPT

Jack: ". . . try to figure out how this thing works. It may take a while and it may not. You've already worked with something very similar to this. . . . Then once you've figured out how the system works—this is the hard part—see if you can write down instructions so that somebody else using it will be able to make it balance (even if they didn't understand how it worked) by following your instructions."

Joe: "Do you want a pictorial view of what it looks like?"

Jack: "Whatever it takes so that you can do that well enough for somebody else to understand it. Finally, see if you can come up with some mathematical expressions which describe how it works. OK? And in particular, looking for things which work by adding together; things that work by multiplying together, or subtracting . . . . OK? . . here's a formula behind this . . . you see if you can invent one, OK? So fiddle around with it; 40 minutes, something like that; I'll come by and check how things are going . . . . There's some paper . . . . got something to write with? Try to do some talking . . . ."

Joe: "First of all, we've got to figure out how this works, right?"

Jack: "Yeah."

George: "How to balance it?"

Jack: ". . . no matter what the combination of weights I put on here, you ought to be able to figure out what weights to put on to make it balance . . . . OK, so if I put one weight on here . . . and another weight on here . . . you ought to be able to figure out directly what weights to put on without doing trial and error; you ought to be able to just look at it and say, 'Oh, that's there, and that's there, and so I need a weight . . .'"

George: ". . . over here . . ." (laughs; some confusion and laughter as he apparently puts the weight in the wrong place)

Jack: "OK. So you gotta have about . . . ."

Joe: "These are all evenly spaced, right?"

Joe is inquiring about the spacing between the hooks.

Jack: "Yeh, these are evenly spaced . . . . The question is, what is *this* like, in terms of those, and so on. OK? And there's a pen—try to talk to each other so there's something on the tape . . . . 'cause otherwise we won't be able to retrace what you were thinking at different times."

George and Joe agree to all this and Jack leaves.

Joe: "This is similar to the other one, only if this string wasn't hanging here (over the pulley)."

George: "Why don't we do it without that . . . that string on that?"

Joe: "OK, just ignore the string, just leave it the way it is without putting any weights . . ."

George: "Shall we take it off?"

Joe: "Yeah, take it off . . . ."

George: "Try to balance it without this here string on."

Joe: "Without the weight it will balance itself because there's an equal amount of distance on both sides . . . divided right in the middle."

Having reduced the complexity of the situation, the next step is to define the problem. In the following section, notice that Joe is very vague about several critical aspects of the problem. He uses "same area" instead of "same place." His explanation of the tilt fails to indicate that he has a scheme for forming causal links; ". . . it would tend to go left because it's unbalanced." On the other hand, his use of the word "compensate" shows that he is consciously searching for reciprocal relationships.

Joe: ". . . both sides equal so they're in equilibrium; so therefore it will stay . . . balanced, OK?"

George: "So the problem is, if I put a weight here . . . where would you put the other one?"   (0000 0100)

Joe: "You put a weight at one end . . . . You should put it in the same area to balance it . . . . the same number of spaces 'cause the weights are the same, so if you put five pounds here, you have to put five pounds in the same area to compensate for that equilibrium, right? . . . so they stay the same. Now, if I was to put *this* weight on the left-hand side at the far end, then there would be a difference; it would tend to go to the left because it's unbalanced."   (1000 0100)

The following section contains a series of experiments that can be viewed as random play directed towards obtaining a greater familiarity with the balance. However, several themes emerge from this activity. George states early that nothing is to be learned from trying obvious arrangements they know will work. Both students naturally try small perturbations rather than large reorderings. They are investigating how such changes affect the balance; in other words they are searching for a causal link between position of the weights and tilt. The session ends with Joe asking, ". . . why does it work?" This question indicates that they are ready to move from play to theorizing about the cause-and-effect relationships.

George: "OK, so we would have to put it to the right in order to balance it . . . ."

Joe: "OK, so now the same thing: . . . all right, you have to put this weight on the right-hand side at the end to balance with the weight on the left-hand side. Try it and see what happens . . . ." (1000 0001)

George: "I know . . . it would balance. What about taking this weight?"

Joe: "OK, take that weight . . . . you could try to add another on and see what happens; then you can add one at the end, but let's try it."

George: "No, no . . . let's try the heavier one on one side.   (1000 0100)

Joe: "No, no . . . we'll try *it* once, just to see what happens, right? So you add one more to the right-hand side: see what happens . . . .

(1000 0110)

It's too much, right? It goes all the way down on the right side So to even it out I would put in the first hook . . . to balance it out:

(1000 1010)

then it works. Now, try to put it in the middle clip on the right-hand side and I'll bet you it will work . . . ."

(1000 0110)

George: "Of course not . . . ."

Joe: "Right . . . . it just goes right down . . . . So how would you balance this now? One weight on the left—on the far left—and two weights in the middle . . . How would you balance that?"

George: (sighs) "I don't know . . . ."

Joe: "Well, where would you put the weight in order to get it to balance? When you have . . . let me draw a picture of this . . . leave it like it is . . . OK? See, you got this, down . . ."

George: "It's gonna have to be the first one . . . ."

George is suggesting adding weight to the first hook on the right to create (*1000 2100*).

Joe: "So, right in the middle . . . and this has gotta be the first one . . . ."

George: "Yeah . . . ."

Joe: "All right: let's see, we have one weight here and . . . two weights here. All right, how would you balance this? Right now it's leaning to the right."

George: "Leans to the right; it's out of balance."

Joe: "OK. How do you balance?"

George: "Put it here."

(1000 2100)

Joe: "OK."

George: "It works!"

Joe: "Good. It works. Now why does it work? You know, why couldn't we put it over here, just by . . . we said, "We'll put it over here; it'll be too much, right?"

(1000 1110)

George: "Uh huh . . . . (affirmative)"

Joe: "It'll just balance off and go back over to the other side . . . . Try it over here and see what happens; on the second one to the left."

(1010 1100)

George: "It's gonna be heavier." (Weights start crashing around.)

Joe now begins a monologue in which he describes the relevant variables and tries to create causal links between weight, position and force.

Joe: "It's gonna lean to the left . . . . so now you have it leaning to the left instead of leaning to the right. You have the same amount of weight, but different displacement, different position, OK? So you have like no force acting down. So, OK, how would you balance this one? On the right-hand side, now? Where would you put it? You could put it on the end. If you put it on the end, it's gonna come down and . . . go up . . . and balance . . . . put it right over here: right at the first paper clip. Now you have three in the middle; well, the first three on the right-hand side and you have two on the left-hand side; but there's one at the end which is a big difference, makes a big difference . . . and there's not, there's nothing at the end of this one. OK, let's balance this."

(1010 1110)

(long pause)"

George: "How do you think that makes them balance?"

Joe: (mumble) "You got one here, and another one here . . . . What was your question?"

George: "Why do you think they balance when there's more to the right than to the left?"

Joe: "Why do I think it's balanced when there's more going to the right than there is to the left?"

George: "Uh huh . . . ."

George is unsatisfied with Joe's vague description of how the system operates. Furthermore, he is struck by the conflict between the visual situation (three weights versus two) and the fact that the system balances. Joe's response is still vague, but it includes two key concepts: ratio and the relevance of his experience with seesaws.

Joe: "Well, right now, OK, the reason why it's balanced: there's three weights on the left, and there's only two weights on the right; wait a second: there's two weights on the left and three weights on the right. The reason it's balanced, given that the ratio is two to one is that the big difference is on the left: you got the weight at the end. It's like, if I was to sit, if you were to sit at the balance on one end, and I was to sit on the other part at the middle, even if I weigh more, you'd still "outweigh" me because you're sitting at the end: . . . there's more distance."

George: "All right, that's right."

Joe: "OK, this distance makes a big difference; this difference here. If I was to put this at the end, if I was to put one of the weights on the right-hand side at the far end, this would just go out of balance."

Having agreed on a rough qualitative model of the system, George now asks the critical question: "How do you know what distance to go?" (How can you tell exactly where to place the weights?) Joe responds by gradually developing a rough model based on the theory that the distance out from the fulcrum is inversely proportional to the weight. However, he only considers the ratio, 1:2. This is a special case and involves for most people an intuitive rather than mathematical calculation.

George: "How would you know by the . . . they would work the same way to somehow . . . distance? How do you know what distance to go?"

Joe: "What distance to put the weights in, you mean?"

George: "Yeah."

Joe: "OK, you have . . . you know that you have to keep the distance; you have to keep the same amount of weights and the same distance on both sides; otherwise you'll go out of balance. You can compensate for that by putting more weights on one side than, than the other side . . . . Wait. In other words, you can put more weights closer to the middle of the balance, while on the other side you can have one weight, say one weight, at the end . . . . Like, one weight here would probably balance two in the middle. Let me try it . . . . One weight at the end would balance (probably) two . . . . Right. Hmmmmmm . . .

(1000 0020)

Guess not. OK, well, how about here? (1000 0200)

There you go . . . . That balances: two weights on one side, on the right side, balancing one weight on the left side. You stay how . . . distance here, right? The big difference is that this is on the right-hand side, the second-to-the-last clip, there's no weight there."

At this point George and Joe seem to decide that they are ready to start looking for a mathematical relation. They differ, however, on the degree to which they understand the balance system.

Joe: "Right? So . . . we have to come up with some sort of a uh, a formula for this . . . ."

George: "OK, let's start from the beginning."

Joe: "Let's hook up the string and see what happens."

George: "No, 'cause we haven't figured out *this* one. Now we're going to try a more complex one . . . we have not solved *this*."

Joe: "We have solved it . . . ."

George: "No we haven't."

Joe: "Well, what's your problem, anyway?"

George: "Like if you . . . like if I put a weight here, right?"   (1000 1110)

Joe: "Right: it's going to go out of balance."

George: "OK. Now why, would you tell me—without any mistake, I mean . . . If I give you one chance, just one chance, would you balance that at that weight?"

Joe: "If I can balance that?"

George: "Yeah. Without any mistakes."

Joe: "Yeah. I can do it."

George: "OK. Where would you put it?"

Joe: "I would put this right here at the end. Cause I have the feeling that's where it's going to go . . . . That will balance it."   (2000 1110)

George: "Well, it didn't."

Joe: "Well, well, well . . . Wait a second."

George laughs.

Joe: "Sorry 'bout that: it didn't. (laughs) All right, hold it, hold it."

George: "See what I mean now?"

Joe: "OK. Didn't balance, My God . . . should have balanced it . . . How about that? Didn't. Wait a second, hold it, hold it; it was close. We have three here, right in a row, right?"

George: "Yeah."

Joe: ". . . the only way we can balance that, is if we have one here, like that . . . ."   (1003 1110)

George: "All right. How do you go about . . . how come . . . you can get . . . ."

Joe: "It's all trial and error . . . ."

George: "No, it's not . . . . What's together about trial and error?"

Joe: "All right, right . . . listen . . . I did it trial and error. I guess, take the distance from here to here . . . take this distance . . ." (long pause)

George: "See what I mean now?"

Joe: "Yeah . . . . All right: there should be a formula which you can just plug in . . . and you can know which side to put . . ."

George: "We gotta *make* the formula then . . . ."

Joe: "So we have to come up with the formula to find . . . so we have to uh . . . using trial and error, we just know exactly where to put the weights regardless . . ."

George: "Yeah . . . . OK, let's start from the beginning . . . ."

George has made his point. They return to the beginning, only this time their approach is much more theoretical and systematic. They start by determining which variables are involved; then they consider how these variables may be related.

Joe: "OK. What we know is involved in that . . . off handedly, What do you think . . . what do we know that is involved . . . in all this? You know, there's certainly distance involved, right?"

George: "It's the length . . ."

Joe: "Length and weight."

George: "Uh-huh . . ."

Joe: "Yeah . . . length and weight involved in this, right?"

George: "Right . . ."

Joe: "So it's gonna have to be a formula which will involve length and weight. OK? Let's figure it out . . ."

George: "Ah . . . . oh (expletive deleted) . . . . should work out . . . . . There . . . . . um hum . . . . ."

Joe: "Hear a good thing about ratio, huh?"

During their physics course, there was a constant emphasis in the importance of ratios.

George: "Yeah, yeah."

Joe: "A ratio-type of problem . . ."

George: "I'm trying to get ahead;[1] let's see . . . . one . . . two . . . six . . . (counts): one, two, three, four, five . . . six . . . hold it . . ."

Joe: "Any ideas?"

George: "Ah . . . getting there . . ."

Joe: "Hey, right two."

George: "Hmmmmmm."

Joe: "What is it?"

George: " 'Alfo!' "

Joe: "What did you find? . . . . . anything with a . . . think I should look at it on paper . . . draw it out, see if I can find something here: looks . . . . "

George: "OK. Got one, two, three, four . . . five."

Joe: "Bells? here . . . . go get a balance, but uh . . ."

George: "Oh."

Joe: "Don't tell me you've got it . . . ."

George: "Seems like I've got it, but . . . ."

Joe: "Why don't you put it together, then . . . . . let me know what you have." (pause).

George: "See these here." (George points to the balance.)

Joe: "OK, I call this one here, like L_1, OK? (Joe is referring to a drawing he has made of the balance) . . . . In order for L_1 to be balanced right, just look at this, just look at the picture now, look at this . . . All right . . . in order for L_1 to be balanced . . . . oh . . . wait a sec. OK. In order for this to be balanced, L_1 has to equal L_2. They have to be equal to each other. They have to be the same equal distance between these two, so I can say that much, that L_2 is equal to L_1."

Joe is inexperienced in the process of inventing abstract names to represent the critical features of a problem. He is imitating what he has seen his professors do without recognizing the function of subscripts or the need to create a simple, unambiguous system. Although the ability to invent and manipulate algebraic

---

1. This comment is interesting in its similarity to the title of a paper by Karmiloff-Smith and In-helder: "If You Want to Get Ahead, Get a Theory." (1975)

symbols may seem abstract, it can in fact be handled quite concretely. Joe continues to do this during the next several pages of transcript.

George: "Just call it side one and side two."

Joe: "Doesn't matter. Call it anything you want."

George: "OK."

Joe: "It's just, this led from the middle of this, from the center of the piece to the . . . OK? And by adding more weight to each one, to either one of them, makes a big difference. Now, the distance of these lengths also, by adding weights at a different distance, also makes a big difference. Now we have to find out—come up with a solution to that . . . . . OK, now start off simple, just by adding one weight, the first one . . ."

George: "We know that already . . ."

Joe is still trying to get his new notational system under control. While his method of proceeding is systematic and sensible, one has the feeling that he believes that this re-representation of the problem is itself a form of solution. Although his actions may appear misguided, it seems to be that Joe is in fact following a well-established scientific tradition. Notational experimentation seems to be a major component of early scientific investigation and is usually coupled with an inability to separate incidental characteristics of the specific formalism from essential features. The effort expended in reconciling the Schrödinger and Heisenberg formulations of quantum mechanics suggests that this can be a nontrivial problem.

Joe: "No, no, just leave it and see what—you know . . . All right: *that* . . . OK . . . that balances out . . . (mumbles) L_1 is equal to L_2 . . . only—"

George: "Put that there; then leave two spaces."

Joe: ". . . . was equal to one *weight* . . . on both sides . . . same length . . .
(0000 0200)

George: "See this?"

Joe: "Yeah."

George: "This is the distance, right? From here to here . . . It's one?"
0200)

Joe: "Un hm . . ."

George: "From here to here it's two."
0200)

Joe: "Distance, two."

George: "You got two segments, two . . . whatever you want to call them . . ."

Joe: "Two partitions, say . . ."

George: "Yeah . . . and two weights. If you multiply the distance times the weights . . . you get four, right?"   (0000 0200)

Joe: "Hmmmm, I don't quite getcha . . . what are you saying now?"

Why is Joe so puzzled by George's relatively simple suggestion? It is because George has made the critical step. He has invented an unnamed concept that is neither length nor weight but an abstraction derived by operating on both of them. Here for the first time we see evidence that George is using **formal operations.** *He has begun to deal with quantities that are not directly observable; he is using ratio reasoning and is therefore able to quantify what he previously could only describe qualitatively.*

George: "We want to count one distance as some kind of weight."

George's statement shows that in multiplying weight and distance he is doing more than inventing an algorithm for dealing with balances. He is trying to construct a new concept; one that is like weight but that is not weight. Joe on the other hand, is still employing concrete reasoning. The difference between George's formal concept and Joe's concrete version is illustrated below in Joe's use of "equal to" as opposed to George's term "equivalent to." Joe still wants to use the directly perceivable quantity distance but George is using something else.

Joe: "Take the first distance, and call it something, say uh . . . first distance, call it XY."

George: "First distance is equivalant to some weight. Right?"

Joe: "D_1 equals, uh . . . . alright. OK, now what are you going to say about D_2, distance two, from the center of the piece?"

George: "At all times we got two distance (two units of distance)."

Joe: "So we're gonna have, distance two equals."

George: "No . . . You got two distances, right? Then you multiply times the amount of weights that you have, which is two."

Joe: "Yeah. Two distances. You multiply times the weights. What does that give you now?"

George: "I get four, right? So I figure . . ."

Joe: "Now wait, whoa . . . whoa . . . whoa . . . No . . . wait a second; OK, let me see what you're saying. Yeah; two distances multiplied times . . ."

George: "Times the amount of weights you have . . . ."

Joe: ". . . So it's distance times the weights?"

George: "Yeah. You get four. Now, I figure if one weight to get the same force on the other side if we get one weight . . . . See what I did? The only problem is, on this side, I didn't multiply. I counted the three . . . spaces and count the wrong way at some other space and got four and it came out to be alright. See what I mean?"

(1000 0200)

Joe: "Yeah."

George: "So we got four on four; one, two, three distance . . ."

Joe: "Right—got three distances out of there, and one weight; then you go two distance, one distance, you add another weight; you had to compensate that. That did not equal . . . sides were equal, right?"

(0100 1100)

George: "Right."

Joe: "Side *A* equals side *B*."

George: "The problem is, what if you move this weight; that would be one, two, three distances." (0100 0110)

Joe: "Yeah, that's putting more weights on the right side . . . that's putting much more weight on the right side."

George: "That's six, so we would have to have six here." (0100 0101)

Joe: "No . . . you wouldn't put it there."

George: "No . . ."

Joe: "You'd put it right in the middle (of the left arm)." (0110 0101)

George: "No, it wouldn't work."

Joe: "Of course it would work; they would balance each other right in the middle. Where else would you—"

Joe is still guided by visual symmetry.

George: "Right here?"

Joe: "Yeah . . ."

George: "I mean . . . some weight . . ." (0110 0101)

Joe: "Oh, no . . . if you put that at the end it might have worked. Try it."

George: "You're saying to put it here? Right? To put this one here?"
(1100 0101)

Joe: "Yeah, right; it might work."

George: "No, it's not gonna work . . . three, four, five . . . No. This side's going to be heavier."

Joe: "No, it works . . . no . . . still heavier . . ." (It doesn't work.)

George: "So we know it came out the way we want, so we might as well try another. Where do you think we have to put this one? In order to balance it?"   (0110 0001)

Joe: "Uh, right behind this here, see being balanced. It's not tipping too much to one side—put it at the far end. I mean right from the start; beginning there." $\wedge$0001)

George: "Why there? Do you have a special reason?"

Joe: "That's just something that came right out of my head: It looks like it should belong there, you know?"

George: "You're telling by the slope of the weight with the balance."

George is trying to tell Joe that he has to stop being guided by directly visible cues.

Joe: "Yeah, I just looked at the slope to see how much weight needs to be put there to balance it. It's just a guess—it might not work."
(0110 1001)

George: "It works."

Joe: "It works. Yeah."

George: "OK, let's see what is the relationship between the two sides. (pause) . . . . Ha . . . Ha . . . Ha . . . Sometimes it works, sometimes it doesn't. I don't know—I'm doing it wrong; or right . . . . anyway . . . ."

Joe: "Probably it's just a simple algebra equation . . . probably . . . ."

George: "What relationship do you see between those two sides?"

Joe: "I think that there's a ratio—that you use that same ratio throughout the . . . the . . . so right now it's one-to-one's going to be balanced . . . all the way to that same exact place on both sides. So . . . L_2 is to L_1

...... same weights ..... I figure you have some type of an equation, simultaneous equation ...."

George: "Why don't we say that?"

Joe: "Cause you have one side equals one and both sides have to equal the same amount of numbers ... No, wait a second, not really because it differs with distance so you can have one side being/weighing five pounds, while the other side can weigh ten pounds 'cause it defers with respect to distance."

George: "What relation do you see here?"

Joe: "To look at it logically, now, if we have ..... now we know that length one doesn't particularly have to equal length two ... you don't have to get three on one side and three piles on the other side. You don't have to do that because of the distance."

George: "OK."

Joe: "You can actually put less weight on one side than you can on the other side, or vice versa ...."

George: "OK. Let's try by using each space equals to one weight, just to see what happens."

In the following section they seem to regress to an older theory: that of adding length to weight. It is not clear whether George is now playing devil's advocate in order to teach Joe or whether he is reconsidering his own solution. He may be doing both.

Joe: "All right."

George: "It might not work. We can always try anyway .... So we put this weight here and that weight there .... And where would you put this one? Just one, on the other side?" (0000 1100)

Joe: "Just one to balance it out?"

George: "Yeah. One weight to balance two weights."

Joe: "One weight to balance two? I'd put this at the end—first, I'd just put it at the end."

George: "Why? We can't do this sloppy thing, now, guessing—OK?"

Joe: "There's no way I can do it now, without guessing. Right now all I can do is guess lucky ...."

George: "Well, we're trying to try an experiment."

Joe: "OK . . . you have weight two . . . . two weights on one side over distance one and two. Where do you put the other one to balance it out?"

George: "We said that we would use the spaces . . . . so there's two spaces and two weights."

Joe: "Right."

George: "Which would be equal to four weights . . . OK? According to what we said—"

Actually, by the addition rule they should have gotten five. This careless error causes some confusion since it allows the rule to work.

Joe: "So the—, one, two, three . . . four . . . two times four . . . Noooo . . . . Wait a second—There's two here; two plus two is four;—there's one, two, three, four space." (1000 1100)

George: "With the weight, that would be five." (This refers to the left side.)

Joe: "With (two and two?) weight, yeah, that would be five. So therefore, it doesn't balance."

George: "Better one lead . . . that type . . ."

Joe: "Right, right, right . . . It doesn't balance, so OK . . . OK, so now we have four here, right? So there's one, two three, *plus* the weight . . . it's going to make it four, OK? It should balance now. It balanced. I still haven't got it; I'm just about to see it." (0100 1100)

George: "OK. Let's see if it's gonna work for some other . . . ."

George is probably worried that the rule has worked.

Joe: "All right. We have three here, right? Three spaces plus the weight. OK? How about taking this out . . . now we have three spaces plus the weight: four. S'gotta be one, two, three, four—OK. This will not work." (0100 1001)

George: "Well, just put it to see if you can get it down to this, right?" (0100 1001)

Joe: "OK . . . . So I'd say that is one, two, three, four, five, six, seven . . ."

George: "Six."

Joe: "If you're going to count this as two——cause the weight as being two. If you're going to count this as one, you get . . . . six."

George: "OK. Six."

Joe: "Six . . . leaves you right at the end there. Which I doubt."

(2000 1001)

George: "Yeah, it's going to work if you put it there."

Joe: "It is?"

George: "Yeah. Now—"

Joe: "No it won't because there's no weight . . ."

George: "Wait a minute: one, two, three, four . . . five, six . . . so we know it's not going to work, OK? . . . . Let's try to multiply the distance times the weight. OK?"

George recognizes that this case proves that the addition rule does not work. In true professorial fashion, he jumps to the next step without giving Joe time to follow his reasoning.

Joe: "Whoa . . . whoa . . . whoa . . . whoa . . ."

George: "We tried adding the distance—"

Joe: "Distance times the weight?"

George: "Yup."

Joe: "That's equal to *force* . . . . Yeah. Isn't it?" (expletive deleted)

George: "Well, let's try it anyway, we'll find out (laughs). OK. So we got one, two, three, four: four spaces—one weight."   (1000

Joe: "Right. Four times one. It's four."

George: "So we got to put it here."   (1000 0001)

Joe: "OK. That balances. OK. So you have L_1 equals that and L_2 equals . . . . alright . . . . so that works, then."

George: "Why don't you use left and right side so we won't get confused?"

Joe: "OK. L_1 equals R_1."

George: "OK. Why don't we put this one here? How would you have done it just using one single weight here?"   (1010)

Joe: "I think I would subtract the difference . . . of the weight . . . this idea . . . OK . . . You put the weight on this side, right? And you get the amount of space; multiply it and find out how much weight you have on this side in respect to the place it's at, OK? Take the difference from that and add it on to this side, here. Naw, that doesn't make sense."

George: "It doesn't, because this one . . . that's as far as it can go . . . we have   (1010 0001)
to add one more weight—"

Joe: "We have to add one more."

George: "Why don't we move this one to—"   (1010 0001)

Joe: "Leave it, leave it, hold it, hold it . . . . So we have an X missing here, right? You see it? We've got an X missing. We have something missing. We have a variable, X missing, OK? We know what the weight on this side is—if we know what the weight on this side is . . . we have X plus the weight on this side . . . we can find out what it is equal to. Do you follow me? Like, in other words, we could have this: $4X$ is equal to 12, right? OK, let's say there's 12 on this side here . . . there's only four on the other side time some other X; so you want to find out what X is equal to, right? So you have X is equal to 12 over 4: X is equal to 3. So, what you do, you go one, two, three . . . no you don't; you go: one, two, three. Try it."

Joe's system is too idiosyncratic to easily decipher. It probably is based on measuring distance from some place other than the fulcrum.

George: "You mean, if you put that there it's gonna balance?"   (0010 0001)

Joe: "Yeah, you put it there, it's gonna balance."

George: "You're crazy: it doesn't balance . . . . it looks unlogical. It's not balancing. It's going to balance less."

Joe: "That will balance it. I'm serious . . . leave it there—leave it there."

George: "Oh, you mean if you add one?"

Joe: "Yeah."

George: "Oh, put it then; I don't think it's gonna work."

Joe: "What do you mean, it's not gonna work: of course it's gonna work. There you go. Because you leave the same amount of space."
                                                                                      (0020 0001)

George: "Oh . . . . Oh!"

Joe: "See it?"

George: "Yeah, but, according to the equation, that . . . you're not using . . ."

Joe: "OK, OK . . . This works. Cause I know how to use it. But I don't think I can explain it. I don't think I can relate it to someone else."

George: "Hm. Hm."

Joe: "Well, maybe it's just lucky. Just luck."

George: (Laughs) "OK . . . . but according to the equation and what you did, they don't relate."

Joe: "OK. They don't relate. Good. Why not?"

George: "You got three there. And you put it to one."

Joe: "All right, but I'm going from this way to here."

George: "Oh. I see what you mean."

Joe: "See? I'm going from where the weight is to the other side."

George: "Why don't you go from here? From the center to the end? . . . or do you have to do it from the weight to the center?"

Joe: "Because, I say this is all the weight you can put four, so I can put all the weight there will not equal to four. You understand? If that's equal to $4X$, if I put another weight there, it wouldn't be equal to $4X$, cause it is 4 times the distance back; it's four times this amount of distance, right? OK, let's see if this works, doing something else."

George: "You used multiplication, right?"

Joe: "Yeah, I am, but I don't think it's right. Let me try another one."

George: "Should be multiplication."

Joe: "OK. We have like this: more complicated. Three weights on right. How do you balance this? Now you have . . . let's say this is equal to . . . who in (expletive deleted) knows what . . . $4X$ is equal to 3, so we have 12 is equal to—$12X$ is equal to . . . oh (expletive deleted). Well, what do you want to say this is equal to?"

George: "Fifteen?"

Joe: "No. Say, 18."

George chuckles.

Joe: "Right? X will just be 18 divided by 12. Nice and even. Nice and easy."

George: "Why don't we do this with less weights?"

Joe: "Alright."

George: "We're using too many weights. It's more confusing . . . OK. Let's put these two there: one, two, three . . . six—two weights. Why don't we put it here closer . . . . . . OK. One, two, four; two weights . . ."

Joe: "Two times two."

At this point, Joe seems to have switched his origin to the center of the balance. Note there is no indication of this change in the dialogue. We can only induce that there must have been a change because Joe's comments start to make more sense to us.

George: "OK, so we have one, two, three, four—"

Joe: "Four, right."

George: "Four times one is four. So it should balance."   (0020 0001)

Joe: "I'll be doggoned; that balances."

George: "We've got that . . . . Now why don't we move the two weights closer to the center."

Joe: "Yeah. Let's move the two weights to the first piece, OK? Now, do the same thing."

George: "Two times one is two."

Joe: "OK. Just add it there. I think it's gonna work there."   (0002 0100)

George: "It does work."

Joe: "We got something here. We got something here."

George chuckles.

Joe: "OK. Let's add this at the end. Oh, well, put it third position at the right. See what happens, OK? Ah . . . what have we got? We got one, two, three—"   (1000 0020)

George: "Six—"

Joe: "Times two: six."

George: "One, two, three, four . . . times four."

Joe: "So we're gonna have to put two—"

George: "Got four?"

Joe: "Yeah, so we got one, two, three, four times one: four."

George: "So we need one weight here."

Joe: "Right. So we need another one here."

George: "Should balance it: it doesn't . . . ."   (1001 0020)

Joe: "It doesn't balance . . . . wait a minute: one, two, three, four, five, *six*."

George: "Now, wait a minute. This is equal to what? One."

Joe: "That's equal to one."

George: "Yah. That's the problem. See: should've been here: this equals to two."

Joe: "Right."   (1010 0020)

George: "But then it's gotta come out two—"

Joe: "Try it—put it there."

George: "Yeah . . . two times two . . . it's two—yeah . . . so it should balance."

Joe: "And it does . . . hold it, hold it, wait—look at this here . . . first of all, now we have to . . . . we got to solve this."

George: "OK. Up to now we know that we can multiply the distance times the weight in order to find out."

Joe: "Yeah let's get down to the facts that we know. We know that the amount of space times the weight—"

George: "equals to the force."

Joe: "—equals the force."

George: "Call it the force."

At this point, I enter the room and note that Joe has written:

"L × W = L × W

Force = weight × distance"

*Based on this, I hastily conclude that George and Joe have solved the problem
and are ready for a lecture on torque.*

Joe: "We found out some things on the table, we're not exactly sure . . . .
See what we're doing?"

Jack: ". . . you got rid of the other thing (points to string over pulley) . . . .
OK . . . . Here we go. Now, you call that thing force?"

Joe: "What do you call it?"

George: "Well, give it a worthy name."

Jack: "Yeah, let's give it a different name. I'll give it a name for you. The
name that it's given in physics. It's called *torque* . . . . it's written
with a funny Greek symbol that looks like a "T" but has a wavy top.
And torque is [a special kind of] force, remember weight is really
gravity's pulling down, so this weight is really a force . . . mass is not
a force, but weight is a force. OK? So force times distance equals
torque, but only in a very special way. Torque has to do with twisting.
If I'm going to put a torque on it, that's a torque. I twist your arm,
and I'm putting a torque on it. Where a force is like a pushing
strength."

Joe: "That's not . . . that's right . . . OK, OK."

Jack: "A torque is a twist. So, the way that it twists, is you see that if I push
down here, I've got a bigger, longer lever on it, if I push, I'm gonna get
more torque than if I push with the same force here. The easy way to
see that is: try opening the door with your finger right there. Very
hard . . . . Whereas *here* I get the same torque with much less force,
because that's got a longer so . . . what the rule is, see, the simplest
rule is that the torque on the left is equal to the torque on the right
. . . . And then the question: What is the torque on the left? If the
torque on the left is the distance times the weight, but if the weight's
scattered in different places then it's the sum of all those different
weights added together. So you have a multiplication rule here, because you multiply this times the weight, but you have an addition
rule because you add up all the different systems."

George: "In other words, this would be like an equation, right?"

Jack: "If there were enough of these, it would be like an equation. What you
would to would be to sum up . . . . If I took a bag of sand here and
filled it up to some weird shape, then the only way you could figure it
out would be to integrate it. But you don't have to worry because
there's just a few of them. Now you can see if you can do the next
step."

George: ". . . if we can write the formula?"

Jack: "Yeah, that's a good idea, why don't you get that down."

George: "OK."

Jack: "You know, with cars, you talk about the torque—how much twist the engine can put in. The difference in horsepower: now if you are really up on cars, and you're really worried about accelerating down to some place as fast as possible, the horsepower isn't as important as the torque."

Joe: "Which is the torque and which 'screw?' "

Jack: "You've got to have so much twist to get it to work. And there, again, one of the things that's bad about a screw usually is that you've got hardly any distance at all. And if you are having trouble with it, if you can take your screwdriver and put an arm out here so you can twist like that . . . then you get tremendously more torque."

George: "Before torque and/or force?"

Jack: "Torque is like a circular force. It's force times distance. But remember, force times distance is work. So it's not the same thing as work. It's a different type."

Note that by trying to be precise and distinguish torque from energy I have only added unnecessary confusion.

George: "OK."

Jack: "It's always at 90°. Force at 90° angle between the force and the arm coming out is always 90°. Now in terms of work, the angle between the force and the direction in which you push it is always zero degrees. It's the same direction. So you get torque when you consider the 90° thing which is making a twisting motion, and you get work when you talk about pushing it. But you don't need to understand all . . . . this is something a grade-school child can figure through: you don't need to know the name torque and all the rest of it . . . you just need to be able to figure out where the balance is."

Once I catch my error I overcompensate and in so doing set up a very threatening situation for George.

George: "Ha, ha, ha . . ."

Jack: "I say a grade-school child can figure it through; I can also say that many college students have a hard time with it too. I mean, there are many things around that very young children can do well, and that adults have trouble with too. Children aren't all that stupid."

George: "Oh my God! That makes me less bright . . . . . (pause) OK, so we can say torque is a force—twisting force?"

Joe: "It's circular . . . ."

Jack: "That's a very good way of saying it, yeah."

Finally I shut up and leave the room, allowing George and Joe to resume learning.

George: "How do you spell 'twist'?"

Joe: "t" "w" "i" "s" "t" . . .

George: "Now you can say . . . . How do you get the twist, the torque: because you talk of multiplying the distance times the weight, right?"

Joe: "Well, ... yeah. The torque would be distance times the weight, yeah. It also has to do with work, right? Different from work."

George: "Equals distance?"

George and Joe now enter what appears to be a highly repetitive dialogue on the subject of summing torques. In fact, the idea that the total torque is equal to the sum of the component torques is far from simple and is well worth the time George and Joe devote to it. It is also gratifying to note that they have learned something from me, the word torque. But note what they learned from me was the **word** *not the* **concept***; that they had to learn on their own.*

George: "What else do we need to say? So that would be the formula: distance times weight equals torque."

Joe: "I feel you can just say also that it's the sum of the torque of the left-hand side equals the sum of the torque on the right-hand side."

George: "Yeah, it could be the sum of the torques."

Joe: "You could do this: you could probably say sum of torque one equals the sum of torque two."

George: "Or you can say: torque A plus torque B equals torque C: same thing, right?"

Joe: "Yeah, that'd be the sum of two torques equals torque C. Yah, you can say that . . . . Yeah, you can put down that torque A plus torque B is equal to torque C."

George: "Yeah, torque A plus torque B is torque C."

Joe: "It'd be the *sum* of the torque: don't forget to put the sum. 'Cause if you know what torque C is, you can find the other one because it's a sum. Or the sum of this is equal to this."

George: "So now, the total torque could be the sum of two other torques."

Joe: "Torque one plus torque two."

George: "The sum of two torques equals a total party. How would we call it?"

Joe: "Well, specify that distance times the weight is equal to torque . . . on one side."

George: "OK. So that's all we need . . . . OK. Now let's try using the string."

They hook up the string to the left-hand side. The string is attached four units out from the center of the beam and pulls at an angle so that when George and Joe place two weights on the hook the string produces a torque of five.

George: "Let's work on it a little bit more and see what happens. Are we supposed to have some weight here?"

Joe: "Try a weight there—yeah . . . . Took it right down, no problem. You always say what's extension? here . . . . That balances it out. Yeah, it does." ↑(5) (1000 0000)

They have not yet determined that the string produces torque in the same direction as the right side of the beam.

George: "It shouldn't."

Joe: "It shouldn't."

George: "Yeah, it should."

Joe: "That'd better balance it . . . 'Course, all it is is these two things hanging on. Wait a second, hold it, from here to here."

George: "I don't think that one's got anything to do with it; not yet."

Joe: "Sure it does. Look at this: tilt it."

George: "Yeah, there's some friction there."

Joe: "Friction coefficient."

George: "What if we put our weight here, then you know . . ."

Joe: "It would take it right down: that would go up."

George: "Yeah, then we would have to try to balance it, right? Let's see what happens. Where would you put it? Here?"

Joe: "Yeah, that'd be a good place. Still not enough weight. Try two at the end. Put one above the oth   that should do it. Makes no difference."

George: "Doesn't, huh? . . . . ." (pause)

Joe: "It balances it up . . ."

George: "What would happen—What if we put a weight here? A-ha!"

Joe: "OK . . . now you increased the *total* sum of the weights. What you have there as of now is two balances . . . . This is just like one balance, and the string on the ropes is another balance."

They seem now to understand in which direction the string pulls.

George: "So where would you put this one? In order to balance it?"

Joe: "We checked with it here—try over here. No—'course not. Put it at the end."

George chuckles.

Joe: "No . . . . OK. Hold it . . . . (pause) still bad . . . ." *Several attempts are made: all 'bad.'* "There we go: no, it's too much."

George: "I think we're going backwards."

Joe: "A-ha!"   ↑(5) (1000 1002)

After a long period of trial and error, Joe finally finds a configuration that works. The dialogue below that follows this discovery, shows how George and Joe have developed an entirely new approach to learning from the one they used at the beginning of the exercise. Rather than experimenting with different set-ups they now experiment in their heads with different explanations, until they find one that works. From here to the end of the dialogue they make no changes to the weights on the balance.

George: "So what do we have there?"

Joe: "To balance, you need . . . four and . . . one . . ."

George: "That's five . . . nine . . . five against nine . . . ah, so if that's four, this should be five."

Joe: "Why don't you move that around?"

George: "It's going to unbalance."

Joe: "Yeah, put it back."

George: "Let's call it quits, then, it's not so good."

The following section is a striking example of the power of student-student dialogue. George is ready to quit; he has lost interest in the problem and he sees no way to make sense out of it. Joe suggests a formula which George apparently has already considered, at least he dismisses it immediately. Then Joe considers the direction of each torque. This seems to be the missing key for suddenly George comes alive. Together they work out a solution which neither would have accomplished alone.

Joe: "Well, you gotta analyze the stage we'll be in . . . there's a lot of friction."

George: "So it's hard to tell, huh? Pretty dead. So what we supposed to find out anyway?"

Joe: "I assume we're supposed to find out another formula for this . . . . So I guess it is that the sum of the torque on the left-hand side plus the sum of the torque on the right-hand side equals that one piece over there, that string."

George: "Nope. That isn't it?"

Joe: "What do you think? What are they doing exactly now? That's pulling down, right? The other one is going up/ the other arm. So . . . ."

George: "Um hum (affirmative)."

Joe: "We could switch sides: try it on the other side."

George: "Wait a minute: what's going on *here*? We know the sum of these two equals—the sum of this one, right? Would you say that's true?"

Joe: "The sum of the one on the string—the one at the end?"

George: "Yeah. The right side. It's pulling the left side up, right? And so is the string."

Joe: "Yeah, that's what's happening. We know that there is some similarity between the right side and the far left side of the (string). They both work in the same way. Only when they're pulling the left side up, that is, 'cause they're both going down, they pull."

George: "Yeah. Now, by knowing that we can find out the force of the torque between the two, right? See what I mean? Like the same formula we made up. So this would be one, two, three, four . . . . nine."

Joe: "Now wait a minute—what's the difference? Instead of adding, you just subtract it."

George: "No . . . we know the total force here is nine, right?"

Joe: "It's anything—call it anything."

George: "No, I mean according to the formula we developed, it's nine, right?"

Joe: "Right."

George: "And that's force—I mean four—there on the other side (?) . . . . So what would you say the force of the string is?"

Joe: "It wouldn't be balanced so it's that minus that: those two minus the other one there."

George: "Or . . . that one minus nine, right? Four minus nine, which is five. So the force of this string is five. Right?"

Joe: "Yeah. That's what I say, you have to subtract to get the minus . . . . the weight is five."

George: "The weight or the force, or whatever it is. The torque?"

Joe: "So what you're saying is: left side minus right side equals X—whatever X is . . . . so distance times weight equals torque here. Torque of the left side minus torque of the right side equals X-torque?"

Looking back over the previous section it is interesting to note the stages through which Joe and George construct their theory:

1. they analyze the direction of each torque;   (qualitative model)
2. they notice the analogous function of the string and the right side of the balance with which they now have practical "hands on" experience;
                                                                       (empirical knowledge)
3. they use their old formula to calculate the magnitude of the string's torque;
                                                                       (algebraic knowledge)
4. they construct a general law abstracting from the procedure they used to solve one specific case.   (mathematical model)

The coordination of these four representational systems seems to be an essential component of physical understanding. (See Clement, 1977, for further explication of this idea.)

SUMMARY

The four kinds of knowledge mentioned above can also be used to describe the progress Joe and George made in their attempt to understand the simple balance. First they built up a rough qualitative understanding, then Joe noted the analogy to seesaws with which he had had practical experience. Next they spent a considerable time trying to find an algebraic rule which would describe how the balance behaved. They were unable to find such a rule until George made the formal breakthrough of creating a new concept, torque, which was not itself directly perceivable. Finally, after I made the "invaluable" contribution of giving George's new concept a name, they stated a general mathematical law: the sum of the left torques equals the sum of the right torques. Of course a great many other things occurred during this hour, not the least of which was the discovery that they could determine a mathematical formula on their own without assistance. Few college students recognize this possibility; and current modes of instruction do little to alter this state of affairs.

Among the many instructional implications that can be drawn from this dialogue I would like to emphasize five:

1. Teachers talk too much. I was shocked and embarrassed the first time I listened to the dialogue, and the experience has had a significant effect on my teaching. Nevertheless I know that I frequently fail to restrict my comments to what students can assimilate. All of us need to heed Arnold Arons when he says "teachers must learn to shut up and listen." One of the best ways to do this is to encourage serious dialogue between students.

2. What at first appears to be poor performance may on closer examination turn out to be impressive work. We frequently claim that students cannot reason because we do not understand what is involved in reasoning and have failed to grasp the nature of the problem which the student is working on.

3. During learning, performance may temporarily decrease; only over the long term is there a steady increase. The cyclical and oscillatory nature of learning is expected in athletic instruction (Austin, 1974) but not in most academic programs.

4. Students can construct mathematical concepts and formulas if given time to do so. Most current instruction systematically denies students this opportunity; consequently we are producing students who are utterly unskilled in representing new situations mathematically.

5. The correct use of terminology is a natural consequence of understanding, but understanding does not result from skill in the correct use of terminology. We introduce students to scientific jargon far too early. For the most part this exercise is totally unnecessary since students can pick up new vocabulary fairly easily as soon as they see the need for it.

Although dialogue between students is not always as productive as this one was, it is one of the most effective forms of instruction. Meaningful learning is a slow and repetitive process; students can be hindered in this process, if there is too much intrusion by the teacher. Promoting constructive dialogue is not easy. It requires a thorough appreciation for the students' stage of development and a deep understanding of the material to be taught. Teachers must be able to distinguish superficial knowledge from well-coordinated conceptions and design tasks that force students to recognize where their knowledge is incomplete. Several techniques have recently been described for inducing such dialogues, for example: Wales and Stager (1978), Monk and Finkel (1978), and Whimbey and Lochhead (1978). But no technique can substitute for the teacher's willingness to give students the time they need to carefully think through ideas which the naive instructor sees as obvious.

ACKNOWLEDGEMENT

I would like to thank John Clement and Fred Byron for many helpful suggestions concerning earlier drafts of this paper. Discussions with Bob Gray have helped me clarify several issues concerning the nature of formal reasoning. Finally I would like to acknowledge my indebtedness to George and Joe, for all they have taught me.

REFERENCES

Arons, A. "A Coordinated Program of Instruction in Physical Sciences and Science Teaching for Pre-Service Elementary School Teachers," Progress Report No. 2. University of Washington, July 1973.

Arons, A. *The Various Language.* New York: Oxford University Press, 1977.

Austin, H. "A Computational View of the Skill of Juggling." MIT Artificial Intelligence Laboratory, 1974.

Clement, J. "Some Types of Knowledge Used in Physics," Technical Report. University of Massachusetts, Amherst, 1977.

Driver, R.P. "The Representation of Conceptual Frameworks in Young Adolescent Science Students," PhD Dissertation. University of Illinois, Champaign-Urbana, 1973.

Finkel, D.L. and Monk, G.S. *Contexts for Learning.* University of Washington, 1978.

Fuller, R. *The ADAPT Book.* ADAPT Program, University of Nebraska, Lincoln, 1977.

Inhelder, B. and Piaget, J. *The Growth of Logical Thinking from Childhood to Adolescence.* New York: Basic Books, 1958.

Karmilov-Smith, A. and Inhelder, B. "If You Want to Get Ahead, Get a Theory." *Cognition,* 3(3), p 195, 1975.

Karplus, R., Karplus, E., Formisano, M., and Paulsen, A. *Proportional Reasoning and Control of Variables in Seven Countries.* This Volume.

Klahr, D. and Siegler, R.S. "The Representation of Children's Knowledge." In *Advances in Child Development,* Vol 12, edited by H. Reese and L.P. Lipsitt, New York: Academic Press, 1977.

Renner, J.W. and Stafford, D.G. *Teaching Science in the Secondary School.* New York: Harper and Row, Inc., 1972.

Wales, C. and Stager, R. *The Guided Design Approach.* Englewood Cliffs, NJ: Educational Technology Publications, 1978.

Whimbey, A. and Lochhead, J. *Problem Solving and Comprehension, A Short Course in Analytical Reasoning.* Philadelphia: The Franklin Institute Press, in press.

SECTION II

New Approaches
to
TEACHING

Introduction: Teaching

The previous section presented a variety of approaches for investigating the nature of cognitive processes. There is even greater variety among the applications of this research to instruction. However, to a large degree these applications do share a common epistemological basis, namely a constructivist theory of learning. In this view significant learning only occurs when the learner actively manipulates or restructures his or her own knowledge. Thus new concepts cannot simply be transferred to a *passive* learner no matter how carefully they are presented. This section opens with a paper by Gene D'Amour which presents an introduction to one version of this epistemology: Popper's critical fallibilism. Readers interested in more radical formulations are referred to Wittgenstein (1953), Feyerabend (1975) and von Glasersfeld (1978). D'Amour ends his paper with an example of how Popper's approach to addressing scientific and philosophical questions can be implemented in the classroom.

Continuing the theme that cognitive process instruction, above all else, depends on a certain attitude towards learning and knowledge, Richard Wertime, Arnold Arons, Ira Goldstein and John S. Brown offer us the subjective testimony of four expert observers. Wertime considers the interaction of thought and attitude in learning as seen from the context of teaching in the humanities. Goldstein and Brown in their paper offer the apparently contrasting vision of the technocrats' dream—a computerized companion or PAL (Personalized Assistant for Learning). Yet both papers are fundamentally about the same thing: creating an environment in which learners feel free to take the risk of constructing new ideas, and in which they can control some aspects of their own learning. Arons' paper considers specific skills we need to teach if we are to produce such students.

While Arons lists a large number of skills, Robert Gray focuses on one skill which he believes is fundamental to all the rest. He reviews the efforts that have been made in teaching abstract reasoning and concludes that most have dealt with only part of the problem. He suggests that a great deal more research is needed before we can really understand the nature of the transition between concrete and formal operations.

We then turn to specific examples of cognitive process instruction in science. First, Lois Greenfield provides an overview of several innovative instructional programs in engineering. Andy diSessa exposes several of the false assumptions made in current instructional programs and shows how these can be overcome in physics and introductory mathematics. Bob Bauman, Ruth Von Blum and Robert Sparks describe new approaches to courses in physics, biology and chemistry ranging from the remedial to the senior undergraduate level.

Mathematics presents an unusually difficult task to the instructor interested in developing cognitive processes. Jim Kaput gives an analysis of why this has been the case, suggesting that a radical reformulation of the mathematics curriculum may be in order. Herb Koplowitz and Arthur Whimbey illustrate how some aspects of this reformulation have been implemented experimentally at the introductory or remedial level and describe problems that the teacher must be prepared to face. Finally, Alan Schoenfeld concludes on an optimistic note by showing that problem-solving heuristics can indeed by taught to mathematics students . . . sometimes.

REFERENCES

Feyerabend, P. *Against Method.* Atlantic Highlands, NJ: Humanities Press, 1975.

von Glasersfeld, E. "Cybernetics, Experience and the Concept of Self." In *Toward the More Human Use of Human Being,* edited by M.N. Ozer, Boulder, CO: Westview Press, 1979.

Wittgenstein, L. *Philosophical Investigations.* New York: MacMillan, 1953.

Problem-Solving Strategies
And the Epistemology of Science

Gene D'Amour

One of the primary aims of cognitive process instruction is to teach students "good habits of thought," because education is more than the simple accumulation of facts: it involves developing the ability to actively and independently learn from experience. Three basic questions, then, face the cognitive process educator:

1. What thought processes are *actually* used by students?

2. What thought processes *ought* students to use—that is, what is a "good habit" of thought?

3. What educational strategies are most likely to help students move from their actual habits to better habits of thought?

Those who make observations of human problem solving or attempt to mechanically simulate human problem solving appear to be trying to answer question *1*. Such efforts give hope that we can unlock puzzles about actual thought habits. However, in this paper I intend to concentrate, not on the first question, but on the latter two.

One approach to the solution sought by question *2* is to study clear models of ideal thought processes—those which characterize the methods of the physical sciences, for example—and then to isolate the essential elements of such processes. Since it is difficult to deny that the "scientific method" illustrates a model of good thought habits, these tactics should be important to the cognitive processes educator. This paper examines two characterizations of the scientific method: *1.* that of the inductivists which, I argue, is incorrect, and *2.* that of the critical fallibilists, which I support. I shall attempt to show that the cognitive process view of learning is compatible with the critical fallibilist view, but not with the inductivist position.

THE INDUCTIVE HABIT OF THOUGHT

The best-known characterization of the scientific method is that of induction: once sufficient facts have been accumulated through observation, an inductive generalization is made—a form of higher knowledge is reached. Many basic science texts present the history of science from this highly organized, rarefied perspective. Newton is said to build upon and supplement Galileo; Einstein erects his new theory on the "foundation" of Newton; old theories are embedded in new theories as science expands and progresses in an orderly fashion (Kuhn, 1962). The inductivist perspective can be analyzed by being broken down into five tenets:

1. Science begins with a solid base of "brute facts," basic statements that are justified by observation.

2. These basic statements are logically *prior to* and *independent of* the theories inferred from them.

3. *Inferences* from such statements are made in accordance with an *ideal* calculus—for example, the probability calculus.

4. Making inferences in accordance with an ideal calculus assures the attainment of the primary aim of science, namely, reliable theories.

5. Science arrives at theories of increased reliability in a step-by-step fashion; the scientific method is cumulative.

These tenets characterize not only science, according to the inductivist, but also learning from experience in general. If students learn to think inductively, they will learn how to make valid empirical inferences and how to justify the products of such inferences. Their knowledge of the world will grow rapidly as they not only rediscover known truths, but go on to discover new truths.

PROBLEMS WITH THE INDUCTIVIST HABIT OF THOUGHT

The inductivist model disguises a formal epistemology in which the student of science is ultimately reduced to a passive role. The senses simply *receive* the "brute data" of reality and, after enough are gathered, an axiomatic inductive inference is performed. Students need develop only the capacity to draw the "correct" inference. The cognitive structures necessary for this are quite simple. The conceptual level and background of the student play, at best, a minor role. Qua creative thinker, the student is reduced to passivity.

Such criticisms are not new. The inductivist model has been under serious attack for over 20 years, though college science texts are just beginning to reflect this. There are many sound criticisms of inductivism, in addition to the foregoing.

For example, tenet 2 has become a focal point of criticism. Its detractors claim that there is no such thing as observation of "brute facts": all observations are really "theory dependent." This dependence is revealed in the situation of physical observation, which is comprised of the state of the observer, the nature and state of the instruments of observation, and the environment in which the observation takes place. All of these observational components themselves involve or are assessed by scientific theories. It is widely known, for example, that Galileo had difficulty explaining why naked-eye observations of the planets appeared to be inconsistent with Copernicus' theory. He was able to refute pre-Copernican "facts" through the use of a "superior and better sense," the telescope. Why was the telescope a better means of observing the stars than the naked eye? Because, claimed Galileo, the construction of the lens was based on Kepler's theory of optics (which, by the way, was incorrect).

The critics of induction claim that not only do theories determine *how* the scientist observes, but also *what* the scientist observes. A scientist doesn't just solely *observe*; he observes selectively. The tool of selection is a theory. It "tells" the scientist what to observe as well as how to describe what is observed. To Ptolemaics, for example, a tower on the earth was seen as standing still. Their assumption that the earth was the sole frame of reference entailed a concept of absolute motion. Galileans, on the other hand, used instruments to alter the raw data of experience, and motion became relative instead of absolute. The tower was *now* seen to be moving, with the earth, relative to the "motionless" air. Thus, what the Ptolemaics and Galileans alike described as the "facts" of observation depended upon their respective theoretical or conceptual frameworks (Feyerabend, 1970).

Criticisms of inductivist tenet 3 have also been made. Some are historical, showing that the major figures in the development of modern scientific theories did not work in an inductive manner. Particularly forceful are those arguments against the combination of tenets 3 and 4, which together hold that induction attains the primary aim of science, i.e. reliable theories. The critics claim that the uniqueness of science is not found in the reliability of its theories, but in their falsifiability. Karl Popper (1959) elucidates this view as he recalls his days in Vienna in the 1920s:

> I found that those of my friends who were admirers of Marx, Freud, and Adler, were impressed by a number of points common to these theories, and especially by their apparent explanatory power. These theories appeared to be able to explain practically everything that happened within the fields to which they referred. The study of any of them seemed to have the effect of an intellectual conversion or revelation, opening your eyes to a new truth hidden from those not yet initiated. Once your eyes were thus opened, you saw confirming instances everywhere: the world was full of verifications of the theory. Whatever happened always confirmed it. Thus its truth appeared manifest; and unbelievers were clearly people who did not want to see the manifest truth; who refused to see it, either because it was against their class interest, or because of their repressions which were still "un-analysed" and crying aloud for treatment.
>
> The most characteristic element in this situation seemed to me the incessant stream of confirmations, of observations which 'verified' the theories in question; and this point was constantly emphasized by their adherents. A Marxist could not open a newspaper without finding on every page confirming evidence for his interpretation of history; not only in the news, but also in its presentation—which revealed the class bias of the paper—and especially of course in what the paper did not say. The Freudian analysis emphasized that their theories were constantly verified by their 'clinical observations.' As for Adler, I was much impressed by a personal experience. Once, in 1919, I reported to him a case which to me did not seem particularly Adlerian, but which he found no difficulty in analyzing in terms of his theory of inferiority feelings, although he had not even seen the child. Slightly shocked, I asked him how he could be so sure. 'Because of my thousandfold experience,' he replied; whereupon I could not help saying: 'And with this new case, I suppose, your experience has become thousand-and-one-fold.'
>
> What I had in mind was that his previous observations may not have been much sounder than this new one; that each in its turn had been interpreted in the light of 'previous experience,' and at the same time counted as additional confirmation. What, I asked myself, did it confirm? No more than that a case could be interpreted in the light of the theory. But this meant very little, I reflected, since every conceivable case could be interpreted in the light of Adler's theory, or equally of Freud's . . . . I may illustrate this by two very different examples of human behaviour: that of a man who pushes a child into the water with the intention of drowning it; and that of a man who sacrifices his life in an attempt to save the child. Each of these two cases can be explained with equal ease in Freudian and in Adlerian Terms. According to Freud the first man suffered from repression (say, of some component of his Oedipus complex), while the second man had achieved sublimation. According to Adler the first

man suffered from feelings of inferiority (producing perhaps the need to prove to himself that he dared to commit some crime), and so did the second man (whose need was to prove to himself that he dared to rescue the child). I could not think of any human behaviour which could not be interpreted in terms of either theory. It was precisely this fact—that they always fitted, that they were always confirmed—which in the eyes of their admirers constituted the strongest argument in favour of these theories. It began to dawn on me that this apparent strength was in fact their weakness.

With Einstein's theory the situation was strikingly different. Take one typical instance—Einstein's prediction, just then confirmed by the findings of Eddington's expedition. Einstein's gravitational theory had led to the result that light must be attracted by heavy bodies (such as the sun), precisely as material bodies were attracted. As a consequence it could be calculated that light from a distant fixed star whose apparent position was close to the sun would reach the earth from such a direction that the star would seem to be slightly shifted away from the sun; or, in other words, that stars close to the sun would look as if they had moved a little away from the sun, and from one another. This is a thing which cannot normally be observed since such stars are rendered invisible in daytime by the sun's overwhelming brightness; but during an eclipse it is possible to take photographs of them. If the same constellation is photographed at night one can measure the distances on the two photographs, and check the predicted effect.

Now the impressive thing about this case is the risk involved in a prediction of this kind. If observation shows that the predicted effect is definitely absent, then the theory is simply refuted. The theory is incompatible with certain possible results of observation—in fact with results which everybody before Einstein would have expected. This is quite different from the situation I have previously described, when it turned out that the theories in question were compatible with the most divergent human behaviour, so that it was practically impossible to describe any human behaviour that might not be claimed to be a verification of these theories.

Popper's arguments about the relationship between observation and theory also hold strong implications for the inductivist construction of the history of science as development which proceeds step-by-step to accumulate increasingly reliable knowledge. Popper maintains first that scientific theories are not arrived at by any sort of inductive process at all. The formation of a hypothesis is, instead, a creative exercise of the imagination; it is not a passive reaction to observed regularities. This opens the way for the interpretation of the history of science as one of a succession of unexpected appearances of bold new theories which, in revolutionary fashion, bring about the overthrow of accepted theories. When this occurs, there is a radical change in the way both the world and, consequently, the scientific enterprise itself are conceived. As Thomas Kuhn and others have noted, the change is so great that those working in the old paradigm and those working in the new one have difficulty finding a common factual base for the purpose of theory comparison (Kuhn, 1962).

THE CRITICAL FALLIBILIST HABIT OF THOUGHT

Popper sums up his views on the methodology of science as follows (1959, p 36):

1. It is easy to obtain confirmations, or verifications, for nearly every theory—if we look for confirmations.

2. Confirmations should count only if they are the result of *risky predictions;* that is to say, if, unenlightened by the theory in question, we should have expected an event which was incompatible with the theory—an event which would have refuted the theory.

3. Every 'good' scientific theory is a prohibition; it forbids certain things to happen. The more a theory forbids, the better it is.

4. A theory which is not refutable by any conceivable event is nonscientific. Irrefutability is not a virtue of a theory (as people often think) but a vice.

5. Every genuine test of a theory is an attempt to falsify it, or to refute it. Testability is falsifiability; but there are degrees of testability; some theories are more testable, more exposed to refutation, than others; they take, as it were, greater risks.

6. Confirming evidence should not count *except when it is the result of a genuine test of the theory;* and this means that it can be presented as a serious but unsuccessful attempt to falsify the theory. (I now speak in such cases of 'corroborating evidence.')

The distinction between Popper's "critical fallibilism" and scientific inductivism should now be obvious. What is crucial, from the standpoint of education, is that Popper's theory of the role of the imagination in scientific theorizing is strongly in accord with findings of those who have studied the processes of cognitive development—and with theories of learning based on those findings. What Popper—and students of cognitive development like Piaget—have in common is a view that encourages multiple trials and a variety of approaches to problem solving.

It is my contention that these approaches are superior to the inductivist theory and method which have lain behind the teaching of introductory science courses in this country for many years, as well as behind our concepts of the activity of science itself.

The student, like the scientist, begins the learning experience by initiating a hypothesis. Next, he or she must attempt to refute this hypothesis by seeking empirical evidence of its untruth—its falsifiability. If the hypothesis stands the test, it may be accepted—but never as "knowledge," only as a hypothesis that has not yet been refuted. In this way, what the student comes to accept as "scientific knowledge" rests upon his or her own efforts to disprove it—a method of instruction far more effective than the traditional memorization of pronouncements offered, *ex cathedra,* from the lectern. We posit, as indeed Piaget seems to have discovered, that the active participation of the student in a process of simultaneous criticism and discovery, produces better students— students with a sure and intuitive grasp of the fundamentals they will need to build upon later. That this accords with what can be shown to have been the actual history of the development of modern scientific theories should help explain its effectiveness as a teaching method. Critical fallibilism, then, has helped us to discover an answer to question 2, "What thought processes *ought* students to use?"—simply put, the method of trial and error (Popper, 1959).

ANSWERING QUESTION 3, BASED ON RESPONSE TO QUESTION 2

Any answer to question 3—"What means do we use to move students from how they do think to how they ought to think?"—must follow from one's answer to question 2. The lecture method, for example, appears to be consistent with

the inductivist position—a lecture might simply present facts with the expectation that the learner will make an apropriate inductive inference. The lecture is
not, on the other hand, consistent with the critical fallibilist position since it
does not directly involve the student in actively reflecting on his own thought
processes while seeking to learn something: it omits providing the student with
opportunities for active hypothesis generation and critical testing. Similarly,
the traditional laboratory section has, more often than not, given students the
opportunity to carry out by rote methods experiments whose results are
precisely known. The pedagogical assumption behind such lab exercises has
been that the student will come abreast of modern developments in the field by
duplicating the experiments which lead up to them. In practice, however, the
student usually ends up checking the results to make sure that they accord with
an "answer"—provided by the laboratory manual or the professor. But this is
not "trial and error." In other words, once it is possible to see whether the results
accord with what is expected, the divergence of the findings from those of the
professor lead the student to conclude that he has made an error in procedure.
Methodical procedure rapidly becomes what counts most, and in this case,
student adherence to the ideal model becomes somewhat too complete. If we are
to meet the dual requirements of intelligent procedure and creative imagination
in students of science, we need to find some middle way in which students can
be helped to discover scientifically appropriate methods within a context sufficiently flexible to permit self-reflection and learning from mistakes: a more
"open" situation than those provided by either the traditional lecture or laboratory exercise. To this end, a discussion of one new instructional method called
Guided Design—more compatible with critical fallibilism than the inductive
theories—follows.[1]

GUIDED DESIGN

The concept of Guided Design is based on the belief that teachers have
something much more important to give their students than simple facts. That
something is a model of how an intelligent human being makes scientific judgments. For this reason, Guided Design is part system, part attitude. It reshapes
the traditional approaches to higher education. Students, who work in small
groups of four to six each, attack open-ended problems rather than masses of
cold information. It is based on the assumption that the student who works
through an ascending order of well-designed problems, actively seeking solutions—instead of passively assimilating knowledge or faithfully adhering to
procedure—will emerge not only better educated, but stronger intellectually.

Guided Design was created by two engineers, Charles Wales of West Virginia
University and Robert Stager of the University of Windsor, who applied the concept of systems design, as it is used in engineering, to problems of education
(Wales, 1978). A Guided Design course not only focuses on developing students'
capacity for sound scientific judgments, but also helps them learn specific concepts and principles. In class, the students solve open-ended problems which require them to think logically, gather information, communicate ideas to one
another, and use ordered steps in a process of making judgments. Among the

---

1. The compatibility of the inductivist or critical fallibilist positions with other instructional
methods like programmed instruction or the Personalized System of Instruction (PSI) may vary depending on the design of the specific instructional unit considered.

judgments they learn to make are to identify the problem, state the basic objective or goal, state the constraints, gather information, generate possible solutions, analyze, synthesize, and evaluate the solution. Students are guided through each problem by a series of printed "instruction-feedback" pages (prepared in advance by the instructor), by their discussion with other students on their respective teams, and by the instructor who acts as a consultant. The students do the thinking, make the value judgments, and play the role of professional decision maker.

In most cases, students in a Guided Design class cover the same material offered in traditional courses on the same level, but there is a difference. Each open-ended problem requires for its solution familiarity with a unit of subject matter which each student is expected to gain independently, outside of the classroom. Although little, if any, class time is devoted to recitation of the reading assignments, the small-group problem-solving sessions accomplish this same goal while providing a meaningful framework for structuring learning. This approach helps students to understand that facts, concepts, principles, and values are all included in the background information upon which making scientific judgments depends. Such organization helps students to establish patterns that will serve them after graduation—where continued independent learning is a prerequisite for success.

In a Guided Design class, the primary roles of the teacher are those of guide, prompter, manager, and consultant. During class the teacher moves from group to group, listening, asking leading questions, and encouraging students to participate critically in the decision making. Thus, the students are active, they learn from the decision-making model developed by the professor (and presented in the instruction-feedback material), and they receive the benefits of personal attention—both from fellow students who are also being critical and may hold different views, and from the instructor, who interacts with each group.

Before the students begin work on each project they are told that it is not necessary that the group agree completely among themselves, nor with the printed instructions, nor with the feedback. The projects do not present dogma under the guise of open-mindedness, but rather present a series of questions with which the students are to grapple. Group discussion has been found to serve, not as a goad to the right answer, but as an invaluable aid to getting the students to think about the question—to explore its nature and its implications more deeply. To provide guidance, the views of a professional regarding each question are provided in the printed feedback, not a statement of the "real" answer, but an additional, informed opinion. Students are told that there is no guarantee that the opinion is correct—the group's answer may even be better. In fact, the students often discover that the professional's answer is reasonable, logical, and quite similar to their own. Work of this sort is important because it introduces students to certain ways of thinking about ideas and methods of problem solving in general, without stressing specific answers as solutions to specific questions. The emphasis has been shifted from the notion that knowledge lies in the acquisition of facts, to the understanding that useful knowledge is the result of certain thoughtful modes of activity.

Guided design, then, is a technique that emphasises "good habits of thought," by habituating students to step-by-step modes of decision making with reference to a model that is widely accepted in decision-making theory. It provides students the opportunity to actively put forth creative hypotheses at each step in the decision-making process. These hypotheses are then criticized through group experimentation until students arrive jointly at a conclusion. After this, they then have yet another chance to criticize their ideas as they compare them to the printed feedback which gives them the opinions of both other students and professionals.

SUMMARY

Cognitive process educators are discovering how students actually think. However, if they are to educate, they must know in addition how students ought to think—what constitutes good habits of thought—as well as how to develop these habits. Identifying good habits can be accomplished by studying "ideal" thought patterns. I have argued that the critical fallibilist, trial-and-error, approach is a more appropriate model of learning from experience than that of the inductivist.

After an ideal pattern of thought is identified, the cognitive process educator should use it as a guide to evaluate possible alternative approaches which might be used in education. I have tried to illustrate how this can be accomplished by focusing on Guided Design. Programs of this nature are consistent with the epistemology of critical fallibilism, and there is some evidence that they can improve students' ability to learn (Wales, 1977, 1978).

REFERENCES

Feyerabend, P. "Problems of Empiricism, Part II." In *The Nature and Function of Scientific Theories,* edited by R. Colodny, Pittsburgh: University of Pittsburgh Press, 1970.

Kuhn, T. *The Structure of Scientific Revolutions.* Chicago: University of Chicago Press, 1962.

Popper, K. *Conjectures and Refutations.* New York: Harper and Row, 1959.

Wales, C.E. and Stager, R.A. *Guided Design.* University of West Virginia, 1977.

Wales, C.E. and Stager, R.A. *The Guided Design Approach.* Englewood Cliffs, NJ: Educational Technology Publications, 1978.

Students, Problems
and
"Courage Spans"

Richard Wertime

Applied cognitive psychology—cognitive process instruction—speaks insistently of problems and of problem-solving strategies. This emerging approach to learning is a profoundly hopeful one, for it embraces the belief that the rational powers of our students are vital and dynamic ones susceptible to discipline and hence continuous growth. At a time when the reasoning skills possessed by many of our students appear to be woefully deficient, such affirmation of the intellect is more than merely welcome; it is downright necessary.

At the same time, it is easy to overlook some basic matters which must be borne in mind when we are speaking of heuristics, problem progressions and the like. Just what *is* a problem? How does a problem come into being and, as it were, live its life? How do students *feel* about problems of an intellectual nature? And I must add this question, perhaps the most elementary of all, since I am offering to look once more at well-trodden ground: just what is a student?

My answers to these questions might serve one of two functions: they might help to revive some old and fairly handy distinctions that have fallen out of use; or they might help to elucidate some of the psychological aspects of problem-solving activity. Problems, despite their being the stuff of everyday life, both academic and otherwise, are neither simple nor homogeneous. Of course, the same holds for students. If we can grasp more clearly the stages and the rhythms which characterize a student's growth as well as his efforts to meet the challenges inherent in all problems, we may well be on the way to becoming better teachers—shrewder, more sympathetic people better equipped to help our students grasp the richness and efficiency which are native to their minds. Students tend, specifically, to misunderstand the various ways in which their courage operates in problem-solving; by introducing the notion of what I call the "courage span," I hope to help alleviate the chronic self-shortchanging which sadly victimizes many of our students.

What is a "student"? We are accustomed, nowadays, to letting the term cover an astonishingly wide variety of meanings. It encompasses those who actively wish to be in our classrooms for instruction in our discipline; it encompasses those in our classrooms who would rather be elsewhere but have requirements to fulfill; it encompasses the reluctant who detest school generally, but feel unable to put it behind themselves for whatever reasons; and it encompasses that handful of devoted excitable people who have attached themselves to us because we teach what we teach and are who we are. It encompasses, in short, anybody who meets his bills and is entitled to be present when the morning buzzer sounds.

Such a generous and democratic use of the term clearly loses some distinctions. The term "student" derives from the Latin word *studere,* "to be eager, zealous, or diligent, to study," and the first meaning cited in the *Shorter Oxford English Dictionary* is that of a "person who is engaged in or addicted to study." This renders a "student" sharply different from a "pupil," which term originally meant "an orphan who is a minor and hence a ward; in Civil and Scottish Law, a person below the age of puberty who is in the care of a guardian." A teacher, then, is likely to have a good many pupils who are scarcely his students in the proper sense of the term; colleges and universities, at least in America, find themselves in the guardian's role with many young people, and it remains an open question whether or not it is socially useful—or intellectually productive— for the role of a keeper to devolve upon those whose proper function has nothing to do with the care of minors.

What counts here, of course, is the essential distinction between the willing and unwilling, between the so-called "motivated" and "unmotivated" student; for without this distinction we cannot talk rightly about the business of solving problems, since a problem, as we will see, requires a passionate engagement of the person who would tackle it. It is one thing to struggle to engage a pupil's attention in the hope of making him over into a bona fide student, and quite another thing to talk of helping students increase their problem-solving powers. Strictly speaking, wooing pupils ought not to be among a teacher's responsibilities, for the burden of desire must rest with the student. At the same time (and this will be important to all that follows), I am convinced that a fair proportion of our reluctant-seeming pupils are in fact crypto-students—that is, would-be students who have been so badly damaged by the educational system that they dare not allow themselves to have intellectual hope. And these are deeply ambivalent people; on the one hand, they yearn to enjoy the sense of full sufficiency and general self-esteem which they glimpse here and there among some of their bolder fellows; and on the other hand they tend to withdraw from combat quickly when confronted with a challenge. They have, in a sense, become indoctrinated weaklings; they are so demoralized that they would rather enjoy the consolation purchased by despair than endure the fruitful stress of confronting the injustice they habitually do themselves. Put most simply: they are terrified of problems—yet desperately wish not to be, for they know, if only subliminally, that the will to face problems is a self-affirming will.

Much remains to be said about the nature of a student, but we can best serve the question by turning, now, to the nature of problems. Again, the etymological basis yields useful clues: the English word "problem" derives from the Greek *ballein,* "to throw," and has for its literal meaning "something thrown or put forward." The English word "project" has a parallel meaning; from the Latin *pro-jacere,* it similarly means "to throw forward." Thus we might say: a problem is, to some extent, *a project for the future we commit ourselves to by an act of the will.* This means by implication that a problem entails risk, since all "future-projects"—to use Hannah Arendt's term—involve uncertainty (Arendt, 1978). There is also an essentially binding or promissory dimension to the act of facing a problem. A problem is not an entity which has its existence independent of a person (regardless of the illusion fostered by textbooks full of problems); it is an intensely personal and passionate affair, one which is deceptively hard to break off at will. Moreover, problems are hard work. They require attention and courage, and they involve, as I will argue, a significant act of self-surrender which can seriously jeopardize the individual's sense of himself. Lastly, a problem is a hopeful enterprise which involves an act of faith. This act of faith bears both upon the student's self-assessment at the moment he acknowledges the existence of a problem, and on how he expects to assess himself at some point in the future. It is hardly any wonder, then, that so many people tend to shy away from problems; the acknowledging of a problem entangles us deeply with ourselves and with the world.

Now, just as there are willing and unwilling students, so likewise there are problems which we seek out voluntarily and those we feel are thrust upon us. Intellectual problems are often ones we seek out, whereas our existential problems are often ones we see as burdens: a chess game is fun, but a broken rear axle is a miserable business. It is worth keeping in mind the simple fact that education— and cultured life in general—is founded on leisure, that margin liberating us from the humdrum necessities. Problems are, in this sense, distinct from dilemmas, which have an ensnaring quality to them which admits of less escape.

But why, if this is so, do so many of our students regard their studies as a burden, and not as a positive opportunity? The truth of it is that, the more we look at it, the harder it becomes to correlate "real-life" problems with a sense of being burdened, and "purely intellectual" problems with a festive sense of choice. We are just as prone to feel burdened by our intellectual problems as we are to seek out existential problems voluntarily—and, of course, vice versa. Artists, for example, who commit suicide over the failure of a work (or, even more curiously, over the completion of a work) illustrate this truth. So do adventurers who place themselves in mortal danger for the excitement it affords them.

The reasons underlying this are pertinent to an understanding of problems in general. As I asserted earlier, the problems which we tackle are deeply involved with our self-esteem. We cannot help but value them; and this value isn't something which exists in isolation. Problems have contexts which anneal them to purposes; purposes necessarily reside in commitments; and commitments are the stuff of which our self-esteem is made. This is another way of saying that *all problems are personal ones* in the sense that they have being in and are entangled with our personhood. Hence, how burdensome a problem is depends on two main factors: *1.* the importance of its context to our general sense of worth; and *2.* the anticipated difficulty we will have in solving it. Our being uncertain about either of these factors can make us less ready to confront a given problem; and our having grave doubts about both of these factors is almost sure to paralyze us. Demoralized students fit this latter category: they doubt the validity of their being in school, and they doubt their capacity to make inroads on the problems which their being in school entails. On the other hand, the scientist who dwells on difficult problems may very well not be intimidated at all, inasmuch as his commitment to the worth of his work—and to the discipline enshrining it—may be so strong as to justify the stress which his work puts him under. I will speak more of this later.

But for the moment, let us turn to how a problem comes into being, and to what it involves for the person who confronts it. Michael Polanyi suggests in *Personal Knowledge* that a problem has its origins in a state of perplexity, and that it necessarily issues into a combined second stage of perceiving and doing in which the individual seeks to dispel his perplexity: "Nothing is a problem or discovery in itself; it can be a problem only if it puzzles and worries somebody, and a discovery [or a solution] only if it relieves somebody from the burden of a problem" (Polanyi, 1962, p 122). At the same time, however, that a problem constitutes an uncertain venture into the future, our perceiving it as a problem entails an essential expectation that it has a solution:

> Though the solution of a problem is something we have never met before, yet in the heuristic process it plays a part similar to the mislaid fountain pen or the forgotten name which we know quite well. We are looking for it as if it were there, pre-existent. Problems set to students are of course known to have a solution; but the belief that there exists a hidden solution which we may be able to find is essential also in envisaging and working at a never yet solved problem. (Polanyi, 1962, pp 126-127)

This intimation of a solution reveals the ingredient of hope which is intrinsic to any problem, and which further serves to distinguish a problem from a dilemma. That the hope may be illusory is quite beside the point; a dying man caught in the grip of an illness may perceive his situation as a "problematic" one, one that is hopeful to the extent that he has so far not despaired of its containing some as-yet-uncovered dimension. Discoveries, indeed, as Polanyi elsewhere says, can never be strictly logical performances, since they involve a leap of imagination and intellect. They rest on "the plunge by which we gain a foothold at another shore of reality"—or what, in our expectancy, we deem to be reality (Polanyi, 1962, p 123).

But, since we live with hosts of problems, and even actively seek them out, there must be a vigorous compensation involved in our enduring the stress which a problem puts us under. Again we come back to the hopes involved in problems, and to the root meaning of "problem" as something thrown or put forward. If a problem, on the one hand, is a mortification of sorts, a tiny death or discomposure, it is also at the same time a healthy self-assertion which involves the same bold gamble that any project does. At the instant that we identify a problem as such, we step back from the world as we had known it the instant before; and this regression into ignorance and relative insufficiency is instantly offset by our perceiving an opportunity to reconstitute ourselves in our relation to the world. Thus, the realizing of a problem entails our making an investment in our future self-esteem, the immediate price of which is the relinquishment of some of our previous self-esteem.

Now obviously this prospect is a more pleasant one when the problem is one which we have actively sought out. We could go so far, in fact, as to say that a problem we seek out—especially an intellectual one of the sort which is confronted in the forefront of science—is an act of admiration in the Kierkegaardian sense of a happy self-surrender. What the scientist surrenders *to* is that aspect of reality which contains the solution to the problem he is pursuing. Or such, at any rate, is his subjective experience: he submits, not invents. And so it is to some extent with any problem we contend with, even an artistic one: it draws us into a dialogue with a reality not of our making to which we wish to submit for the purpose of strengthening ourselves. Or, as Martin Buber puts it,

> This is the eternal source of art: a man is faced by a form which desires to be made through him into a work. This form is no offspring of his soul, but is an appearance which steps up to it and demands of it the effective power. The man is

concerned with an act of his being. If he carries it through, if he speaks the primary word out of his being to the form which appears, then the effective power streams out, and the work arises. (Buber, 1958, pp 9-10)

What is happy about the encounter is its power to affirm our sense of our personal sufficiency, our initiative, our courage: we have entered into a battle which we might well lose. But, especially at the highest levels of problematic activity, as, say, in advanced research, victory and defeat are almost equally honorable. The scientist, very possibly, may even revere his most arduous problems inasmuch as these problems constitute, for him, tokens of the world's mystery and beauty.

This is all well and good for the practicing professional. However, in our teaching, we are largely concerned with students who are unable to solve their problems as successfully as they might, or who are reluctant to wrestle with them with their full native vigor. Here, the questions of courage and personal pride come in, as well as the more embracing question of how we, as teachers, can foster problem-solving courage while helping our students learn to think. I have stated my belief that morale in many students is low; and I have said that hope is always involved in problem-solving activities. How, then, do we keep our students from lapsing into despair? How to shore up their dignity while inviting them to persist in enduring the tensions that go hand-in-hand with problems?

Students fail to understand the indirect ways in which the mind gets things done. This failure of understanding is both a sincere and costly one, for it effectively shuts them off from a just appreciation of, and easier access to, the full range of their mental powers. If they lack persistence, it is not because they are lazy, or cowardly, or docile; it is much rather because they have never had a knowledge of their persistence revealed to them. Of course they have been told about deliberate will-power and self-reliance and such good things; what they have not been told about is the "trained unconscious" which builds up in people as they work to acquire skills. *Incubation* and *inspiration* are the major mental powers least known by our students, as well as the least trusted by those who do know of them. This fault is largely ours; we have not, as educators, taught our students very well how to activate these powers. Nor have we by and large shown them how their subliminal cognitive processes relate to their fully conscious ones.

We need only look at the assumptions underlying our students' views of problems to see that this is so. When some problem stumps them, they are likely as not to classify themselves as being incapable of solving such problems: "Gee, well, you know, *I'm* just not the kind of person who can do that." For some, to be sure, such a statement is a perfunctory act of self-consolation, not much more credible to them than to us. But many students honestly feel that there are certain kinds of problems—such as mathematical ones, or literary ones—which require certain temperaments, certain inborn qualities that they simply lack. Such a belief is a mockery of the truth in these matters, a wild and grand distortion of the natural-talent thesis, which, to be sure, has some validity. Again, the distortion is sincere as well as costly. Or they will soberly think that problems are do-or-die ventures at which one gets a single try, like certain games on the carnival fairway. This belief—that they have only a single chance to get it right—is perhaps the most demoralizing of all their errors, and it bears witness to the prevalence in our unhappy classrooms of "right answer" approaches. It is also a belief which bears witness to their ignorance of how the mind sleeps on things. The educational system which has nurtured most of them has not, until now, encouraged lengthy meditation as a problem-solving strategy.

Why are our powers of incubation and inspiration so little understood? In part, I would think, because our culture is generally clumsy in its understanding of the will; in part because the bureaucracy of short-term instruction militates so strongly against their being rewarded within the academic establishment. The first of these failures is the much deeper one, the one that merits closer inspection than space permits my giving it here; but it is also the one we can counter more effectively in our classroom instruction with a major renovation of our educational structure. I would like, here, to introduce my notion of the "courage span" as a possible device for helping students better grasp their latent problem-solving powers. It is a term I derive by analogy from the familiar "attention span." Very simply, the courage span is the time which elapses between the taking on of a problem and the abandonment of that problem.

The term may be useful if only for forcing into the open what a central and complex role the will plays in problem-solving. As every teacher knows, there is abandonment and abandonment. There is ultimate abandonment, which engenders self-deprecating misery in the abandoner, even when the misery is mollified by a sense of relief; and there is temporary abandonment, which makes itself felt as an abiding frustration. The purposive tension which is involved in the latter is often unpleasant—indeed, often aggravating—but it does not entail the same sort of fundamental misery which a final abandonment does. The point of this, of course, is that to put a problem aside so that the mind might sleep upon it is a radically different thing from heaving a problem overboard. In truth, we can abandon fewer problems than we might wish to, for our wills are tenacious—and this for the very good reason that a problem, once realized, and once pursued, no matter how little, has *become a part of us;* and to relinquish it eternally is to give up part of ourselves.

But stress and discouragement do go hand-in-hand with problems, and students should be prompted to appreciate this fact. Moreover, we should help them learn to husband their courage wisely so that they might not mistake fatigue—which is natural enough—for permanent abandonment. How long their courage span will be will depend on those two factors which I cited somewhat earlier as governing a student's sense of the burdensomeness of a problem—namely, the importance of its context to his general sense of worth; and the anticipated difficulty he will have in solving it. The first of these two factors is vital to any serious theory of learning, and it is the reason I spent time defining such terms as "pupil" and "student." There is, unavoidably, a basic correlation between the length of a student's courage span and the degree of his commitment to the work he is undertaking. Commitment mobilizes all of our intellectual powers, both the active and the passive, the short-term and the long-term. As Polanyi says again,

> Obsession with one's problem is in fact the mainspring of all inventive power.
> . . . It is this unremitting preoccupation with his problem that lends to genius
> its proverbial capacity for taking infinite pains. And the intensity of our preoc-
> cupation with a problem generates also our power for reorganizing our thoughts
> successfully, both during the hours of search and afterwards, during a period of
> rest. (Polanyi, 1967, p 127)

But *how* does a student coordinate the active parts of his faculties with the more passive ones? Does "sleeping" on a problem guarantee a solution? Very obviously not. We must clarify here what it means to let the mind work by itself upon a problem. The prior requirement for authentic "passive work" is the person's having disposed himself to a *general life of pursuit* with regard to a given field of inquiry. Leslie Farber, perhaps our most astute thinker on the issue of will, has located this kind of self-disposing activity in the "primary" realm of will, which makes itself felt as the deep drift of intention underlying the person's life. The "secondary" will, according to Farber, is the will which we consciously exercise in our day-to-day lives, and it is this latter will that we are most aware of when we take up a problem, or put one down (Farber, 1976, pp 4-8). So there are, in a sense, two distinct courage spans, each one pertaining to these different realms of will. *Short-term* courage spans are what we strengthen most readily through the teaching of heuristics; *long-term* courage spans are the ones which we nurture when we uplift a student by whatever legitimate means—such as by taking him into apprenticeship, or by otherwise celebrating his investment in his work. Apprenticeship especially (itself a passionate relation between a student and a teacher) is conducive both to a student's developing his incubatory powers and to his readying himself for true inspiration. The "trained unconscious" which typifies a full practitioner or expert in a discipline begins to consolidate during the apprenticeship phase, which is the last phase, really, of life as a pre-professional student. It is also during this phase that a student begins to grasp surely the hierarchical nature and the profound interconnectedness of all the problems in his discipline.

Once we have grasped the contributions of the will to our natural problem-solving rhythms, we are no longer quite so puzzled by the inconvenient fact of our repeatedly abandoning and resuming difficult problems. There is, quite simply, no other way to tackle them but to parse them into units fit for our talents and our courage. Interestingly, the succession of abandonments and resumptions which typifies our work on a very difficult problem sometimes yields, at the last, a relaxed and fully-formed solution which seems to come upon us from nowhere—"an illustration of that strange law," as John Updike says, "whereby, like Orpheus leading Eurydice, we achieve our desire by turning our backs on it" (Updike, 1962, p 236). Moreover, some of us are more or less intelligent cowards in the face of massive problems. Some students appear to give up easily, only to prove more willing than their diligent-seeming fellow students to return to a problem again and again. Such "goof-offs" often prove serious in the long run: they have learned the limitations of their short-term courage, and they are wisely protecting their long-term courage against erosion. Such students are often maddeningly indifferent to our proddings, as well they ought to be (Perry, 1963, pp 135-145).

On the other hand, those students who apprentice themselves to us very often allow us to foist problems upon them—and do so gladly in the convivial expectation that their growth will enable them to use such problems as the foundation on which to build problems of *their* devising. Thus, acquiescence becomes the training-ground for independent initiation and authentic new beginnings. Were this not so, the enterprise of high culture—especially advanced science—would grind to a halt.

Indeed, conviviality is the key to much of this. For as we get individually older, our intellectual lives get harder and more problematic—though by no means necessarily worse. The progressive accumulation of legitimate problems and projects is never offset by a correspondingly large number of solutions or successes. And yet the increasing severity of our intellectual responsibilities does not destroy our hope, or courage for that matter either, since both hope and courage are intrinsic to the fabric of problems. What happens, I suspect, is that we displace our hope from the egocentric realm into a more universal one: although we can't solve all of our problems because of our mortal limitations, they may be soluble, and others may find solutions for them. The early hope for self becomes a hope for mankind. Hence the deepening humility and expanding conviviality which typifies the intellectual as he moves toward death.

The foregoing discussion is a sketchy presentation of the problem of a problem. At the least it might serve to raise questions which are pertinent to cognitive psychology and to education in general. I am reminded here, in closing, of a remarkable assertion which G. Spencer Brown makes in his *Laws of Form*. A somewhat crusty and even, possibly, reckless statement, it may stand as a challenge to those who would put all of their faith in articulate thought and logical method:

> To arrive at the simplest truth, as Newton knew and practiced, requires *years of contemplation*. Not activity. Not reasoning. Not calculating. Not busy behaviour of any kind. Not reading. Not talking. Not making an effort. Not thinking. Simply *bearing in mind* what it is one needs to know. And yet those with the courage to tread this path to real discovery are not only offered practically no guidance on how to do so, they are actively discouraged and have to set about it in secret. . . .
>
> In these circumstances, the discoveries that any person is able to undertake represent the places where, in the face of induced psychosis, he has, by his own faltering and unaided efforts, returned to sanity. Painfully, and even dangerously, maybe. But nonetheless returned, however furtively. (Brown, 1972, p110)

These are strong words from a strong individual. My own hope is that the foregoing pages will make them seem less rash than they otherwise might seem, less at the antipodes from the notions of us teachers who believe that conscious thought and the systematic pursuit of problems are the main considerations requiring attention in education. They may indeed be; but they are not the only ones.

REFERENCES

Arendt, H. *Willing,* Vol. II of *The Life of the Mind.* New York: Harcourt, Brace and Jovanovich, 1978, pp 13-14.

Brown, G.S. *Laws of Form.* New York: Bantam Books, Inc., 1972.

Buber, M. *I and Thou,* second edition. New York: Charles Scribner's Sons, 1958.

Farber, L.H. *Lying, Despair, Jealousy, Envy, Sex, Suicide, Drugs, and the Good Life.* New York: Basic Books, Inc., 1976.

Perry, Jr., W.G. "Examsmanship and the Liberal Arts: A Study in Educational Epistemology." In *Examining in Harvard College: A Collection of Essays,* Cambridge, MA: Harvard University Press, 1963.

Polanyi, M. *Personal Knowledge: Towards a Post-Critical Philosophy.* Chicago: University of Chicago Press, 1962.

Updike, J. *Pigeon Feathers and Other Stories.* New York: Alfred A. Knopf, Inc., 1962, p 236.

The Computer as a Personal Assistant for Learning

Ira P. Goldstein and John Seely Brown

Following is a revised version of testimony given before the House Science and Technology Subcommittee on Domestic and International Scientific Planning, Analysis, and Cooperation, October 12, 1977.

We believe that a revolution will occur over the next decade that will transform learning in our society, radically altering the methods and the content of education and deeply affecting the individual at work, at home and at school. This revolution will result from harnessing tomorrow's powerful computer technology to serve as *personal assistants for learning.* We foresee personal computers performing as assistants, coaches and consultants, becoming, in time, an intellectual resource as ubiquitous but more powerful than today's primary learning technology—the book, and as captivating but more active than today's primary leisure technology—the television. In this paper, we sketch briefly several paradigmatic examples of learning assistants and indicate how they derive from a synergistic union of computer science, information processing psychology and artificial intelligence research. This sketch is intended to provide an overview of the potential role of computers as learning assistants, rather than provide a detailed account of the underlying technology.

WE ARE AT THE BEGINNING OF AN INFORMATION REVOLUTION

While it is no longer revolutionary to assert that powerful personal computers will become widespread over the next decade (see *Scientific American,* September, 1977), it may seem impossible that they can be insightful teachers, responding appropriately to a wide range of unanticipated situations. Page turners, drill and practice monitors—of course. Troubleshooting consultant, mathematical assistant, intellectual coach—incredible! It is our objective here to demonstrate that what would have been an incredible role for computers in the 1960's is an inevitable one for the 1980's. We shall do so by describing three prototypes of the computer as consultant, assistant and coach, respectively.

These prototypes are built upon an emerging cognitive technology, the central concern of which is to provide computers with an ability to *understand the learner*—that is, understand both his task knowledge and his learning style. Its approach is:

- to use techniques of artificial intelligence to represent problem-solving expertise within the computer, thereby escaping the limitations of traditional, frame-based computer assisted instruction (CAI).

- to use techniques of information-processing psychology to build models of the learner's skills, thereby being responsive to the idiosyncratic needs of the individual
- to use techniques of computational linguistics to provide natural language capabilities, thereby escaping the straightjacket of computer jargon

This technology confronts complex problems regarding the representation of knowledge, the nature of learning, and the theory of language. But solutions are gradually being developed for bridging the gap between a learner's needs and the machine's capabilities; a technology is emerging that promises to make the computer a truly personal tool.

The revolution in personal computing is in fact of even larger dimensions than is developed in this paper. We shall emphasize the impact of personal assistants. Of equal importance is the new set of activities the computer makes possible as a programmable musical instrument, a dynamic artistic medium, an interactive novel, a simulated physics laboratory, and many as yet undreamt of uses. Papert (1971) and Kay (1977) have eloquently described its potential as a creative medium. Our goal here is to describe the complementary objective of making the machine an even more powerful learning tool by supplying it with a tutoring ability. While we believe this to be of enormous importance in avoiding potential alienation, it is nevertheless the case that the set of intellectually exciting computer activities is far larger than the subset which is understood well enough to allow the construction of artificial-intelligence-based assistants. Even in such cases, however, the effort to construct such personal assistants is a powerful means of understanding more deeply the cognitive nature of these computer activities.

COMPUTERS AS CONSULTANTS

Learning assistants can have enormous impact on *technical* education. Training competent technicians to repair the ever-changing number of devices and technologies on which our society depends is an important educational goal. The mass dissemination of tomorrow's powerful computers makes possible the widespread use of simulations in technical training. Just as fliight simulators have long been important in trainng pilots, we believe that electronic simulations, for example, will be equally basic to technical training. These simulations provide inexpensive and safe opportunities for students to explore the complexities of a device.

The critical contribution of the learning assistant is to monitor and critique simulated tests and repairs made by the student. SOPHIE, *Sophisticated Instructional Environment* (Brown, 1975), is a prototype for a limited part of the electronics domain. SOPHIE presents the user with a simulated circuit to be fixed. The user can make any measurements he wishes, replace any parts. SOPHIE observes these measurements and employs a deep understanding of electronics to decide whether a given measurement is needed or a given part replacement justified. Its tutorial function is to discuss these observations with the novice technician. In essence, it is a troubleshooting consultant. The student can explore the device with no possibility of harm, in a private setting.

The availability of such consultants also solves another critical educational problem: that of continuing education. A technician faces the difficulty that the devices he is expected to repair are constantly changing. Written material is a clumsy means to document these changes, for it is a difficult medium in which to describe the dynamic nature of repair procedures. We envision SOPHIE-like consultants as being the basic medium by which technicians receive ongoing education. They are dynamic, they are organized around experience, yet they are easily modified to serve as an educational environment for a changing technology.

There is still another role such consultants can play. We believe citizens themselves can employ computer consultants to reduce the alienation engendered by modern technology. SOPHIE-like environments are fun: it is quite enjoyable to take apart a simulated television to see how it works. The embedded computer consultant can provide a guided tour of the device by embodying a deep theory of the domain along with a cognitive theory of what constitutes commonsensible understanding. Even if citizens do not actually repair their television sets, they will have gained a sense of command, of personal power through a better understanding of the devices which they employ.

Indeed this sense of power and command derived from the exploration of computer simulations can potentially extend beyond technical repair to a large number of otherwise unapproachable activities. It is quite conceivable to design intelligent consultants for flying a plane, building a bridge, or even piloting a moon shuttle. Such systems may do much to ameliorate a general alienation felt by people faced with the enormous complexity of modern science and technology.

COMPUTERS AS ASSISTANTS

The second prototype BUGGY is a teacher's assistant for diagnosing the underlying cause of errors in a student's arithmetic skills (Brown, 1977b). Table 1 is a set of problems that Johnny, a young student, was given in a screening test. All of the answers are wrong. Not surprisingly, the teacher concluded that Johnny could not add.

But the teacher was not correct. For Johnny had a perfectly reasonable procedure for addition; it just had one small bug. When adding two digits with a carry, he wrote down the carry and threw away the units digit. Johnny was only one small step away from a correct procedure, a fact that the teacher failed to observe for most of the entire school year. The failing grade produced by traditional evaluation neither diagnosed the nature of Johnny's misunderstanding nor helped him to debug it.

Table 1. Johnny's Test Problems

87	365	679	923	27,493	797
+93	+574	+794	+481	+1,509	+48,632
11	819	111	114	28,991	48,119

BUGGY is a *diagnostic* assistant that constructs procedural models of a student's arithmetic skills. It examines a student's answers and automatically grows a *diagnostic model* that best explains the observed errors. In a recent experiment, BUGGY analyzed over 10,000 problems done by 1300 students. The results were procedural models for each student which identified the bugs in that student's arithmetic skills. In the case of our young student Johnny,

BUGGY would have inferred "why Johnny can't add."

The consequences for standardized tests of the 1980's are enormous. It is entirely reasonable to conceive of the current national tests being replaced by student interaction with systems like BUGGY. The result would be evaluations that are less susceptible to the uninformative right or wrong quality of mechanized scoring, and that would be far more useful to the educator and to the student.

In concluding our discussion of this prototype, we must stress that before BUGGY could be built we needed to have both a deep procedural theory of the basic skills of arithmetic as well as a set of diagnostic heuristics. Before such systems could be built for significantly more complicated domains—such as reading comprehension—equivalent theories must be constructed for those domains. In other words, dramatic advances in computer technology alone will not suffice.

COMPUTERS AS COACHES

The third prototype envisions a renaissance of learning in the home. We see the television set, as we know it, disappearing in 10 years. It will disappear by becoming a part of the home computer installation. Of course, the TV/Computer will continue to receive commercial programs. But this will no longer be its central role. With far more frequency, the American family of the 1980's will be using its TV/Computer to play games; to watch interactive films in which the viewer can affect the outcome; and, most important of all, to access the treasure house of human knowledge that will be stored in the memory banks of these computers.

We believe that the TV/Computer will revolutionize learning in the home by uniting recreation with education into a single self-motivated enterprise. To explore the possibility, we have been studying the marriage of personal tutors to intellectually challenging games. We are looking at computer games because their popularity will provide a crucial point of educational leverage.

The idea is this: the home TV/Computer makes possible a class of intellectually substantial, but extremely enjoyable games. These games can involve logical deduction, probabilistic reasoning, hypothesis formation, and scientific induction. Hence, the game provides a challenging educational environment. However, as with any game, the player may reach a skill plateau. Good coaching is then required to move the student off the plateau. A computer-based learning assistant provides this coaching by modeling the player's skills, diagnosing plateaus, and supplying appropriate explanations to encourage further discovery.

For example, one class of games we are studying involves the exploration of mazes (Goldstein, 1977). The player must find the princess, slay the Minotaur or discover the treasure, while avoiding the many dangers of the maze, pits, bats, gremlins and other fairy tale nemeses. Each move yields information about that region of the labyrinth. The skilled player uses this information to make logical deductions about the location of dangers when possible, probabilistic inferences that make relative judgments of danger when no certain conclusion is possible, and strategic decisions balancing the goals of gaining information about the maze with the risk of a given course of action.

Such games are exciting for both adult and child. Indeed, we have observed players as diverse as first graders and Texas Instruments executives enjoying them. And the current excitement they engender is as nothing compared to the games of the 1980's when a six-square-foot screen and video disc memory reveals the misty caverns, pits and shadowy form of the dread Minotaur.

It is important that these games provide an active environment that captivates children and adults, and that exercises important intellectual skills. But their revolutionary impact derives from our ability to add a *cognitive component* to these systems, a personal coach serving to advise the players about what is intellectually significant about their current game situations.

To illustrate our prototype coach, suppose Mary, a young player, has reached a skill plateau. For example, she is not understanding the common-sense heuristic that multiple evidence carries more weight than single evidence for competing hypotheses. Our prototype coach can observe whether Mary is consistently failing to apply this heuristic in her play. If so, the coach waits for an appropriate situation and offers advice. For instance, the coach might say:

> Mary, it isn't necessary to take such large risks with the Minotaur. You have multiple evidence that the Minotaur is in cave 14 (where you want to go) which makes it quite likely that the beast is there. It is less likely that cave 0 contains the Minotaur. Hence, Mary, we might want to explore cave 0 instead.

In the absence of such advice, Mary might be long delayed in acquiring this heuristic and the other basic skills exercised by the game. In a natural way, the computer coach marries education to recreation.

Of course, a human coach could also perform this function. But this is not practical when one realizes that these games will be in the home, played as recreation by any member of the family at any time.

Naturally, there are many subtleties in creating a successful computer coach: the coach must not interrupt too often, it must not give explanations that are too lengthy, and it must retain a rapport with the student. To this end, we are constructing a *procedural theory of the teacher* that takes account of such considerations. It is a difficult enterprise and much work remains to be done. But it offers the possibility of making the computer the ultimate in congenial tools, one that is sensitive and responsive to its user.

"Wumpus" is only one game among many that we could have discussed. For example, we have constructed computer coaches for the PLATO project's arithmetic game "How the West was Won" (Burton, 1976) and the math educator's game of "Attribute Blocks" (Brown, 1977a). Rather than relying on canned responses, all of these coaches embody an ability to construct diagnostic models of the strengths and weaknesses of the student's play, and techniques for generating explanations for any situation that may arise in the game.

The potential here is enormous. Imagine, for a moment, 10,000 computer activities—all intellectually challenging, all different. Now imagine tens of millions of citizens engaging in these activities as recreation. And finally, add to this vision the availability of the coaches. Explicitly, the coaches provide advice to improve play; they serve a recreational purpose. But, implicitly, an educational purpose of the most profound kind is being served. In essence, we are creating a new environment for learning, centered in the home and equipped with personal tutors available on demand. This hands-on environment is useful for all citizens and may be ideal for reaching that segment of the population for which the classical schoolroom situation has failed. We believe that such personal computers will have a major impact in the sociology of learning by enabling the *home* to assume a new importance as a center of education.

A NEW SCIENCE OF LEARNING IS EMERGING

We have organized our discussion around examples to give a concrete picture of the future we envision. But such an organization conveys only implicitly our belief that a new science of learning is emerging. It goes beyond traditional psychology in its pervasive use of the procedural metaphor, beyond traditional education in its employment of new learning environments, and beyond traditional computer science in its focus on personal computing. Already it holds the promise of reformulating technical curricula, of transforming educational evaluation, and of diminishing the difference between recreation and education. Its fundamental contribution is the construction of procedural models of the syllabus, of the learner and of the teacher. It is these models that guided the repair advice of our technical consultant, the diagnoses of BUGGY, and the choice of explanations made by our computer game coach.

The continued development of this procedural theory of learning is the critical goal for future research. The availability of powerful computer technology is not by itself sufficient to insure a revolution in learning in our society. It would not revolutionize education, for example, to place the *Encyclopedia Britannica* within the memory banks of every home computer. While this may be worthwhile, it would not be qualitatively different from placing the books themselves in the home. The unique quality of the computer that does make possible a revolution is that it can serve as an active agent—a servant, assistant, consultant or coach—in a way that books and television cannot.

On the other hand, the mass dissemination of powerful computer technology does hold out the possibility of preserving an important strength of books and television, namely their ubiquity. Traditionally, computers in education has been an expensive activity: students are given limited access and their use is divorced from the family and the home. Over the next decade, the phenomenal drop in cost that we have seen for calculators will recur for personal computers. The result will be for the home to assume a new importance as a center of education. We foresee learning assistants as household appliances, as useful to mom and dad as to the kids. Surely our west coast friends will be using such personal learning environments to understand the latest results in earthquake prediction, while we New Englanders are studying with equal interest the latest advances in solar heating technology.

This is happening none too soon. Given the dramatic pace of progress in our society, education must be a lifelong experience. But for an adult, (indeed perhaps for the child) the classroom, where one sits among 20 to 30 others, receiving limited personal attention, surely is not acceptable. The home-based learning assistant is an attractive alternative. Learning again becomes centered in the home—an activity for all family members, motivated by interest, not arbitrary schoolroom demands.

A FRONTIER OF THE MIND IS BEING OPENED

We believe that society is on the brink of a true revolution, as far reaching in its consequences as the first industrial revolution. It is a revolution that holds the promise of opening new frontiers of the mind. Arthur C. Clarke describes, more poetically than we are able, the possibilities inherent in the personal computer.

> Duncan walked to the (console), and the screen became alive as his fingers brushed the ON pad. Now it was a miracle beyond the dreams of any poet, a charmed magic casement, opening on all seas, all lands. Through this window could flow everything that Man had ever learned about his universe, and every work of art that he had saved from the dominion of Time. All the libraries and museums that had ever existed could be funneled throught this screen and the millions like it scattered over the face of the Earth. Even the least sensitive of men could be overwhelmed by the thought that one could operate a (console) for a thousand lifetimes—and barely sample the knowledge stored within its memory banks. (Arthur Clarke, 1976, p 149)

These are the thoughts of a fictional character who addresses the Congress of the United States on July 4, 2276, the 500th anniversary of the republic. Clarke is right that the personal computer will have these qualities. But it need not take 300 years. With vision, planning, and dedicated research, we believe this new frontier of the mind can be crossed within a decade.

REFERENCES

Brown, J.S. and Burton, R. "Multiple Representations of Knowledge for Tutorial Reasoning." In *Representations and Understanding Studies in Cognitive Science,* edited by D. Bobrow and A. Collins, New York: Academic Press, 1975. 1975.

Brown, J.S. and Burton, R. "A Paradigmatic Example of an Artificially Intelligent Instructional System." *Proceedings of the First International Conference on Applied General Systems Research: Recent Developments and Trends,* Binghamton, NY, August 1977a.

Brown, J.S. and Burton, R. "Diagnostic Models for Procedural Bugs in Basic Mathematical Skills." ICAI No. 10, Bolt, Beranek and Newman, August 1977b.

Burton, R. and Brown, J.S. "A Tutoring and Student Modelling Paradigm for Gaming Environments." In *Computer Science and Education,* edited by R. Coleman and P. Lorton, Jr., ACM SIGCSE Bulletin, 8(1) pp 236-246, February 1976.

Clarke, A. *Imperial Earth.* New York: Harcourt Brace Jovanovich, 1976.

Goldstein, I. and Carr, B. "The Computer as Coach: An Athletic Paradigm for Intellectual Education." Proceedings of 1977 Annual Conference, Association for Computing Machinery, Seattle, October 1977, pp 227-233.

Some Thoughts on Reasoning Capacities Implicitly Expected of College Students[1]

A. B. Arons

INTRODUCTION

In most text materials, homework problems, and lecture presentations encountered in college-level study of natural sciences, social sciences, and humanities, it is tacitly assumed that the students are already in command of a variety of thinking, reasoning, and linguistic processes. An attempt is made in this paper to identify and make explicit some of the more important tacit assumptions. Among the assumed capacities are reasoning patterns characterizing Piaget's category of formal operations, but there are also expected patterns of still higher complexity and sophistication.

In recent years the administration of Piagetian tasks in logical reasoning has revealed that a very large proportion of college students tend to use predominantly concrete as opposed to formal patterns of reasoning. This observation points to a profound discrepancy between most secondary-school- and college-level course content on the one hand, and the actual student reasoning patterns on the other: most course presentations assume that students are generally prepared to utilize formal reasoning processes.

Efforts are currently being made to devise ways of enhancing formal reasoning skills and to reduce this discrepancy. This suggests that one should examine in greater detail the common assumptions about modes and processes of student reasoning—beyond the Piagetian examples—which are implicit in college-level course materials. Bloom (1956) defined a taxonomy of educational objectives which encompasses some general skills and defines a very broad framework for curriculum design. My objective is different; it is to help isolate those reasoning abilities which are commonly and often inappropriately assumed to exist in all college students. What varieties of thinking, reasoning, and linguistic skills are tacitly assumed to be already available to the students? What further skills can be identified as implied objectives in undergraduate instruction?

Putting aside questions concerning the extent to which such reasoning capacities may or may not be developed in the student population, and also deferring questions about possible methods of instruction that might foster the development of such skills, this paper concerns itself only with identifying what

---

1. Some of the content of this paper was originally developed in a discussion group which was part of the Symposium on Learning in Adolescence, held in connection with the bicentennial celebration of Phillips Academy, Andover, Massachusetts in February 1978.

appear to be some of the most important assumptions about student reasoning processes and levels of preparedness. No pretense is made that the listing is complete or exhaustive; no hierarchical ordering is intended. The goal, rather, is to examine our tacit assumptions and develop a higher degree of awareness of them. It should be clear from what follows that the reasoning processes identified are not confined to physics or the natural sciences exclusively, but arise in every area of study requiring logical thought.

RATIONAL AND LINGUISTIC PROCESSES IMPLICITLY EXPECTED OF COLLEGE STUDENTS

Recognition, Identification, and Control of Variables

It is generally expected that students can recognize and control, or weigh variables in setting up experiments or in solving verbal problems in the natural and social sciences. For example, they should be able to account for the possible influence of length, diameter, and composition in the well-known Piagetian task of the bending of rods, then control two of these variables in establishing the effect of the third. Similarly, the interpretation of historical phenomena requires recognition and sorting of political and social factors within the information available, following which the student must decide whether cause-and-effect relationships are discernable or whether the variables are so confounded as to preclude a reliable inference. Economics, political science, and experimental psychology all depend upon student facility in dealing with similar considerations of variables and their relations.

Arithmetical Reasoning

Ratio reasoning is required for predicting the alteration of gravitational or electric forces between point objects with changes of mass, charge, or spacing. Similarly, the use of scale ratios is necessary for interpreting maps, determining sizes of objects viewed under a microscope, or for comparing relative changes in volumes and surface (or cross-sectional) areas with changes of scale. Further, it is useful to relate such arithmetical changes to, say, the disproportionately large cross-sections of bones in larger animals, or to the rapidity of the dissolving of a material under finer subdivision. The use of similar kinds of ratio reasoning in connection with demographic data may be required in political science or sociology, and in connection with scaling of factors or rates in economic problems.

Interpretation of division is needed in dealing with concepts of density, velocity, or acceleration in physics; moles, gas behavior in chemistry; and in the study of population and other growth rates in biology. Combinatorial problems arise in elementary genetics, in probability considerations, in control of variables, and in design of experiments.

Forming and Comprehending Propositional Statements

Formation of intelligible propositional statements requires an intuitive grasp of the rules of logic, and of the grammar in which such statements are to be made. For example, forming or understanding verbal statements involves inclusion, exclusion, and serial ordering. In addition, one must grasp syntactical constructions such as double negatives, subjunctive mood, and the capacity to deal with elementary one- or two-step syllogistic reasoning. This is *not* meant to include involved propositional logic in which one is forced towards symbols or Boolean algebra for elucidation. The *basic* skill indicated, however, applies to all areas of study which require the use of language.

Ability to Paraphrase Paragraph of Text in One's Own Words

This expectation is applicable to all areas of study. A word of warning is needed here; students may be able to rephrase a paragraph using language similar to that in the text without understanding its content. Thus the insistence that they put it in their own words is of critical importance.

Awareness of Gaps in Knowledge or Information

This problem has two dimensions—gaps in the student's own knowledge, or incompleteness of known information in a given area of study. In the former case, it is expected that when a student fails to recognize the meaning of a word or symbol used in an oral presentation or a passage of a text, he or she will sense the need for establishing its meaning, and have the motivation to do whatever is necessary to establish it.

When the problem is incompleteness of information in a particular context, the student should realize that a definite conclusion cannot be reached; or should note that conclusions or decisions are being reached in the face of incomplete data and hence that such conclusions must be qualified accordingly. This overlaps with problems of psychological maturity, upon which depends the capacity to recognize and tolerate ambiguity in the material under study.

In a sequence of development of a given subject matter, the student is expected to gradually distinguish what has become known or clearly established at any particular point from what has *not* been so established. This implies learning to anticipate some of the questions still to be asked.

Understanding the Need for Operational Definitions

In general, students are expected to learn the criteria by which it is possible to determine whether or not a definition is operational. These criteria include: realizing when a concept in a passage of text has not been clearly defined; recognizing the necessity of using only words of prior definition in forming a new definition; becoming aware of the appeal to shared experience in forming operational definitions.

Translating Words into Written Symbols and Written Symbols into Words

The skills necessary for such operations are more rigorous than those needed for paraphrasing textual passages. Examples include: transforming a verbal statement into its equivalent arithmetical, algebraic, or graphical form in any of the natural or social sciences; interpreting a graphical presentation or the results of a symbolic problem solution in words, extracting its content while expressing the relevant qualifications and restrictions.

Discriminating between Observation and Inference

Students in all academic disciplines must learn to recognize the observational, empirical, or experimental facts that are available in a text presentation or in laboratory work. The next step is to separate these clearly from the concepts that may be formed or the inferences that may be drawn from them. An example of this would be identifying observed facts concerning the extent of illumination of the moon relative to its position with respect to the sun and then separating these from the inference that moonlight is reflected sunlight. Another example would be distinguishing the observed behavior of electrical circuits (that bulbs get dimmer as more are put in series at the same source) from the concept of "resistance" which is induced from the observed behavior. In this particular case, further distinctions would then need to be made between inferences concerning the nature of electrical resistance, and the predictions that can be made concerning phenomena in more complicated circuits.

Other examples from quite different areas of inquiry include: separating Mendel's observations of nearly integral ratios of population members having different color or size characteristics from the inference of discrete elements controlling inheritance; the distinction—common in the study of literature—between analysis of the *structure* of a novel, or poem and an *interpretation* of the work; and the historian's task of recognizing the distinction between primary historical data and his own interpretation of such data.

Analyzing a Line of Reasoning in Terms of Underlying Assumptions

Every line of reasoning has an underlying set of assumptions separate from the factual data it may utilize. Students need to develop the capacity first to discover and second to distinguish among assumptions, assertions, and reasoned conclusions.

Drawing Inferences from Data and Evidence, Including Correlational Reasoning

Separate from the analysis of another's line of reasoning comes the formulation of one's own. For example, given the observation that the spot formed on the screen by the cathode beam (Thompson's experiment) remains coherent (i.e. does not smear out) under both electric and magnetic deflection, what inferences can be drawn concerning the origin and character of the beam? Or: given the code of laws of Hammurabi, what—if anything—can be inferred about how the people subject to it lived, and what they held to be of value? Yet another example is the problem of recognizing possible functional or cause-and-effect relationships in either positive or negative correlations in the face of statistical scatter or uncertainty; for example, discerning relative or competing effects of light, heat, or moisture on a biological population (with a simultaneous awareness of whether or not the variables have been adequately controlled).

Ability to Discriminate Between Inductive and Deductive Reasoning

Students should be able to follow inductive reasoning used in generating a concept, model, or theory, and use deductive reasoning in testing the validity of a construct. They should perceive the analogous patterns of scientific thought arising in such broadly diverse areas as the Newtonian Synthesis, wave versus particle models of light, atomic-molecular theory, gene theory, theory of evolution, economic or sociological models of society or its parts, and so on.

Performing Hypothetico-deductive Reasoning

Students should be able to visualize, in the abstract, outcomes that might stem from changes imposed on a given system or situation, whether it be in scientific, literary, historical, economic, or political contexts, and effect such visualizations through reasoning within the basic principles or other rational constraints applicable to the system.

Performing Qualitative, Phenomenological Reasoning or Thinking

In science and mathematics, the ability to recall formulas and manipulate them algebraically does not by itself indicate complete understanding of a subject area. Students should also be able to give *qualitative* explanations of principles and make direct inferences from them without referring to the results of numerical calculations. They should be able to apply phenomenological reasoning without relying on mathematical formalism.

Checking Inferences, Conclusions, or Results

Skills in this category include: testing for internal consistency; using alternative paths of reasoning; examining extreme, limiting or special cases.

In some instances, only initial or preliminary levels of the skills listed in the preceding section are actually presupposed in college work at introductory levels, while enhancement and further development of such skills are often implicit objectives of the courses of instruction. In addition to these objectives, others are at least implied, when not explicitly articulated, in most statements of the cognitive goals of higher education. Two of these more general goals which subsume many of the preceding objectives can be articulated as follows.

Developing Self-consciousness Concerning One's Own Thinking and Reasoning Processes

It is generally desired that students learn to become explicitly aware of the mode of reasoning being used in particular situations. This provides the basis for consciously seeking to transfer familiar modes of reasoning to new and unfamiliar situations. In general, students should learn to attack a new problem or unfamiliar situation by first forming very simple related problems, by asking themselves questions derived from the simplest and most concrete aspects that seem to underlie the given situation.

Developing the Skills of One's Discipline

Finally, students are expected to combine the preceding modes and processes into the general skills of problem solving as practiced by the discipline(s) of choice.

COMMENTS AND CONCLUSION

In forming the preceding list, an effort has been made to isolate and describe reasoning modes and processes—together with levels of awareness that appear to be bound up with clear thinking and genuine understanding in various disciplines; and to indicate that such processes are common to very different subject-matter areas. However, this description of assumed student capacities (or course objectives) does *not* constitute a prescription for *how* such capacities are to be attained.

In designing procedures for assisting student development, we need to be sensitive to styles of learning; balance of verbal, visual, and auditory presentation; the question of which subject-matter areas may be the most suitable starting points for different individuals; and in general, the volume, pace and level of materials presented. It must not be forgotten, however, that the reasoning modes and processes proper to academic inquiry are skills that must be mastered—they are not a matter of style or preference. I have recently attempted to design a course in physical science which encourages student growth along many of these lines (Arons, 1977); readers interested in instructional techniques are referred to this work as one example of a curriculum oriented toward some of the preceding goals.

In recent years, the administration to college students of typical Piagetian tasks characterizing formal patterns of reasoning has revealed that at least 50 percent in the broad cross section completely fail to perform the tasks or perform them with only partial success. In general, these tests probe only a few items among the capacities listed in this paper and do not probe the more sophisticated capacities or those involving verbal-linguistic aspects. Inquiry among secondary school and college teachers usually yields quick, subjective denial that an appreciable fraction of the students exhibit these capacities. Yet much of the material and instruction with which students are confronted implicitly assume that the reasoning capacities are either already developed, or that they will automatically evolve with maturation and the study of the subject matter.

It is becoming embarrassingly clear that such automatic development actually occurs in only a relatively small proportion of our students—perhaps the upper 25 percent, those whom we characterize as among the "brightest." The remainder, who might well develop such intellectual capacities at a slower pace, are not afforded the opportunity to do so. They need the time to reason slowly, to make mistakes and retrace their steps without being crushed or punished; to revise their thinking and test it themselves for internal consistency. Under pressure of the volume and pace of material with which they are deluged, many students seek refuge in blind memorization and completely lose sight of the intellectual processes through which they could bring order into the chaos that seems to surround them. As a matter of fact they never develop any sense of their own intellect or of the deeply satisfying feeling that emerges when one recognizes that he understands something.

In order to cultivate the development of the reasoning capacities listed above, it is necessary to give students time, explicit help and encouragement, and repeated practice in all subject-matter areas. It is also necessary to enhance their consciousness of thought processes by urging them to stand back and examine the reasoning processes in which they have engaged, and to express them in their own words. When this is conscientiously done, even if it be at the expense of "coverage" of subject matter, students develop an entirely new intellectual stance, characterized by heightened respect for their own intellect and by pride of achievement. In consequence they begin to be conscious of the speciousness of the rewards they received in circumstances in which they were driven to memorize without understanding, and they are motivated to attack still more demanding inquiry without giving up readily on encountering difficulty or temporary frustration. In other words, to quote Justice Learned Hand's ironic phrases, they become more "willing to engage in the intolerable labor of thought—that most distasteful of all our activities."

Justice Hand used this line in a discussion of the circumstances under which liberties might be preserved in a democratic society. His basic thesis was the absolute necessity of a thinking, reasoning citizenry. What capacities characterize such a citizenry? The sophisticated distinction between enlightened self-interest as opposed to short-range self-interest must be based on hypothetico-deductive reasoning. Such reasoning is also inevitably involved in visualizing possible outcomes of different policies and decisions in economic and political contexts concerning which one must exercise a vote. There is a repeated need to discriminate between facts and inferences in the contentions with which one is surrounded. There is the necessity for making tentative judgments or decisions, and it is better that this be done in full awareness of gaps in existing available data, knowledge, or information. There is the highly desirable capacity to ask critical, probing, fruitful questions concerning situations in which one has no initial expertise. There is the need of being explicitly conscious of the momentary boundaries of one's own knowledge and understanding of a particular problem. Every one of these capacities appears in our list and can be cultivated and enhanced, at least to some degree, in the great majority of college students through properly designed programs embracing a wide variety of subject matter. Thus there is reason to hope that an educational framework, consciously designed to develop the reasoning capacities of college students, can enhance their professional careers and make them more responsible citizens of a democratic society.

REFERENCES

Arons, A.B. "Cultivating the Capacity for Formal Reasoning: Objectives and Procedures in an Introductory Physical Science Course." *American Journal of Physics,* 44, 1976, pp 834-838.

Bloom, B.S. *Taxonomy of Educational Objectives. Handbook I: Cognitive Domain.* New York: Longmans, 1956.

Toward Observing
That Which Is Not
Directly Observable

Robert L. Gray

INTRODUCTION

Instructional strategies in physics have over the past decade been influenced by the developmental theory of Jean Piaget. In general, curriculum revision and design have been directed toward the incorporation of concrete materials in laboratory settings requiring active involvement on the part of students. Given the nature of Piaget's research methodology, it is no surprise to find exploratory manipulation the cornerstone of these newer instructional strategies. From kindergarten through freshman and sophomore college classes one finds example after example of action-oriented instruction (Arons, 1973) in which concreteness is rigorously pursued. The theoretical rationale for this emphasis upon concrete objects derives from the unique and crucial role played by the object prior to the stage of formal operations (Piaget, 1969). The interplay between sense data derived from objects and overt actions performed on them forms the intellect in the Piagetian model.

The educational community ought now to seek a theoretical base by which to guide the growth of formal operations. That part of an overall theory which connects the manipulation of real objects to the formation of concrete operations has fulfilled its educational promise. There seems, however, to be little evidence supporting the notion that formal operations grow solely out of the debris of paper, glue, test tubes, inclined planes, litmus paper or any other laboratory apparatus. The metaphor of a proportionality algorithm (a favorite Piagetian task indicative of formal operations) succinctly states the issue urgently requiring attention.

"Objects are to concrete operations as _____ is/are to formal operations."

The metaphor needs somehow to be completed. Piaget might insert "the performance of operations upon operations." (Nodine, 1971) But what does that mean in an instructional setting?

The experiences of the author over the past few years in physics courses for both science and nonscience majors suggest the need for an instructionally useful, articulated distinction between reasoning at the concrete stage and reasoning at the formal stage. The notion of abstraction is in this paper the distinguishing feature, and the relationship between abstraction and concreteness is considered in the context of instructionally assisted promotion from concrete to formal operations.

PERCEPTUAL UNDERSTANDING

A concrete mode of interaction with the physical world enables an individual to "make sense" of his surroundings in a very literal fashion. Muscle actions play an important role in developing notions of weight. Sensitivity to texture accompanies the formulation of the concept of surface friction with attendant influences upon mechanical motion. A number of misconceptions can be traced to concrete roots. Sinking or floating derive their early and incorrect causal descriptions by differentiating floating cork as weighing less than sinking stone. A causal link between force and velocity contrary to the Newtonian model is clearly associated with an egocentric description of how individuals place and relocate the objects in their environment.

A perceptually based strategy for understanding has a pernicious nature. It works and works well when by understanding we mean "know about" or "repair" or "be familiar with." Dealing successfully with the world in this manner may establish a predisposition toward modes of understanding not thought of as abstract. A very simple example would be a technical understanding of automobiles or television. A majority of auto or television repairs are effected as the direct result of observation—listening for sticking valves, looking for broken parts, smelling burned resistors or feeling worn gears. This concrete level of understanding is obviously useful. At the same time perceptually based understanding, derived as it is from specific experimental episodes highlighted by ends rather than means, has limited transfer potential. In terms of transfer the analogy between automobiles and television sets provides a striking example. The physics of a fly-back transformer in a Zenith television set and an ignition coil in a John Deere tractor are conceptually identical. Abstract notions of magnetic field, flux changes, and induced currents are identically comparable, but descriptions of this type are not sensible in the manner of burnt contact points or charred transformers. Instructional programs will have come a long way when a mechanic trained by General Motors can deal effectively with a Sony product.

ABSTRACTION AND FORMAL REASONING

Lest the description of concrete technical understanding appear as a harsh judgment upon those who provide useful, and, in the categories of television and automobile repairs, nearly indispensible public services, the point is not to downgrade the utility of observation and detailed acquaintance. Rather the intent is to develop by example the distinction between knowledge of the physical world (sense data) and logico-mathematical knowledge (abstractions imposed by individuals upon sense data). Curriculum development and instruction, inspired by the theories of Piaget, require equal emphasis on both.

Explanations derived solely from sense data tend to establish a pattern of thought which inhibits development of the notion of abstraction and hence of formal reasoning. From the point of view of education at the university and college level, the cognitive awareness which prefers sense data may be at best reluctant and at worst unable to entertain concepts not readily and directly based on the senses. Even though a purely constructionist theory argues (von Glasersfeld, 1974) that objects—indeed, an ontological reality per se—are mental constructs and therefore abstract, it does appear that our understanding is all too often thought of as contained within the barrage of sense data which originates outside the individual. Thus, much current classroom use of the Piagetian teaching and learning strategy consists of a reduction of his theories to a simple "tune up your senses and understanding will follow."

A primary characteristic of formal reasoning, then, is the abstraction imposed upon perceptual data by the perceiving subject. Abstraction (Langer, 1957) requires the recognition of a relational structure apart from the specific thing, event, fact, or image in which it is exemplified. To bring this definition of abstraction into sharper focus an example in medical diagnosis is appropriate.

Consider as an "object" a recorder tracing of the lead I electrocardiogram (EKG). Figure 1 depicts the essential features of the EKG. The "QRS complex" represents the main contraction of the heart muscle. The diagnostician, particularly if an inexpensive and reliable survey of heart conditions among members of large populations is desirable, needs to know which features could

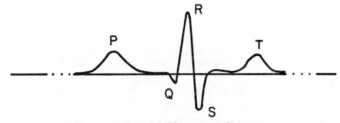

Figure 1. Lead I Electrocardiogram

be utilized as a tool in the treatment and prevention of heart disease. An observer in the habit of preferentially attending to observable features might attempt to correlate heart condition with the shape and/or size of the actual tracing. The size of the large wiggle, the distance between wiggles, the shape of the pulse—long and skinny or short and fat—are possibilities linking EKG with the condition of the heart. The question arises whether the useful abstraction is directly available from the object or apart from it. Is the search essentially the art of seeing (explicitly in the form of sense data) or does the search require the observer to add features which are not seen? Does the object alone contain the means for understanding or is understanding brought to the object by a cognitive act on the part of the observer?

The example of the EKG tracing gives us an example of what is meant by features apart from the object. A recent scientific article (Djalaly, 1977) describes the "T-intercept" of a lead I EKG as a useful diagnostic tool in treatment of heart disease. The construction of the T-intercept is shown in figure 2. Note that the T- intercept is not part of the original object. It is not anywhere on the

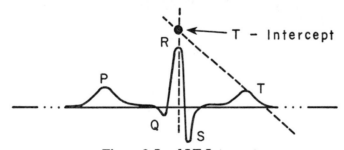

Figure 2. Lead I T-Intercept

original tracing. It was added by the investigators through the rather simple, albeit not obvious, expedient of extending two straight lines with subsequent identification of their intersection. Of course the dotted lines are determined by features of the object—the down slope of the R pulse and the center line of the T pulse—but the important correlating feature is a combination of the two. For a number of common heart diseases the T-intercept is a more reliable index of heart condition than either the R or T pulse. This technique of operating on objects to produce abstract characteristics useful as organizational means is a distinguishing feature of concrete and formal operations and further differentiates active from passive problem solving. A passive problem solver is one who permits the problem to dictate actions rather than initiating the type of action to be taken.

COMBINATIONS OF OBSERVED PROPERTIES

At the elementary-school-level curriculum materials (Karplus, 1974) consistent with stage theory provide ample opportunity for the invention of organizing classification schemes based upon perceptual data. Sorting buttons, identifying whales as mammals rather than fish, serial ordering according to size, and classification of electrical conductors and insulators are examples of the process. Such activities are important, but probably depend more upon selective provisioning and opportunity than instruction. What does require instruction, both in its own right as an intellectual activity and as an important precursor to abstract reasoning, are the utility and necessity for classification schemes which are *not* based on a directly observable principal index, schemes that go beyond mere sense perception.

The EKG example is a more abstract classification scheme than sorting buttons according to size. While each involves sense data the T-intercept concept *follows from an active treatment of the data,* in other words, from the invention of new data. In the case of circular buttons an analysis cognitively analogous would be to classify buttons as those objects whose circumference to diameter ratio is approximately three. The "threeness" of buttons is a property quite apart from those directly perceivable.

Geometrical optics beginning at the concrete level lends itself to a useful discussion of object classification and identification which becomes increasingly removed from concrete sense data. Cylindrically-shaped thin plastic containers of water exhibit many of the commonplace optical phenomena of magnification, image reversal, real image formation and so on. (Elementary Science Study, 1968) With minimum direction and suitable provisioning students can, using containers of different sizes filled with transparent liquids of various types, sort out the role of "curvedness" and "liquid type" as important variables which characterize a lens. The more careful observer will also note a third feature of importance having to do with the spatial organization of the object being viewed, the lens and the eye. From the point of view of concept formation this spatial organization is described by attributing to the lens a focal-length distance measured from the center of the lens to a point external to the lens. Eyeglasses have focal lengths of a few to tens of centimeters. Here is an example of an abstract property of an object apart (and here in a very literal sense) from the object.

Defining focal point as a point in space a fixed distance from the lens is cognition fundamentally different from cognition of radius of curvature or transparent fluids. It is this pocess of conceptualizing via the invention of abstract properties not sensibly perceived, as one does not see the focal point, which is a necessary precondition for the transition to formal operations. Thus, abstract properties of this type are often linked to perceptions but they do not have an analogous concrete character. Their acquisition, description, deployment and comprehension are not available through passive sense detection. Piaget's clarion call (1971) for active methods of learning and instruction has been heard concerning acquisition of concrete knowledge and firm development of concrete operations. But the educational community needs now to address the equally active but cognitively distinct task of acquiring abstract, as opposed to sensual, modes of thinking. This is especially so if the purpose of higher education is to promote transition from concrete to formal operations.

RATIO AND ABSTRACT PROPERTIES

Those familiar with the physics of lenses will recognize focal length as being proportional to the ratio of index of refraction ("liquid type") to radius of curvature ("curvedness"). Looking still deeper, index of refraction is the ratio of the sine of two angles defining air to have an index of refraction of one. It is clear how quickly a useful classification scheme becomes uncoupled from the perceptual properties of the thing called a lens.

The construction and use of ratios is a particular operation occurring frequently in quite diverse disciplines and circumstances. Unit pricing, density, slopes of linear graphs, force constant, cost per gallon and inflation rates are examples. Karplus (1970) has made a detailed study of the use of proportional reasoning which involves the concept of ratio with his well-known problem of Mr. Short and Mr. Tall. These studies have considered Piagetian concepts of developmental level as well as cognitive styles among diverse populations and age groups. Fuller at the University of Nebraska-Lincoln wrote a variation of Karplus' problem. Part *A* of the problem is essentially the original version. Part *B* is a related question. Fuller's complete problem is appended to this article.

The Fuller problem was administered to 100 students at the University of Massachusetts in Amherst. Approximately 70 percent were junior or senior nonscience majors enrolled in a course concerning the physics of sound. The results are shown in Table 1.

Results of part *A* of this problem are as reported by Karplus and Peterson with nearly identical data, approximately 80 percent responding correctly. The more intriguing aspect of the data reported here is the fact that considerably fewer than half answer both *A* and *B* correctly. The determination of the width of Mr. Tall's car, measured in buttons, is obviously a more difficult task. It would seem that stating a problem without explicitly placing it in the format of a proportionality, with three knowns and one unknown, not only fails to prompt solution skills apparently present, but, more seriously, calls into question the cognitive elements actually present among those who can use the algorithm.

It is the author's contention that instructors who present proportionality algorithms in a very concrete setting fail to take account of their truly abstract character. The invariant paper clip/button ratio, which is neither paper clip nor

Table I. Responses to Part A and B of the Fuller Problem

Response Categories, Part A	Percent Occurrence	Response Categories, Part B	Percent Occurrence
Wrong guess or wrong answer with no or incorrect reason	7	Correctly answered parts A and B	39
Proportion indicated but answer wrong	3	Correctly answered A but not B even though recognized	27
Use difference reasoning rather than proportion reasoning	7	Correctly answered A but not B and used different or no reasoning	17
Proportional reasoning with correct answer	83	Incorrectly answered A and B	17

button nor anything else, is the concept which allows for solution of part B. That ratio, initially determined by the height of Mr. Short, has nothing in particular to do with Mr. Short. The ratio is an abstract organizational scheme which relates the linear dimensions of two quite distinct objects. Thus we find—as was the case with the example from geometrical optics—that the actual content of instruction is an abstraction completely distinct from the concrete properties of the examples used to illustrate it.

ABSTRACTION AND RELATIONSHIPS

Previous sections have considered abstract properties which are derived from an *active* treatment of concrete data; hence they have their roots in the manipulation of real objects but are never concretely attached to objects. Von Glasersfeld (1977) describes the act of constructing abstract properties as belonging not to the perceived object but rather to the active perceiver. Abstract construction is thus a performed operation, and the term "abstraction" is here meant to imply the conception of formal properties of concrete objects which are not given exclusively through empirical observation of such objects. This section considers abstraction in the context of mathematical relationships with an emphasis on their meaning consistent with intellectual development.

The necessity for abstraction at the formal operational level emerges as interest shifts from the identification or description of discrete static states of objects (including in large measure their functional use) to the transformations associated with changes from state to state. Attending to the nature of change requires what Piaget (1964) has called production and conservation. Production derives new states from old via transformation operations. Conservation insures a permanent conceptual foundation independent of both specific static state details and specific changes that may result from the transformation operations.

In physics a mathematical model is often used as the means to describe transformational relationships. Such mathematical models combine the production and conservation elements of the transformation process. The production function finds expression in the mathematical function, such as additive, multiplicative, and derivative, while the conservation element appears either explicitly in the function itself or implicitly in the range and domain of the function.

Instruction which focuses attention upon the production and conservation nature of the transformation process, and emphasizes its nonconcrete features, would appear to be a necessary stepping stone toward formal operations. In physics there is opportunity to contrast two types of production function which, developmentally speaking, represent the shift from concrete, static states to the more formal notion of transformation.

The new length L of a metal rod whose temperature is increased by an amount T is a well-known phenomenon. In terms of its original length L_0 and a coefficient of expansion α the relationship is written:

$$L = L_0 (1 + \alpha T) = L_0 + L_0 \alpha T$$

Here is seen the production of the new length L as an increase of length $L_0 \alpha T$ added to the original length L_0. A production function of this first type represents a transformation in which a single attribute of the rod is quantitatively changed: one starts with a metal bar and ends up with a longer metal bar.

An example of the second type of production function is:

$$\frac{E}{R} = \frac{1}{R} \frac{d\phi}{dt}$$

To the concrete learner, this and the earlier expression are algebraic formulae without distinction apart from the obvious one of different letters. (Students show a remarkably blind allegiance to an egalitarian democracy of formulas.) The relationship above says that the time rate of change of the magnetic flux threading a closed circuit induces a current in that circuit. Here the production starts with a time rate of change in the magnetic flux and ends up with an induced current. Thus, unlike our earlier example, where the relationship to be understood was one of change with respect to a single attribute of otherwise identical things, the present case concerns transformations in nonidentical domains of dissimilar "things." The student must note, not only that a change is taking place, but also what aspects of the situation are not changing. The second function, then, expresses a relationship between qualitatively quite dissimilar "things."

The notion of a transformation process of the second type is not exclusively within the province of an abstract discipline such as physics. Consider the production function:

$$\text{Total dollars} = \frac{\text{dollars}}{\text{gallon}} \times \text{number of gallons}$$

This production function transforms gallons into dollars.

Or another example:

$$\text{Mr. Tall's height} = \frac{3 \ \text{paper clip}}{2 \ \text{button}} \times \text{Mr. Short's height}$$

This production function transforms Mr. Short into Mr. Tall. The conservation element of the transformation process is explicitly included in the form of the constant ratio $\frac{3}{2}$. Were it not for the fact that the ratio is constant, or conserved, it would be impossible to infer the height of Mr. Tall from knowledge of Mr. Short. All relationships imply an invariant and it is precisely such invariant concepts which make it possible to attend to change as the principal focus of formal operations.

As abstractions, ratios have two features which, in the context of instruction designed to promote formal reasoning, deserve investigation and discussion. First, these ratios are never perceptions. They are always conceptions and as such must be dealt with in nonconcrete terms. Yet, the ratio representing unit cost is typically not described as a concept independent of cost or quantity. It is usually described as what must be paid to buy one unit.

The second feature of ratios can be discovered through recognizing that a ratio is the outcome of a mental act of construction on the part of the student. Renner and McKinnon (1971) employed five Piagetian tasks (Piaget, 1958) to assess the logical thought processes of college freshmen. These authors found two thirds of their respondents unable to solve two of five problems involving reciprocal implication and the relationship between sinking and floating. Piaget (1958) cites these two problems as particularly appropriate indicators of the level of development of intellectual capacity for formal operations. In both instances, successful solution requires the student to perform a constructive act. The reciprocal implication problem is manageable when a line perpendicular to a reflecting barrier is added; floating or sinking requires the density ratio for solution.

Karplus (1974) in the fourth of his articles on proportional reasoning speculates concerning the relative level of abstraction between the explicit construction of a ratio (the paper clip/button ratio), and the use of a direct statement of proportion as means toward problem solution. Citing anecdotal evidence from scientifically trained adults Karplus suggests the use of the paper clip/button ratio, although formal, may be at a lower level of abstraction. In a more recent paper on teaching for proportional reasoning, Kurtz and Karplus (1977) report delayed posttest results showing substantial regression from using an algebraic equation to a multiplicative proportional statement. These authors interpret the regression to imply that additional insight into the constant ratio is not provided by algebraic equations.

The disparity between parts *A* and *B* of the Fuller problem and the findings of Kurtz and Karplus concerning algebra suggest to this author that Piaget's formulation of the transformation process involving production and conservation might provide a theoretical base for instruction in abstract reasoning. The well-known formula syndrome among students of physics encourages and sustains a static view of physical phenomena thereby inhibiting the growth of formal operations. Mathematical modeling interpreted in terms of production (with emphasis upon the dynamic character of the transformation process) and conservation (with emphasis upon invariance) holds not a little promise for instructionally assisted promotion from concrete to formal operations. The blank space of the proportion metaphor in the introduction to this paper could usefully be completed by "abstraction *brought to* objects."

SUMMARY

For the better part of a decade curriculum revision within science generally and physics in particular has been strongly influenced by Piaget's developmental theory of intellectual growth. The psychobiological connection between overt physical actions performed upon objects and the resultant internalized schemes coordinating these actions has been the principal theoretical tenet guiding instructional innovation to date. The enhancement of intellectual skills among students exposed to the active or manipulative methods is well documented. If one includes anecdotal evidence of the positive change in attitudes toward learning achieved through the active methods, then those who have been engaged in the "Piagetian movement" can with confidence assert that a very good beginning has been established.

It has been the premise of this paper that stage promotion, not just growth, is the essential goal of education at the university and college level. Concrete operations are by definition bound to situational specifics and concrete properties. Educational institutions for students beyond adolescence must, given the explosive and unpredictable nature of information and facts, give preference to instruction and learning which in the final analysis enables individuals to deal logically, rationally and creatively, regardless of the context.

To the above end this paper identifies one area worthy of greater investigation than is reflected by the literature. The notion of abstraction, especially abstraction quite apart from objects, and its implications both for the intellectualization of a discipline and the intellectual development of an individual require attention. Most of the data available at the present time places 20 percent to 50 percent of lower-level college and university students in the stage of concrete operations. These data would be considerably different if the key to the transition were to be found among the directly sensible properties of the objects of the world.

THE FULLER PROBLEM

Mr. Short is shown below. Mr. Tall is *not shown.*

Mr. Short's height has been measured with large buttons one above the other. Four buttons reached from the floor to the top of his head. Thus:

Mr. Short is 4 buttons high.

Mr. Tall measured with the same buttons was 6 buttons high. Thus:

Mr. Tall is 6 buttons high.

Mr. Short, as shown, has also been measured with paper clips. We find:

Mr. Short is 6 paper clips high.

Part A. What is the height of Mr. Tall, measured in paper clips? _____

Please explain carefully how you found this answer.

Part B. Mr. Tall's car is 14 paper clips wide. How wide is Mr. Tall's car, measured in buttons? _____

Please explain carefully how you found this answer.

REFERENCES

Arons, A. "Toward Wider Public Understanding of Science." *American Journal of Physics,* 41 (6), pp 769-782, 1973.

Djalaly, A., Hedayatti, H.A., and Zeighami, E. "Screening Capabilities of the Lead I Electrocardiogram." *Journal of Electrocardiology,* 10 (3), pp 245-250, 1977.

Elementary Science Study, Educational Development Center, Webster Division, St. Louis: McGraw-Hill, 1968.

von Glasersfeld, E. "Piaget and the Radical Constructivist Epistemology." Presented at the Third Southeastern Conference for Research on Child Development, 1974.

von Glasersfeld, E. Private communication, Fall 1977.

Karplus, R. *Science Curriculum Improvement Study: Teachers Handbook.* University of California, Berkeley, 1974.

Karplus R. and Kurtz, B. "Intellectual Development Beyond Elementary School IV: Ratio, the Influence of Cognitive Style." *School Science and Mathematics,* 74 (6), p. 476, Oct 1974.

Karplus, R. and Kurtz, B. "Intellectual Development Beyond Elementary School VII: Teaching for Proportional Reasoning." SESAME and AESOP, University of California, Berkeley, 1977.

Karplus, R. and Peterson, R.W. "Intellectual Development Beyond Elementary School II: Ratio, A Survey." *School Science and Mathematics,* LXX, Dec 1970, p 813.

Langer, S.K. *Problem of Art.* New York: Charles Scribner's Sons, 1957.

Nodine, C.F., Gallagher, J.M., and Humphreys, R.D. *Piaget and Inhelder: On Equilibration,* Proceedings of the First Annual Symposium of the Jean Piaget Society. Philadelphia: Jean Piaget Society, 1972, pp 17-18.

Piaget, J. and Inhelder, B. *Growth of Logical Thinking.* New York: Basic Books, 1958.

Piaget, J. *Science of Education and the Psychology of the Child.* New York: The Viking Press, 1971, p 68.

Piaget, J. "Development and Learning." *Journal of Research in Science Teaching,* 2, p 176, 1964.

Renner, J.W. and McKinnon, J.W. "Are Colleges Concerned with Intellectual Development?" *American Journal of Physics,* 39, p 1047, 1971.

Engineering Student Problem Solving

Lois B. Greenfield

The engineer has earned his reputation as a problem solver. Where does he learn this skill? Surely if one were to look in on most engineering classrooms in this country, one would find both students and teachers concerned with the solution of problems. In what ways do engineering educators teach problem-solving skills to their students? Do they, in fact, make special efforts to teach such skills?

First, I would like to emphasize the difference between the product of problem solving, i.e. the answer or solution to a problem, and the process of problem solving, or the method of attack on a problem. The answer or solution to a problem is readily observed, and can be quantified. The engineering student's homework solutions can be graded, and marked right or wrong. Emphasis is placed on accuracy of the answer, but the method of solution may be equally important. In the "real world" a variety of answers may satisfy the problem conditions—indeed, two engineers may look at a problem and develop two completely different solutions to what they have seen and identified as two completely different problems.

Although it is possible to infer the process or method of attack used in solving a problem from the product or answer, the conclusion may be misleading. For example, if an engineering student gets the wrong answer to a problem on a test, can we determine the reason for the error? Do we know whether the student has used the wrong formula, made an error in arithmetic, neglected an important bit of data, lacked knowledge of necessary facts, or completely misinterpreted the nature of the problem he was asked to solve? The reason for error may be none of these; or the student may have been so upset by the examination and all that hinged on his performance, that he was unable to demonstrate what he knew.

The instructor may try to infer the process of problem solution from the product obtained by means of statistical techniques. It is possible to determine which questions on the test are most difficult, as well as the order of difficulty. Particularly in objective kinds of tests, it is possible to determine the number of students who get a problem wrong if it is presented in one fashion, and the number who get it wrong when it is presented differently. It is possible to ascertain the wrong answer most frequently given to a particular problem. From such data, the instructor can attempt to figure out why a particular question is difficult, or why he thinks it would be difficult for him if he didn't know how to solve it. The instructor cannot, however, be sure that he is not going beyond the implications of his data. He cannot know with certainty why one answer was preferred above another, where the student might have gone astray in his reasoning, whether the student misinterpreted the question he was being asked, or

229

even whether the student was offering an entirely correct answer to the question he was answering, although it was a question different from the one posed by the instructor.

Another approach to the study of the processes of problem solving is to infer them from the rules of logic, assuming that one step in reasoning follows another in logical sequence. For example, Polya's *How to Solve It* (1945) outlines a procedure for solving problems. "First, you have to understand the problem . . . Second, find the connection between the data and the unknown. You may be obliged to consider auxiliary problems if an immediate connection cannot be found. You should obtain eventually a plan of solution . . . Third, carry out your plan . . . Fourth, examine the solution obtained." In this analysis, much depends on the luck of the problem solver in coming up with "bright ideas" and on good guesses.

The literature of experimental psychology contains a number of studies of problem solving, many of which deal with children solving puzzle-type problems, which have little applicability here, and many of which infer problem-solving process from the product obtained.

It is also possible to investigate the processes of thought by asking students to report how they have solved a problem immediately after they have given an answer to it. Such analyses offer clues to the problem-solving processes employed, but they have the disadvantage of being edited in the reporting, since there seems to be a tendency on the part of problem solvers to arrange their reports in a neat and orderly fashion to conform to an ideal of "good" problem solving.

Another way to ascertain the process of problem solving is to ask the problem solver to think aloud as he solves the problem. This solution, recorded as it is developed, may not be totally complete, but it is more likely to reflect the actual process used than if a later, retrospective account is obtained. The protocols obtained by having students think aloud as they solve problems offer a close approximation to the problem solver's method of attack on problems (Bloom and Broder, 1950).

When students are asked to "think aloud" as they solve problems, it becomes possible to study the students' methods of attack on problems. If, for example, students were asked to do this problem:

$$\begin{array}{r} 834 \\ -415 \\ \hline \end{array}$$

Thinking aloud as they performed this simple operation, different students would use different methods. Some have learned to subtract by "taking away," some subtract by a process of addition, some might be observed to produce the answer as soon as they perceive the problem, unaware of the separate operations, and some may add rather than subtract, having failed to follow directions. The correct answer (product) may be achieved through *varying* problem-solving processes, or even as a result of chance. The method of attack used on such a problem is most completely revealed only by a study of the ongoing process.

An illustration of this is offered in the sample problem:[1]

I bought an evergreen that was 8 inches tall. At the end of the first year it was

1. Paraphrased by permission of Houghton Mifflin Company, Boston. ©1936 Stanford-Binet Intelligence Scale.

12 inches tall; at the end of the second year it was 18 inches tall; and at the end of the third year it was 27 inches tall. How tall was it at the end of the fourth year?

The manner in which several high school students tackled the problem, as they tried to "think out loud" is presented below. The students first read the problem, then responded as follows:

Student A: "First year grew four inches. Second year grew 12, five be 15 inches. Grew four inches first so third year grew about four times that, four, seven, 108 inches."

When asked how he had arrived at the figure of 108, the student answered, "four times 27."

Student B: "Let's see, four the first year, six the second year, nine the third year. Hm, let's see, four inches, six inches, nine inches, four, six, nine grew probably 11 inches, be 28, I think."

When asked about this answer, the student said, "On thinking, that eight inches, four the first year, two, nine, be, cycle would equal out, two more the second year than the third, be two more the fourth year."

Student C: "Well first year grew four inches, second year grew six inches, third year grew nine inches. Let me see. Oh four, six, nine (pause) Rereads problem. "Six add two, add three, add four so know how—two-thirds as tall as was year after—be 27 is two-thirds of be, holy cow." When asked to continue, the student said, "I don't know how to do it. I forgot."

Student D: "Eight inches increased by four (pause), increased by six, increased by nine is (pause) 12 is four subtract is four, six in between, nine in between, relation four, six, nine to x. 27." (pause) Rereads problem. "Six add two, add three, add four so would be four to nine be 13, be 40 inches." When asked to explain further the student said "Seems that 12 from eight is four, 12 from 18 is six and 18 from 27 is nine. Make it four, six, nine. Two to the four. Three to the six, four get nine, 13 is added to 27."

Student E: "Let's see, four inches first year, six inches next year, nine inches next year. Let's see four, six by two. Three think to be 31 inches. Just a minute. Grew four, six, then nine, oh let's see. First up four, then six inches. Nine probably 35, 36 inches. About a year, 36 inches. Increased each year. Oh, I see now, one-half the height larger each year. End of next year. Let's see one-half 27 is 13½, be 40½ inches tall next year."

From this very brief example, it can be seen that students arrive at the solution to a problem by diverse methods. The students differ in their ability to generalize, to apply the knowledge they have to a solution, to reorganize the information given in the problem. They differ in their attitude toward the solution of the problem, in their ability to break the problem into parts, in the manner in which they pay attention to the product of their own reasoning.

What has all this to do with the engineering educator? What should be the approach to teaching problem-solving skills to engineering students?

In the current literature, there have been reports of attempts to teach problem-solving procedures, as well as descriptions of remedial efforts made to teach engineering students more effective methods of solving problems. These paths and precepts have been developed logically. Most have been shown to be helpful. A brief description of some of these studies presents the varied approaches.

Rubenstein (1975) offers these general precepts of problem solving:

- Get the total picture
- Withhold your judgment
- Use models
- Change representation
- Ask the right questions
- Have a will to doubt

He further suggests paths to generating a solution which include:

- Work backwards
- Generalize or specialize
- Explore directions when they appear plausible
- Use stable, substructures in the solution process (modules)
- Use analogies and metaphors
- Be guided by emotional signs of success

Liebold, et al. (1976) describe an adaptation of Polya's approach to teaching problem solving to freshmen engineering students. They divided the definition of the problem step into "define" and "think about it." The "define" step is broken down thus:

- Define the Unknown
- Define the System
- List knowns, concepts, and choose symbols
- Define the constraints
- Define the criteria

The students have reported some difficulty in applying the skills learned in this special class to their regular homework assignments.

As part of the same program, Woods et al. (1975) observed the problem-solving training to which engineering students were exposed in their classes, attempted to identify major difficulties the students were having in solving problems, and to identify necessary problem-solving skills and teach these skills to the students. They identified a set of steps combining creative and analytical thinking, then used these as a basis for teaching a strategy of problem solving to their students in a tutorial program. The strategy is outlined below (Woods et al., 1975, p 239):

Define   Identify the actual problem

Think   What are the attributes?
About it   Identify area of knowledge
                   Collect information
                   Flowchart solution

Plan	Think up alternative plans
	Translate

Carry Out Plan	Solve

Look Back ⎰ Check reasonableness & math
Check criteria & constraints
Study related problems
Identify applications in engineering, everyday
   behaviour & deserted island
Identify & memorize order-of-magnitude numbers
Study problem-solving skills learned
Communicate results

The strategy was devised rationally. The procedure for improving student skills requires that students in a class focus on a particular step simultaneously, and then discuss each other's ideas. This method, although developed through rational analysis, does place emphasis on the actual process of problem solving. An interesting observation made by the professors assigned to sit in on all the required courses which the students were taking, was the discovery that the lecturers actually had presented a number of hints for solving problems, and numerous examples of solutions, yet the students did not utilize these hints or examples. They apparently failed to note them in class presumably because they were given verbally and not written on the board.

Stonewater (1977) points out that, although engineering instructors are able to specify the problem-solving processes they use, they may have difficulty specifying the processes a student should use, since through experience and practice the instructors have internalized so much of what they do and are not conscious of the methods they use. He points out that step-by-step processes that instructors offer their students as they demonstrate problems may not reflect the ways that the problems were solved by the instructor initially, but rather the solutions may have been edited to appear more elegant. Again, the strategies used for problem solution may not be identified.

Stonewater developed a course called Introduction to Reasoning and Problem Solving using a task analysis to develop strategies for problem solving. In the eight-module course, three modules were devoted to the preparation phase:

1. Preparation for problem solving: discriminating between relevant and irrelevant information, specifying solution derived, visualizing the problem;
2. Drawing diagrams;
3. Organizing data tables

Five gave strategies for solving problems:

1. Subproblem strategy: identify unknowns and sequence the order in which they must be solved.
2. Subproblem strategy: develop an organized method to solve the problem.
3. Contradiction strategy: state an assumption which is the logical negation of what is to be proved and use this to contradict given information.
4. Interference: infer additional information from what is given.
5. Working backwards: start at solution, rather than with givens.

Stonewater reports that three elements of concern in teaching problem solving are *1.* finding techniques to help students improve their organization l ability, *2.* increasing abstract reasoning skills, and *3.* teaching them to transfer learning to other courses.

Stonewater used a diagnostic test to determine which students had mastered a particular strategy. He designed materials that allowed for self-pacing to implement learning, and paired students to study particular materials.

The method of pairing students to teach problem-solving skills has been described and used successfully by Whimbey (1978) as adapted from a method described by Bloom and Broder (1950). Whimbey developed a program where nonengineering students, working with each other on a one-to-one basis, think out loud as they solve problems to try to figure out what is interfering with their success as problem solvers. The students work exercises designed to increase their ability to read and understand technical and scientific writing, as well as to solve mathematically based reading problems. Initially, students contrast their methods of solving problems with the methods used on the same problems by experts. Whimbey points out that good problem solvers differ from poor problem solvers in these characteristics:

1. Motivation and attitude toward problem solving
2. Concern for accuracy
3. Breaking problems into parts
4. Amount of guessing
5. Activeness in problem solution

Researchers at the University of Massachusetts in the Department of Physics and in the School of Engineering have been cooperating in investigating a variety of instructional techniques for developing students' cognitive skills and for teaching analytical reasoning (Clement and Lochhead, 1976), (Clement, 1976), (Lochhead, 1976). By asking students to think out loud, Lochhead and Clement can study the cognitive processes of individual students in order to determine the learning strategies they employ, the basic concepts from which they operate, and the techniques they use in solving problems. This information is then applied to the task of improving the problem-solving skills of the students. Lochhead (1976) reports:

> We find that what students usually learn from a physics course is not at all what we believed. For example, a few months ago I gave a student a ride into the University. He asked me what I did and I told him I taught problem solving to introductory physics students. He replied that he had taken a physics course the previous year (Physics for Biologists): it had been OK but of no lasting value. He had tried to understand the material but that took too much time and wasn't any use as far as the grade was concerned. So after a couple of weeks he settled in on memorizing formulas and found that the homework and exam problems could always be solved by plugging the given variables into whatever equation happened to involve those variables. He got some practice in algebra and also in trigonometry but the physics he learned was just rote formulas which less than a year later he had completely forgotten. This approach to learning and problem solving is an example of a syndrome we call "formula-fixation."
>
> The student is neither allegorical nor unusual. He is perhaps more perceptive than many of his classmates, but by no means unique. Last winter Robert Gray,

who is directing the experimental physics course, was visited by an angry student who had obtained a B+ in introductory physics. The problem was that she had understood none of it. The purpose of her visit was to set up an independent study course for the January term in which she could try to understand the material she had mastered. . . .

There is a popular myth that students cannot understand physics because they are weak in mathematics. The above examples show that the inverse is often the case, namely; an ability to do mathematics makes understanding the physics unnecessary. But students are not the only people skilled at the use of algebra to avoid thinking. We all do it most of the time; and we regularly continue the practice when we teach. With rare exception textbooks and teachers emphasize the mathematical manipulations and spend little effort on explaining the physical concepts or on explaining why the mathematics is an appropriate representation of those concepts.

Clement and Lochhead (1976) comment:

> It is easy for a student to memorize a law and to recite it faithfully when given the appropriate prompt. With a little practice the student may also be able to apply the law in a limited class of problems such as those typically found on tests. However, these abilities in no way imply the reshaping of contradictory intuitions. In fact, a careful investigation of how students solve problems shows that in most cases they are operating with an inconsistent system—a collage of newly learned principles and old intuitive concepts. The old intuitive concepts are remarkably resistant to change and this presents a difficult challenge for teachers.

> One area of physics where these intuitions are particularly strong is Newton's first law. The first law states that a body in motion will remain in motion unless acted on by a force. It is a strange law because it directly contradicts our own perception. Our everyday experience shows that bodies in motion come to rest without the application of a *visible* force. Furthermore they show that to keep a body in motion requires the application of a visible force. Thus learning Newton's first law implies the reshaping of certain intuitive concepts.

These points are illustrated with student protocols, generated as the students think aloud. Programs to teach analytical reasoning fashioned after those of Whimbey (1978) focus on five aspects of analytical reasoning:

1. Critical thinking applied to understanding complex instructions
2. Analyzing of errors in reasoning
3. Solving word problems
4. Analyzing trends and patterns
5. Using analogies in formal contexts

Based on this, faculty members at the University of Massachusetts proposed the development of an Analytical Skills Center, headquartered in the Department of Rhetoric, which would attempt to diagnose the causes of students' weaknesses in analytical reasoning ability, and provide instructional programs to teach these skills. The project would be a joint effort of people from the departments of engineering, physics, mathematics and rhetoric.

In a different but related program, Larkin (1977) and Reif (1977) and others at the University of California Department of Physics, and Group in Science and Mathematics Education have contrasted the method of solution of experts (professor of physics) as they solve physics problems while thinking aloud, with the method of solution of a novice (a student who had completed one quarter of physics). The record of the novice shows a direct approach, simply applying various physical principles to the problem in order to produce equations. The equations are then combined to produce the desired quantitative solution.

In contrast, the experts do not jump directly into a quantitative solution but first restate the problem in qualitative terms. The qualitative description is similar to the quantitative equations which are finally employed in the solution. The qualitative analysis seemingly reduces the chances for error since it can be easily checked against the original problem statement. It also outlines an easily remembered description of the global features of the original problem.

In addition to the interpolation of the qualitative description, novice and expert seem to differ in the way they store physical principles in their memory. The novice seems to store such principles individually while the expert groups principles which are connected, and stores them as "chunks." Thus, when the expert accesses one principle from memory, the other associated principles become available.

Larkin (1977) applied this research to her teaching of a calculus-based physics course. Ten students were trained to apply separately each of seven physical principles needed to solve a DC circuit problem. After this, the students worked three problems which could be solved by systematically applying the learned principles. Then five of the students were given additional training in qualitative analysis and "chunking." All the students were then presented with three additional problems which they worked out individually, thinking aloud as they solved the problems. In the experimental group three students solved all three problems, and two solved two. In the other group four of the five students solved at most one problem. Larkin stresses that "if one is serious about trying to enable students to solve problems in physics more effectively, the following procedures seem promising: *1.* observe in detail what experts do in solving problems; *2.* abstract from these observations the processes which seem most helpful; *3.* teach these processes explicitly to students."

Now, based on this diversity of approaches, what can an engineering educator do to enhance the problem-solving capabilities of engineering students?

It is possible to continue to do as has been done in the past to assign problems and either correct them or provide the correct solution. One could focus on the answer, the product of thought, and assume that, by this method, students are learning to be problem solvers. And indeed, many of them do learn engineering problem-solving skills in this way.

Or, it is possible to present and discuss a logical strategy for problem solving, similar to that described earlier that is being used at McMaster University (Liebold et al., 1976, Woods et al., 1975) with special guidance on problem-solving aspects of homework assignments carried on independently of the classroom presentation. In the classroom it is possible to present hints for solving problems, and to offer solutions to sample problems, calling attention to these hints or examples, or writing them on the board, so that the students will attend to them.

The problem-solving course developed by Stonewater (1977) at Michigan State offers a different technique, a separate mastery learning self-paced model course which teaches students such things as to draw diagrams, develop data tables, and use problem strategies. The special course developed by Woods et al. (1975) at McMaster also uses a rational approach to the teaching of problem-solving skills.

An inexpensive and nondisruptive technique would be to reduce emphasis on the products of problem solving, on always getting the "right answer," and to stress the process of problem solving as material is presented in the classroom. Let the students watch and listen as the instructor, the expert, tackles a problem. Let students watch the way discrimination between the relevant and irrelevant information takes place in a problem-solving process, the way translation from the given problem description into a more workable form occurs, how the problem is redefined into terms for which equations can be developed, and how drawing diagrams, data tables or graphs helps in this process. Call the student's attention to the process by which a problem can be broken into more manageable parts. Permit the students to listen to the entire process of developing a plan, even when it leads to false starts, and to learn how it is recognized that the wrong problem-solving path has been chosen and how to check for consistency. In other words, let the students see the scratch paper which was discarded, rather than just the elegant final solution! In elementary courses, try, as Lochhead and Clement have done, to make such students understand the relationships in equations and not merely plug in numbers.

According to Reif (1977) and Larkin (1977), it is helpful to students if they can organize their knowledge base qualitatively, using verbal descriptions to group principles, to "chunk" relationships. Point out that students should examine problems for such relationships rather than immediately spewing forth equations which may or may not be relevant.

Reif (1977) points out that some common teaching practices used in engineering courses, such as emphasis on mathematical formations, may be deleterious to students' skill at problem solving; avoiding verbal or pictorial descriptions may actually hinder the students. Stressing linear procedures such as combining equations should be downplayed, while hierarchical relations should be stressed.

The educators cited above have provided a variety of models which emphasize the importance of teaching problem-solving processes to engineering students. Deemphasizing "the answer" or product of thought, and emphasizing the process involved in reaching a problem solution will provide a first step in this procedure. In classroom presentations, searching for the problem-solving methods used by students, and providing details of expert or model solutions, will direct the students' attention to the importance of the process of problem solving. This should increase the effectiveness of the students as problem solvers, should provide insight to the instructor and help determine why students are having difficulties with assignments. It may also help evaluate the effectiveness of instructional procedures.

Educators say they are concerned with the process of problem solving, yet often pay undue attention to the product. Students are told to 'solve these problems,' but no one tries to find out the manner in which they have solved the problem, the method of attack they have used. There is a story about a student who consults his professor on an assigned problem, saying, "Professor, I know the correct answer to the problem, but I'm not sure how I got it." The professor reads the problem, thinking a bit, then writes on the paper, "264." The student says "Yes, I know that's the answer, but I don't see how you got it." The professor works at the problem again, and again writes down "264." The student repeats his plea for an explanation. Again the professor cogitates and writes down "264." The students says "But I still don't get it." The professor then states "What's the matter with you? I've already shown you three different ways to work the problem!"

REFERENCES

Bloom, S. and Broder, L.J. "Problem-Solving Processes of College Students." In *Supplementary Educational Monographs* No. 73, Chicago: University of Chicago Press, 1950.

Clement, J. "Seven Laboratories on (1) Qualitative Physics, (2) The Concept of Function," Technical Report. University of Massachusetts, Amherst, 1976.

Clement, J. and Lochhead, J. "On Hobgoblins and Physics." University of Massachusetts, Amherst, 1976.

Larkin, J.H. "Processing Information for Effective Problem Solving." University of California, Berkeley, 1977.

Leibold, B.G., Moreland, J.L.C., Ross, D.C., and Butko, J.A. "Problem Solving; A Freshman Experience." *Engineering Education,* 67, Nov 1976, pp 172-176.

Lochhead, J. "The Heuristics Laboratory Dialogue Groups." University of Massachusetts, Amherst, 1976.

Polya, G. *How to Solve It.* Princeton: Princeton University Press, 1945.

Reif, F. "Problem-Solving Skills and Human Information Processing: Some Basic Issues and Practical Teaching Suggestions." Presented at ASEE 85th Annual Conference, University of North Dakota, June 1977.

Rubenstein, M.F. *Patterns of Problem Solving.* Englewood Cliffs, NJ: Prentice-Hall, Inc., 1975.

Stonewater, J.K. "A System for Teaching Problem Solving." Presented at the ASEE 85th Annual Conference, University of North Dakota, June 1977.

Whimbey, A. and Lochhead, J. *Problem Solving and Comprehension, A Short Course in Analytical Reasoning.* Philadelphia: The Franklin Institute Press, in press.

Woods, D.R., Wright, J.D., Hoffman, T.W., Swartman, R.K., and Doig, I.D. "Teaching Problem-Solving Skills." *Engineering Education,* 66(3), pp 238-243, 1975.

On 'Learnable' Representations of Knowledge: A Meaning For The Computational Metaphor

Andrea A. diSessa

INTRODUCTION

> The true meaning of a term is to be found by observing what a man does with it, not what he says about it.
>
> —P. W. Bridgman

> S understands knowledge K if S uses K whenever appropriate.
>
> —J. Moore and A. Newell

It is now widely agreed, at least in principle, that the educational task is now well modeled as a process of transmission of knowledge. Especially following Piaget, emphasis has shifted away from learning models like Skinner's which assume a great and universal simplicity in the structures of assimilation of knowledge, toward models with a very careful concern for the richness and complexity of the student's "initial state" in terms of capacity to assimilate.

A similar enrichment has taken place in the very notion of knowledge itself. Indeed, the currently prominent concept of *representations* of knowledge symbolizes the rather recent awareness of the manifold ways of encoding and distributing knowledge in an "intelligent system."

Psychology and artificial intelligence have made a great deal of these parallel and vitally interrelated "booms" in complexity. Unfortunately, comparatively little effort has been expanded in education to exploit the rich variety of concepts and theories becoming available concerning knowledge encoding and assimilation. In the central notion of this paper, that of "learnable representations," I am proposing a program of research and development in pedagogical material which can, I hope, lend to education in usable form some important theoretical insights.

The program is, briefly stated, to transform old or invent new representations of physics, mathematics or whatever subject, which do justice to the powerful logical structure of the subject, but which at the same time mesh properly with the cognitive reality of human beings. This task is made particularly difficult by popular, unstated epistemological assumptions about the simplicity of knowing, for example, knowing "mathematical truths." These assumptions make inventing new mathematics and physics seem a dubious enterprise at best, even (especially!) for pedagogical purposes. I wish to argue against those assumptions.

Further, I propose a zero-order theory of what makes a representation learnable, together with an extensive collection of examples from high-school- and college-level physics and mathematics. Important concerns in the discussion are procedural modes of knowledge representation and a particular projection of knowing, "control knowledge," which directs personal activity.

This paper is an elaboration of some simple ideas—that thinking is a complex but understandable *process*, and that education is an interaction with the thinking process and can influence it best by respecting the present structure of the process. My intention is to elaborate some implications of those oft-avowed principles within a perspective provided mainly by the science of intelligent processes, artificial intelligence. The image of teaching and education research which derives from that perspective and its methods of analysis is quite different from most current and past trends.

The crux of my argument is this: since Euclid, axiomatic-deductive systems have, principally by default, served as model representations of knowledge for pedagogical purposes. But while such systems which stress internal simplicity and coherence may serve useful roles for some purposes, they are not good models for understanding the learning process, much less for suggesting how to enhance it. Instead we must stress simplicity and coherence *in relation to the student's prior experience and knowledge.* We must take into account intuitive and other kinds of knowledge and knowledge processing that do not fit any known formal descriptions, let alone an axiomatic-deductive format. Furthermore, we must learn how to bring to the surface procedural and organizational aspects of knowledge which relate to the student's specific thinking process.

But how can we expect to flaunt the "natural formal structure" of subjects like Euclidean geometry and physics? My answer is that even for the expert scientist, formal structure is only a small and sometimes superficial part of what he knows and what we must teach. The appropriate natural structure is not a formal skeleton, but the richer structure of the *functioning* of that skeleton in an individual.

I conclude the paper with a collection of examples from high school and college physics and mathematics that illustrate in detail some ways in which formalism can be put in its proper place.

EPISTEMOLOGY: A PROCEDURAL VIEW OF KNOWLEDGE

The central organizing theme of this exposition is what has come to be called "the computational metaphor." The rise of computer technology has given great impetus to the study of process in the abstract and in particular to the study of the kinds of processes that may be called intelligent. Natural language understanding and production, vision, problem solving and informal inference are examples. Out of this study naturally came a concerned look at perhaps the only appropriate "natural" model for intelligent processing, human intelligence. Thus a symbiosis of natural and artificial intelligence is begun. It is not at all surprising that concepts and theories invented to illuminate and precipitate machine intelligence seem to have a great deal to say about psychology, particularly in the areas of the representations of knowledge and learning. As human thinking serves as a model for intelligent process, so too do theories of abstract processing and machine intelligence serve as rich sources of language and ideas for drawing implications about human processing. Thought and learning seem strongly analogous to the activities in complex computational processing systems. This is the root of the computational metaphor.

One very important branch of the inquiry into the computational metaphor involves the formal modeling of human learning, language abilities or other such processes. I will not be concerned with that here. Instead I wish to pursue a looser but perhaps more immediately applicable vein. I will discuss implications of some of the most crude, but in my opinion, most robust ideas which arise by considering the representations of knowledge and the styles of pedagogy appropriate for human learning in the light of computational concerns. My interest here is with structuring particular domains of knowledge such as physics and mathematics so as to be maximally comprehensible and "learnable." I begin with a crude but telling procedural epistemology.

Classification by Purpose

One may initially classify knowledge by its purpose. I will call knowledge that is directed externally, toward the structure of physical events or abstract relations, *material knowledge*. Knowledge that is directed internally toward personal functioning and the structure of thinking itself, I will call *control knowledge*. One of the prime contributions of the computational metaphor is to call detailed attention to this latter aspect of knowledge which directs organization of the thinking process. The former, material knowledge, is what one conventionally thinks of as curriculum material, the mathematics and physics itself.

Let me use a computational analogy. Consider two different programs performing the same task, say playing chess. The control structures of the two may be quite different. For example, the decision on the legality of a move may be handled in varying ways; one program may check a legal move list while the other may involve context-sensitive productions which can only produce legal moves. Despite the fact that in some sense the programs both know how to play chess, i.e. have the same material knowledge (in fact, there is no a priori reason that their external behavior need be distinguishable), the organization of the process by which they exhibit that knowledge may be fundamentally different. The issue of control is when, why and how that knowledge is used. In practical terms, "teaching" each program to castle may require radically different representations of the notion of castling.

In human beings, of course, the organization of the process of thinking is not as rigid as is popularly expected of computer programs. One may well speak of this control structure as knowledge—some parts of which may be learned or forgotten, some may be quite conscious, some unconscious, some very general and others specific.

My purpose in making the distinction between control and material knowledge is to point out the importance of the *control* aspect of knowledge. In fact, I intend to emphasize the importance of the close interrelationship of these two facets; in organizing information for any processor one must take careful heed of the character of its capabilities. The computational metaphor leads us to look carefully at the character of human thought when constructing the form of knowledge to be presented.

Let me begin exhibiting the important but subtle nature of control knowledge with a rather simple example. Is there any difference between the following versions of Ohm's Law?

$$I = E/R \qquad E = IR$$

Formally the two are the same, differing only by a trivial algebraic transformation, but anyone who has worked with such equations will admit a different feeling toward the two.

The difference is control structure: how the parts are treated and when the whole is evoked. The formal symmetry of A = B $\Longleftrightarrow$ B = A is broken by a usually unspoken control convention that the symbol on the left is the "unknown" and those on the right are the determiners of the unknown. Part of the implied control structure is that one searches for the determiners' values in order to evaluate the unknown. Note that even the word "equal" can have the same control asymmetry as the symbolic equation. Consider the clash of control structures in "let 5 equal x."

The control implications in lexical ordering are clearly almost causal. The symbols on the right taking on the values that they do *causes* the symbol on the left to have its value. I = E/R is a particularly felicitous representation in this respect as it meshes well with the common causal interpretation of E as an externally established *impetus*, R as a given *obstructor* and I as the *resultant* caused by E acting against R. In E = IR, the interpretation of E as a result of a controlling impetus I is somewhat harder to make. R = E/I has the causal interpretation of Ohm's Law directly opposed by the lexical control convention, and not unexpectedly is usually relegated to the status of a "derived" equation.[1]

Anyone who has the algebraic facility to use these relations can realize their identity. Yet they are sometimes taught separately and, more importantly, are often evoked separately, even by experts, for the practical reason that their control structures are functionally different. An important implication of this fact is that the transformation from one to another of these forms in a problem-solving situation may signify more than a trivial change of mind state in the solver.

To better appreciate the importance of the unity of control and material knowledge, perform a rhetorical experiment. Consider an expert's understanding of, say, physics. Ask him about a concept and observe the form of the response. The paradigm is that he generates a situation in which to observe the action of the concept, or generates a process which involves the concept. For example, the structure of the concept of force usually entails an agent, a form of interaction, and a recipient. The expert may not *explain* that, but he *exhibits* that understanding. He says, "If you push on a rock . . . " (agent = you, interaction = push, recipient = rock). In the same way, our expert does not often reply in terms of a formal nature; he neither explains from what the idea can be deduced, nor what follows deductively. Force is more often explained by its function as "the interaction between particles which, if known, allows us to compute motion." Less often (except, unfortunately, in the context of a "standard"

---

1. It seems quite plausible to regard even the causal interpretation of Ohm's Law not so much as material knowledge but as a model abstracted from usual applications to represent control aspects of the law, i.e. what one does with the entities involved. If common sources of electricity were constant current rather than constant voltage, E = IR might assume the role occupied conventionally by I = E/R.

physics course) does one hear precise but formal declarations such as "Force is (mass) x (acceleration)."

The reason for this behavior is more than pedagogy on the expert's part. The way in which *he uses a concept, the control structure needed to use the idea, is as much a part of his understanding as the detailed formal structure of the idea.* In other words, the disposition to embed the concept in a useful context is not accidental, but an expression of the kind of knowledge which is vital for functional knowing of the idea. It's the kind of knowing which causes the concept to be appropriately remembered and allows the expert to solve new problems with it. For teaching purposes it is unfortunate that much of this control structure is implicit.

Classification by Form

The computational metaphor brings to light another frequently ignored fact— that *the structure of process itself can be a mode for knowledge representation.* Lack of attention to this has led to a skewing of educational materials toward the classical mode of knowledge, deductive or syllogistic logic, and away from more process-oriented representations. Many of the examples I will discuss later are attempts to illustrate how process can be an effective knowledge carrier.

This paper is primarily concerned with these process or procedural forms of knowledge representation. Within that broad domain let me pick out two, in some sense antithetical forms which effect an orthogonal cut across the classification by purpose discussed in the last section. The first, *knowledge-of-procedure,* is little more than the name implies. It is characterized by an explicit surface structure which is step-by-step procedure. One expects that knowledge-of-procedure contains explicit reference to purpose and to what circumstances make the procedure useable. An example might be arithmetic which many consider (perhaps incorrectly) to be simply a step-by-step algorithm applied in appropriate circumstances to achieve an established aim.

In contrast one can imagine a deeper and more subtle form, which I call *knowledge-within-process,* in which the surface structure is not necessarily procedure, in which step-by-step analysis may not be appropriate (hence the use of the "process" rather than "procedure" in the name), in which the purpose of the knowledge is only evident in the control structure which evokes the process, or in the function it serves; indeed, the actual subject of the knowledge may be quite invisible.

Gilbert Ryle in his essay on "Knowing How Versus Knowing That" picks a keen example of this kind of embedding of knowledge. In what sense does a hero possess moral knowledge? It is certainly not in the procedure he takes to rescue the maiden in distress. One might be tempted to say that the moral knowledge resides in a list of imperative maxims which the hero consults. But this line is hard to defend. At the very least, the control structure which evokes the "morality list" in response to the maiden's cries has qualities quite different from "knowing *that*" I must do such or other. Ryle resorts to a form of knowledge he calls "knowing how" similar to the idea of knowledge-within-process. In essence it is a blend of *disposition, style of action,* and *capability.* The hero knows his morality in his capability and disposition to act morally. The knowledge lies in the *style* of behavior.

I have chosen a more scientific example of material knowledge-within-process which also emphasizes the fact that such knowledge is of vital importance to the class of knowing generally referred to as intuition.

Through one's experience with moving physical objects one acquires a good deal of knowledge about how much and what kind of force it takes to achieve a certain result. One constantly estimates or remembers the weight of objects and applies forces appropriate for what one wants done. The process is unconscious or often seems so, but vividly makes itself known when a wrong conclusion is drawn. For example, everyone has had the experience of nearly throwing an empty container into the air because it was presumed full. This exhibits knowledge-within-process dealing with the relations of force, weight, and motion.

If you are posed an abstract question—"could you deflect or stop a 20-pound pendulum about to crash into you at three miles per hour?"—you presumably use that same knowledge-within-process. You do no particular analysis or calculation beyond that necessary to make the numbers more meaningful (say, replacing the pendulum with a bowling ball and three miles per hour by five feet per second). It seems reasonable to describe your thinking as imagining the situation and observing your own disposition in such circumstances. Thus this kind of intuition is the accessing of knowledge-within-process by observing oneself.

Such knowledge about moving bodies and forces appreciates a qualitative side of the physical concept of momentum, its relation to mass and velocity, and its conservation (to the extent that inertia represents conservation). These structures undoubtedly play an important role in learning the more formal and precise physics of momentum. Later I will discuss how particularly the control structure of formal physical ideas may be inherited almost entirely from intuition. For example, the conception of *force as a cause* is a vital part of the knowledge-within-process I'm talking about, and I will discuss how this involves a control structure which is sometimes appropriate for doing physics.

In a less positive vein, procedural understanding of this sort can be the source of unfortunate confusion viewed within a concept of pushing to achieve "more motion," changing energy is indistinguishable from changing momentum. That fact causes much confusion for elementary physics students (and did for Galileo as well!) In either case, productive or counterproductive, one should be aware of the possible help or hindrance from these rough but insistent dispositions.

In viewing knowledge-within-process as knowledge we are committing ourselves to a much richer epistemology than might otherwise be acceptable. On the surface such knowledge does not resemble the knowing of a fact in any standard sense. Its evocation may be harder to see and its influence more subtle and more context dependent. More importantly, the ways in which past experience can serve as "knowledge" depend heavily on mechanisms available for invoking and applying it. Consequently a complete epistemology must be procedural in the sense of dealing with the processes of interpretation, analysis, transformation, etc., which can cause knowledge, particularly rather invisible forms like knowledge-within-process, to have significant influence in learning or problem solving. I turn now briefly to this area, accessing knowledge, specifically in the context of accessing knowledge-within-process as in the above example of physical intuition.

Accessibility

One reason that knowledge-within-process is so frequently ignored is that it is generally not verbalized. Furthermore, it may be "inaccessible" in the sense that the student himself does not realize that it is knowledge and can be put to use. It is still more devastating when the teacher fails to recognize such knowledge. Consider: An instructor stands before the blackboard and declares that some problem obviously should be approached using Newton's Second Law. A student says that he did not think to do it that way; he says he doesn't understand. The teacher again declares his approach obvious and as justification proceeds with the details of the solution. But the student is no better off than before. The student was asking for the control knowledge which brought Newton's Second Law to the teacher's mind, not the post hoc verification "see, it works." There is something that the teacher "knows" about F = ma which the student does not know, something which evokes the law in the face of a certain class of problems. Optimally the teacher should know why he thought of that method and be willing and able to discuss it. Otherwise the student may or *may not* generate the appropriate understanding. This is a clearcut case of control knowledge encoded within process (in the teacher, not encoded at all in the student).

Annotation

A fairly general process for accessing[2] knowledge-within-process is observation, analysis, and annotation (hereafter referred to as annotation). With help from a teacher or other source a student can study his own behavior in a particular situation, trying to understand its detailed purpose, what it produces and how it succeeds. Clearly selection of the situation and guidance in deciding what is relevant are important functions of the teacher. The student identifies functional parts of the process, naming them appropriately, making connections and adaptations to a formal scheme. The end result may well be a completely annotated version of the original process, the top level of which may indeed be the sort of knowledge I have called knowledge-of-procedure. In other cases that kind of product is not relevant; one may merely be striving to make an explicit and useful connection between formal learning and experience.

It is worth remarking that a fully annotated version of some knowledge-within-process can be invaluable in teaching. If Ryle's hero were to teach morality, he would undoubtedly compile a list of maxims which annotate his dispositions in imperatives. To cite a different example of the possible roles of annotation, the learning of noun and verb classification should be a process of annotating one's dispositions toward word use rather than learning to apply an abstract criterion. The description of a noun as "a person, place or thing" does not capture any reference to students' own knowledge-within-process about generating or understanding grammatical sentences. The meaning of "noun" includes the *way* one uses words like "cow" or "apple" in speaking, with respect to concerns like word order and the possibility of acquiring modifiers in a certain way. In a parallel vein, defining words—especially determining shades of meaning between similar words—is very often achieved by annotation, using ques-

2. With a more restricted sense of the word *knowledge*, one would almost have to say "generating knowledge" rather than accessing it. Moore and Newell, quoted in the introduction to this paper, rather forcefully refuse to recognize the difference.

tions such as: "How would I use this word?" "What would I think of if someone said the word to me?" and "Can I describe the image the word conjures up?"

Table 1 is a schematic of a possible means of contact connecting knowledge-within-process and more explicit (propositional) knowledge. The dotted line indicates the likelihood of direct inheritance of control knowledge by procedural formulations. See the discussion of force as regards causal syntax and momentum flow in Examples. The Examples section will also carry the burden of conveying the notion of procedural formulation.

**Table I. A Summary of the Procedural Epistemology
With Further Examples of the Classification**

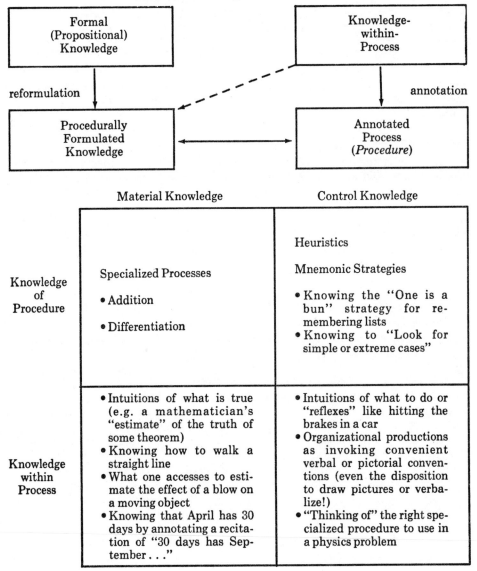

	Material Knowledge	Control Knowledge
Knowledge of Procedure	Specialized Processes • Addition • Differentiation	Heuristics Mnemonic Strategies • Knowing the "One is a bun" strategy for remembering lists • Knowing to "Look for simple or extreme cases"
Knowledge within Process	• Intuitions of what is true (e.g. a mathematician's "estimate" of the truth of some theorem) • Knowing how to walk a straight line • What one accesses to estimate the effect of a blow on a moving object • Knowing that April has 30 days by annotating a recitation of "30 days has September . . ."	• Intuitions of what to do or "reflexes" like hitting the brakes in a car • Organizational productions as invoking convenient verbal or pictorial conventions (even the disposition to draw pictures or verbalize!) • "Thinking of" the right specialized procedure to use in a physics problem

AGAINST AXIOMATICS

To emphasize the import of the computational metaphor, I wish to develop contrasting implications of another, much more common image of knowledge representation, and by extension, another metaphor for the educational process. At the risk of oversimplification, one can call this other image the *logical formalist* (axiomatics and deduction) metaphor.

In what follows it must be understood that I am using "formal" in a special way. In particular I am using "axiomatics" to represent the class of formal representations which strive to remove the subject material as far as possible from the confusions and errors that result from using unannotated structures like intuition. Unfortunately such abstract presentations frequently underestimate the amount and kind of knowledge, particularly in the domain of control knowledge, needed to make the formal system function in the student. "Informal" is by contrast an attempt to use structures the student *has,* annotated or not, and change or debug them rather than write them off as a lost cause.

A prevalent and influential view of the educational task involves dividing the learning process into a large number of discrete steps, the logic for each of which is impeccable. This image closely parallels and perhaps even derives from the image of mathematics (or science) as deductive systems based on a small set of axioms or laws. As each step in the construction of a mathematical theory is to be infallible in its careful and precise deduction from previous levels, so each educational step is simple and easily learned by the student. The knowledge acquired from this step-by-step edification is supposedly as secure as the collection of theorems comprising the finished mathematical theory.

There are several inadequacies of this view which stem mainly from implicit assumptions about the nature of human processing.

Misconception 1: Science is Deductive

The emphasis on the axiomatic system as an end result of mathematical development unfortunately has led some to pick this out as a principal characteristic of mathematics. But what mathematicians *do* does not have the character of the formal end product. Below, R. Courant speaks to this point in the context of 18th and 19th century mathematics history.

> In Greek mathematics we find an extensive working-out of the principle that all theorems are to be proved in a logically coherent way by reducing them to a system of axioms, as few in number and not themselves to be proved. This axiomatic method of presentation, which at the same time served as a test for the accuracy of the investigation, was ... regarded (then and still today) as a model for other branches of knowledge ...
>
> But it was a different matter with modern mathematics ... In mathematics the principle of reduction of the material to axioms was frequently abandoned. Intuitive evidence in each separate case became a favorite method of proof ... Blind faith in the omnipotence of the new methods carried the investigator away along paths which he could never have travelled if subject to the limitations of complete rigour ...
>
> In spite of all its defects, intuition still remains the most important driving force for mathematical discovery, and intuition alone can bridge the gap between theory and practice. (Courant, 1964, p 79)

Morris Kline (1962, p vii) continues the argument.

> Neither Euler nor Gauss could have defined a real number, and it is unlikely
> that they would have enjoyed the gory details. But both managed to understand
> mathematics and to make a "fair" number of contributions to the subject. . . .
> most teachers, instead of being concerned about their failure to be sufficiently
> rigorous, should really be concerned about their failure to provide a truly intui-
> tive approach.

Even in physics it is tempting to regard the deductive and analytic development of mechanics from the cornerstone F = ma as archetypal physics. Yet this
is hardly the case. Certainly Newton in his *Principia* does not follow a coherent
linear deduction from secure basic principles. Historically in almost all physical
theories which had more than private and personal (hence not very accessible)
developments, it is clear that a major part of the effort required to build them
was spent in struggling on a heuristic, imprecise level. Let me cite the case of
perhaps the greatest achievement of 20th century physics. Though quantum
mechanics is formally a theory of vector spaces and linear transformations,
Heisenberg did not even know what a linear operator or matrix was when he
began the theory! But he knew that he needed to know and searched out
mathematicians to find out more.

It should hardly be necessary to point out that the historical lesson applies to
the present. Even in this day when the image of science is precision, there are no
precise, deductive or axiomatic landmarks at any of the frontiers. It is as evident
that future mathematical developments will be based more on new definitions
and axioms than on any current deductive scheme.

The intellectual mechanism of scientific creativity and discovery is structured
differently from a formal synopsis of results. It is a mixture of guessing and
heuristic refabrication, built on established, but *very different* ideas than those
ideas which constitute the end results. I suggest that this is much closer to what
the process of learning in general must be about than the simple model of feed-
ing the student the fundamental principles which contain via a transparent
mechanism (say, deduction) the whole "knowledge."

Is it not reasonable to assume, therefore, that a major part of schooling must
be concerned with involving students in situations where they must carve out
larger chunks of knowledge, perhaps developing a personal mini-theory, rather
than merely proving sublemma x, given y, or applying physical law z to this
situation? If we do not go that far, should we not at least present material in a
nonaxiomatic form, so that the student can develop tolerance for and the ability
to use ambiguities, gain appreciation for guiding principles which are not infallible and other skills which are not usually addressed in "closed" situations?
Shouldn't one spend effort on knowledge and skills at a level above the intellectual mechanism of deduction, the level where the important first steps in
problem solving take place?

Misconception 2: Deduction is Simple

It is a great temptation to assume that axiomatics and deductive processes
are the simplest building blocks of general intellectual skills. Axiomatics do
have a simple step-by-step nature but only within a formal logical framework
and assuming an overview which provides a general direction toward what are to
be considered the final results of the system. Regarding the logical framework, it
must be understood that early students must work as much at understanding

and appreciating that framework and its intent as at learning specific mathematics such as geometry. Theorems, axioms, lemmas, even definitions along with countless proof strategies all have intent and meaning which is invisible to the beginner in the subject. All too often the result of ignoring this is alienation caused by feeling pressed into rules which are arbitrarily made up by someone else and which are not at all subject to discussion or argument.

Furthermore, the understanding of the "natural" flow from axioms to theorems is entirely obscured by the aura of formal infallibility if the axioms are not felt to be secure, or alternatively, if the theorems seem as obvious as the axioms[3]—two phenomena which are common particularly in elementary geometry. I think it is unfair to insist that students follow this flow without the feelings of necessity and security concerning the axioms which were generated by the person who made up the deductive system only after considerable experience in the domain on a tentative, heuristic, and perhaps even playful level. (Euclid's accomplishment was after all organizational, the "content" was previously established, mostly by others, outside of the axiomatic form.)

The mismatch between deductive systems and the character of human thought has surfaced strikingly in current artificial intelligence work. As it has become more and more clear to artificial intelligence researchers that deduction is a poor model of human thinking, so should it become clearer to educators that deduction is inappropriate as a general model for pedagogy (Minsky 1974).

I am not arguing that axiomatics should not be taught, but that they should be approached in somewhat the same way that the active mathematician or scientist approaches them. Axioms and other summary organizations are in order after intuitive and informal understandings have been reached.

A MORE HUMAN STYLE OF PEDAGOGY

I am advocating that abstract formalism be avoided as a model for curriculum, especially on an elementary and introductory level. Instead we need to find a more human style, one that requires a more intuitive way of accessing and generating nonformal knowledge; one that is more heuristic in that it has specific concern for control knowledge; one that is more informal in appealing to other levels of justification than just axioms and deduction.[4]

Teaching What To Do: Material Knowledge

When one asks the fundamental question, "what is it you want to teach?" the best answer is not "science" or "math" but "what scientists (or mathematicians) *do*." (Papert, 1972) The point is to shift emphasis to activity and away

---

3. One finds in some axiomatic treatments that one must prove all right angles are equal. Straight angles by nature of the axioms don't need such a proof. Frighteningly one does not go on to prove the seemingly obvious successor theorem that two angles of any particular measure are equal. The logic is of course there, but in the organization of the axiomatic structure—not in the geometry. Consider further: Is the triangle inequality less fundamental or less "obvious" than some of the axioms found 100 pages earlier than the proof of this "theorem?"

4. One should remark that the "genetic" school of pedagogy shares many of the humanizing motivations presented here. But, though history can teach a good lesson, to mark it the model of "cognitively correct" is making a mistake on the same order as assuming the axioms are the proper synopsis of the history "with the mistakes taken out."

from facts. After all, it is no doubt more correct to say mathematicians know how to generate proofs for x or y than that they know the proofs. Physicists generate solutions to problems; they don't *know* them. It is not that facts (or even calculational algorithms) are irrelevant, but that the higher level activity of deciding when to use a fact or invoke an algorithm is more characteristic of scientific knowlege than the retention of facts involved. This of course is not surprising as *doing new things* is precisely the raison d'etre for scientific knowledge. In other fields even as far removed as the arts one sees clearly that the ability to *reshape* the old in the face of a new context better characterizes the successful practitioner than the ability to recall the old.

The implications of the above are twofold. First is the direct realization that *much of what is to be taught is procedural in quality*. On a rather primitive level this is recognized in current curricula at all levels. Algorithms from reading to adding are stressed. Unfortunately this does not extend to higher level understandings which might teach children, for example, to reinvent "carrying" if they forget the precise mechanism.

Secondly, *facts should be taught with connectives to procedures;* what a fact means must be functionally clear in how it can be used as an imput to established or even possible procedures. The lesson taught by Einstein in the early part of this century is a model for this declaration. Numbers are meaningful only as the product of a particular measuring process, not in their interpretation within an a priori scheme. Concepts are as subject to this mandate as are facts. It is vital to know the proper contexts and functioning of a concept to appreciate its relevance and to use it effectively. Of course, one may hope to have a propositional framing of a concept which makes explicit usually tacit or implicit implications about context, etc. But more often one must allow Bridgman's insight as quoted at the beginning of this paper; that what it is, is how you treat it.

Multiple Representations

There is implicit in this stress on connectivity of knowledge an urge toward multiple representations of the "same" knowledge. The meaning of concepts is very sensitive to the procedural context of their use. Accordingly, one expects that the general rule will be a multiple-faceted approach rather than an attempt to capture all context-related possibilities in a single definition or axiom. It may seem that this means much more to teach. But in the end a coherent but large structure will almost certainly win out over a small but obscurely dense presentation. A poem may easily be memorized while a terse encoding of the same ideas in a sentence of nonsense syllables can be "unrememberable."

The fundamental assumption behind this idea of multiple representations is that a rich, overlapping collection of different views and considerations is much more characteristic of preciseness in human knowledge than a small, tight system. In terms of problem solving the claim is that the parity of restatement or translation is as or more important to problem solving itself than the hierarchy of deduction.

Micro-skills

I do not wish to be misunderstood as saying that high-level, large-scale procedural knowledge is all that is important. There are certain "micro-skills" which simply must be mastered in whatever way the student selects. Perhaps the most vital "knowledge" of trigonometry is the simple skill of quickly and correctly identifying the components of vectors. A student who can prove angle addition formulas and all the rest will truly have gained little from trigonometry if he takes several minutes to find the hypotenuse of a right triangle given an angle and one of its sides. Typical of knowledge-within-process in general, one does not often find educators cataloging or doing analysis of the many sorts of "compiled" micro-skills which are associated with specific curriculum.

Teaching What to Do: Control Knowledge

The recognition that the activities and procedures we are discussing take place within intellectual structures has other implications. First, teaching students to direct their own activity must be a prime target of the educational task, even if it is not explicitly addressed. One must provide an answer to the question "What do I do now?" Secondly, the procedural content and procedural relations of facts and concepts must be taught with respect for the intricacies of personal functioning. They must contain information, implicitly or explicitly, about what is easy and what is hard to do with human mental machinery: how to use that machinery efficiently. This returns us to heuristic and intuitive levels of understanding, for there can be no doubt that these are effective yet thoroughly human and personalized levels of processing. The feelings that many teachers and educators have that such concerns are imprecise and irrelevant to the "real" material must be overcome.

Taking Advantage of What is Known

The movement away from axiomatics has not only the bonus of developing knowledge and skills which are more "human," it allows one to tap knowledge and skills which already exist but are usually unused in formal contexts. Two are: practical language skills and the active, intuitive geometry which allows one to navigate and physically manipulate the world. Both are ignored in most explicit formal treatments. These do not have deductive support but nonetheless are secure due to the rich experience gained in dealing with the practicalities of the world. Being able to choose and walk a straight path across the room is every bit as much knowledge (albeit knowledge-within-process) about geometry as Euclid's axioms. I will have much more to say about this personal and active geometry. For now suffice it to note that it is quite far from the static geometry which is usually taught as the first step into mathematics after arithmetic.

Beyond such specific knowledge that a less formal approach is meant to tap, I would like to argue that axiomatics cannot engage the more general style by which people quickly and effectively learn about the world. *People are more fundamentally model builders than they are formal system builders.* They reason by analogy. They induce. They formulate heuristics and develop dispositions to act in certain ways in certain circumstances. Their views are as much conflicting patchworks as they are coherent systems. Yet, despite lack of rigor, they learn a great deal about the world and they learn it well in a functional sense. A junior-high-school student learning axiomatic geometry must throw all this well-practiced methodology out and restrict himself to the most meager of learning styles.

Developing in the style of and from the contents of heuristic and intuitive world models is a very robust structuring of knowledge. Though personal "world models" may be mistaken in many details and may seem on the surface quite imprecise, they are secure in that they do not arise from abstraction but from procedures which work. Not only is it very likely that these ideas generated from experience contain germs of truth or senses of interpretation which are valuable, but the student knows this. He can have a confidence in ideas which come from knowing that he can navigate the world though he does not know theorems in Euclidean geometry. Again, geometry will be a major example when we turn to explicit examples.

A Dynamic Curriculum

Experience-based knowledge is also robust in relation to its extensibility. Here the diffuseness in the statement of questions and problems characteristic of informal understanding again pays the same dividends to a student seeking to extend and explore on his own. It is much more likely that one can dig out meaningful and precise understanding from imprecise ideas suggested by informal inference from real-world experience than to expect a beginning student to be able to suggest productive new lines of enquiry from formal similarity or other such operations inside a deductive system. A well-structured curriculum allows students to ask interesting and productive questions rather than waiting for the next exercise. It is a rare child who will propose, let alone prove his own theorems in Euclidean geometry. To return to an earlier argument, if the aim of mathematical education is to involve the child in mathematician-like activities, it surely seems reasonable to provide him an environment much richer in suggestions and pointers to new enquiries than the sparse combinatories of axiom juggling. A rich yet fuzzy intellectual environment can in itself return full circle to provide the appreciation and feeling for the proper use of axiomatics and thus provide motivation and sustenance for the time when careful formalism is necessary.

This richness we speak of in an informal environment should be considered a goal in itself. One wants a world as rich and ambiguous in the sense of many possible connections between parts as an erector set or a good set of blocks, not the dull and sometimes frustrating efficiency of a prefabricated toy model which has been cleverly designed to go together in only one way.

Heuristics

There are three final notes to make before turning to examples. First of all, though there has been much talk of heuristics in this paper I do not mean to say that the program here is merely to present students with a thorough and complete list of general guidelines for solving problems. Such an approach as taken by Polya may be useful, but it cannot be the whole story. What I am arguing for is that curriculum material be organized with great concern for the control structure in the student. Heuristics, informal statement, and analogies, should all be an *integral part* of the material taught. These sources of control knowledge should be grounded as far as possible in the knowledge already present in the students.

To be sure, one will have to tolerate in this curriculum a great number of "holes." The student will have to deal much earlier with the realization that he (and his teacher!) has only partial understanding. But the illusion of perfect knowledge fostered by axiomatic presentations is well buried. Although there are inevitable gaps in the students' knowledge there are windows through which further and perhaps more profound understanding may be seen.

An Intuition-Formalism Truce

Again, I have not written symbolic or any other kind of formalism out of this pedagogical theory. Certainly such intellectual machines play an important role in summarizing material knowledge and insuring a uniform precision. But I have argued that it is a great mistake to identify knowing a field with knowing a formalism.

The examples which follow are intended to point the way toward formalisms which are good physics and mathematics as much as they are intended to reflect cognitive realities. There is no reason that a curriculum cannot allow students thorough precision when that precision is the content of the matter.

EXAMPLES

The problem of developing a curriculum which reflects the above concerns for intuition, control knowledge, and experiential support, is not trivial. At the very least, the question of inculcating in the student a particular disposition (the inverse operation to making knowledge-within-process explicit) is problematic. We cannot merely insert into the student's mind a "demon," as one may figuratively do in a computer program, which activates itself in appropriate circumstances. Nor can we expect practice as the mindless rehearsal of musical scales or arithmetical operations to succeed very well in such complex and constantly changing situations as doing physics problems (even if we could set meaningful examples to practice). But we should explain carefully to the student what we understand of the disposition, its cues and modes of operation. Furthermore, and probably more importantly, we can organize the material around and incorporate into it the relevant features to the final dispositional encoding. It seems likely the student will always have to make that last step for himself.

To aid in this broad program one has the "new" outlook of searching for procedural formulations rather than air-tight formal ones, but the way still is by no means clear. This section sketches some examples of the kind of reorganization envisaged. These can fill out and support the rather abstract discussion in earlier sections.

Physics is a protean ground for exploitation of procedure and intuition. It is by its nature mechanism. There is, furthermore, prima facie evidence that children have the sort of procedural-intuitive understanding of physics that we can rely on to replace deduction and propositional security. A child can catch a ball thrown in the air and knows roughly what will happen when you push on things in most circumstances. Below is a sketch of some ideas concerning mechanics which embodies the concerns we have been discussing.

The Concept of Force

What is the concept of force as usually taught in elementary physics courses? After a few philosphical remarks about pushes and pulls, there appears $F = ma$. From there on, by and large, discussion of force revolves around this analytic representation of the idea. Operational understanding is for the most part left to the student to be constructed from many workbook problems and examples of expert solutions. I would argue for explicit treatment of the precise mechanism that force is and for greater effort to bring to bear intuitive knowledge.

What does force do to a body? Clearly it changes something—not the color nor the taste. To zero in on what force changes, it makes sense to consider a situation where there is only one force of significance and one simple object. Perhaps a hockey puck on ice is a good example. What happens when you push on it? If it is standing still initially, the puck goes in the direction of the push. Does force then change position? Certainly not—at least not exactly. A moving hockey puck subject to a push does not simply go in the direction of the push.

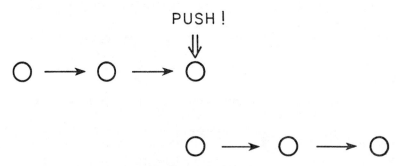

Figure 1. Force Changing Position

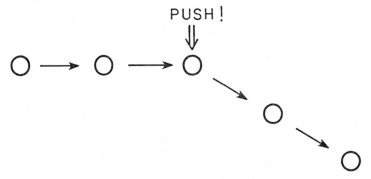

Figure 2. Force Changing Velocity

Force deflects, it does not *move*. A more precise way to say that is: force changes the velocity (direction included), not position. It is the sole function of force to change the velocity. This parallels the function of velocity which is a "changer" in its own right. Its responsibility is to change the position of an object.

Thus we have arrived at a remarkably simple yet precise picture of the mechanism of motion. Velocity is the sole changer of the position, and force is the sole changer of velocity. That is not obvious, but it is true and is only a small (but important) step away from the naive "push or pull."

There are immediate consequences for these ideas. For example:

1. The "counter-intuitive" action of a gyroscope can easily be understood if one looks to see how the *velocity* of the wheel parts is changed rather than

assuming pushing something makes it go in the direction of the push (diSessa, 1972). "Push or pull" is not adequate qualitative knowledge about force to account for gyroscopic action.

2. Given an analytic representation like vectors for velocities and changes in velocities, an algorithm for generating particle motion is immediate. Roughly speaking, if $\vec{v}$ = velocity, $\vec{F}$ = force and $\vec{x}$ = position, each particle is continually computing its next position and velocity by

$$\vec{v} \leftarrow \vec{v} + \vec{F}$$

$$\vec{x} \leftarrow \vec{x} + \vec{v}.$$

That is essentially all there is to a very concrete manifestation of this understanding, a computer program to simulate any "forced" motion. It is a formalism in the best sense which stands in the proper stance with respect to intuition and qualitative understanding, flowing naturally from them but at the same time refining them.

3. A particularly good example of the simplicity of this point of view is gravity. All those complicated trajectories are the result of the simples of all possible forces. One that acts equally on all objects all of the time. In the above equations, $\vec{F}$ is a "universal gravitational constant."

Now of course there is some refining to do. If $\vec{F}$ is not an impulse force, but some continuous "pouring of velocity" into an object, one should write

$$\vec{v} \leftarrow \vec{v} + \vec{F} \Delta t$$

where $\vec{F}$ is the amount of velocity per unit time which is added to the initial velocity. This matches the more explicit

$$\vec{x} \leftarrow \vec{x} + \vec{v} \Delta t$$

Furthermore, to make this notion of $\overline{F}$ coincide more closely with the intuition of force as effort one must realize that heavier objects require more "force" to make a given change of v, thus one has

$$\vec{v} \leftarrow \vec{v} + (\vec{F}/m) \Delta t.$$

This refining is not a disadvantage. Heuristic information is often characterized by a hierarchy of ideas with the key ideas on the top and successive lower levels of warning, restrictions, corrections, just as any procedure has a global plan but also many conditional and contextural parts. "Oversimplication" with successive corrections is, I think, a good mode for presenting much curriculum material as well as a general workhorse method for computer programming. It is not in any way intended to be a sloppy understanding, but it is meant to be an organization which allows one to 1. keep the information most necessary for actions such as problem solving on top, and 2. choose the top level in such a way as to be tied to an appropriate intuition. By these statements I do, in fact, mean to imply that force as a velocity changer is both a key idea in problem solving *and* an idea with considerable intuitive support. On the other end, the notion of "refinability," that the top-level heuristic and qualitative understandings must develop naturally into more precise and careful treatments, is important to the approach.

Causal Syntax and Force

Let me take another look at what may be happening in the student's head when we introduce Newtonian Mechanics. I implied earlier in an example of knowledge-within-process that the causal structure—agent, interaction, recipient—is important to the concept of force. For reference I will call this triad *causal syntax*. It is important in the first place because it is a naturally occurring structure in terms of which humans will interpret similar structures such as the notion of force. The personal context of force is almost always a causal situation: I cause something to move, or it causes me to recoil, etc.

The recognizable form, causal syntax, gives an interpretation of the newer formal concept of force in more primitive terms. At a deeper level there is a control structure implicitly attached to the causal syntax which is appropriate for force. One always assumes a reasonable connection between the causing agent and the resulting actions of the recipient. It is therefore appropriate to try to establish the result of the agent's initiative given a suitable description of it. It is also appropriate to infer specifics about the cause, given the causal mechanism and some specifics about the effect. These translate into dispositions to calculate motion from force and infer force from motion.

There are also negative dispositions obtained from the natural identification of causal syntax in the concept of force. The syntax is directed, not symmetric. One is not disposed to treat the agent in the same way that one treats the recipient. The earth causes a ball to fall, but one does not think about what effect this has on the earth. Elementary physics students are notorious for thinking long and hard about the force B exerts on A having just pronounced the force A exerts on B. Though the symmetry is always taught in Newton's Third Law, "equal and opposite reaction," almost never does one see a careful, explicit confrontation of that proposition with the intuitive causal structure of force. As a consequence, the proposition of symmetry languishes as a formal idea devoid of appropriate control structure until the student has many times been chided to remark on its implications.

Multiple Representations: Force as Momentum Flow

When one looks at statics problems instead of dynamics, the concept of force as velocity changer is almost empty of intuitive content. We must restructure the concept of force unless we intend to let formal analytic implications $F = ma$ carry the burden (as they are capable of doing, *formally*). To this end one can introduce the concept of momentum, mv, and force becomes a changer of momentum. The reason for this is that one has more structure; momentum is a conserved quantity. Thus force is not just a change in momentum of some particle, it is a *transfer* from one particle to another. Providing one can invoke strongly enough the image of momentum as a conserved entity, just as water in everyday experience, a whole host of intuitive knowledge (such as "what goes in must either come out, or it collects!") becomes available for use in problem solving. This image of force introduces an activity into the world of statics which is much closer to a physicist's view of a constantly working and processing world than the name statics implies. It is an image much closer to common experience than formal equations of equilibrium.

Gravity is a force, hence it is continuously pouring momentum into the ball in my hand. The momentum must be going somewhere; I do not see it collecting in the ball (you can always see momentum collect). The momentum must be leaving the ball, flowing through my body and into the earth. All along the way there are forces, stresses, expressing this flow of momentum. Flow of momentum is a precise and correct replacement for the "common sense" feeling for "transfer" of force through a static member. Again I am claiming the intuitively tied image will help in problem solving. Consider the following problem:

> A train consisting of an engine and several identical cars is accelerating at a constant rate on a flat stretch of track. Neglecting friction, what is the tension in the linkages between cars?

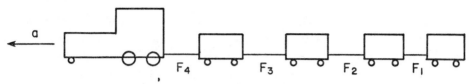

Figure 3. An Accelerating Train

The standard approach to the problem directs isolating subsystems and writing down $F = ma$ for each. The control knowledge of how one selects an appropriate subsystem is rarely if ever discussed, but let us suppose the student does the obvious, writes $F = ma$ for each car.

$$F_i - F_{i-1} = ma, \qquad F_0 = 0$$

At this point for most students the equations stand bare—the physics is done and the question of how to solve them is an algebraic one. The clever student notices quickly that the equations can be solved recursively from $F_1 = ma$; one less clever muddles through.

I have frequently encountered students not quite steeped in the system ($F=ma$ process) whose intuitions suggest a rather different analysis. They explain a feeling that each car is "absorbing" a force (of ma), leaving the car to the rear with "less force." Unfortunately the analysis cannot proceed if the only Newtonian paradigm is the $F = ma$ one. I am about to point out how force as momentum flow can provide a precise Newtonian frame which allows this usually disallowed intuition.

The momentum-flow analysis says that each car is receiving momentum from the car in front, collecting momentum at the rate of ma, and losing some momentum to the car to the rear.

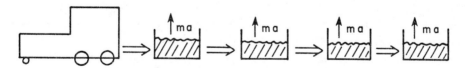

Figure 4. A Momentum Flow View

The equations corresponding to this view are identical, but the sense of mechanism provided makes for an enriched possibility for attaching an intuitive control structure to them.

The recursive solution method is natural in this framework. The linkage to the last car must be carrying the whole of the momentum collecting there, i.e. ma. Knowing the flow out of the next-to-last car, the flow into it is determined, and so on.

But an equally transparent solution exists. Each link is providing all of the momentum collecting in the cars down further. Thus the flow through the nth link is nma. Notice that this solution corresponds to selecting a new system decomposition in $F = ma$ terms, one which in the absence of a sense of the mechanism often poses difficulties such as: How can you just declare a collection of things to be a system?

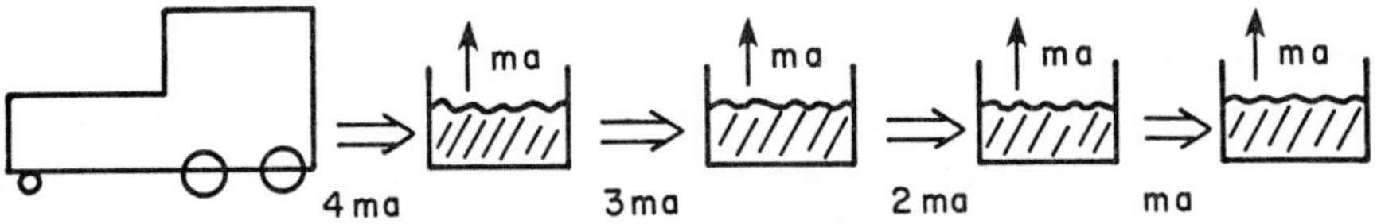

Figure 5. The Flow Solution

The criteria for flexibility of control structure and "naturalness" of solution method are not abstract, but rely in two ways on students' previously acquired knowledge. First, we are relying on students' abilities to easily provide control analysis within a flow interpretation. Heuristics like "look upstream or downstream to see if there is a better place to measure flow," are engaged. Secondly we are relying on the fact that momentum flow is a precise Newtonian annotation for the previously noted intuitions about "absorbing" and "transferring" forces. Thus it will be naturally evoked in problem-solving situations.

The difficulties of introducing the flow of a vector quantity in analogy to fluid or other flow in order to tap experiential control knowledge do not make it a clear pedagogical winner over pure $F = ma$. But I argue that these concerns suggest a more coherent experimental effort in this direction than is evident in the vast numbers of "standard approach" textbooks.

Causal Syntax Revisited

I conclude the example of the concept of force by returning to the role of causal syntax. I previously pointed out that the difficulty of attaching causal syntax to the notion of force is that the syntax is directed whereas there is no physical way of separating or distinguishing action from reaction. It is facile to point out that force as a transfer of momentum is a much more symmetric image (the recipient gaining p is the same as the agent gaining −p) and to thereby conclude that it "fixes" the asymmetric disposition of the causal syntax. Indeed what is more likely is that students will find their whole feeling for the notion of force sliding away from their grasp when an essential (to intuitive understanding) primitive structure as the causal structure is threatened. Again it seems the most rational course to confront the problem directly with a discussion centering on metamorphosis of causality from the simplistic causal syntax to a more appropriate notion.

Turtle Geometry

Standard Euclidean geometry approaches start with objects, points, lines and angles with which students have some familiarity, but immediately and usually quite thoroughly cut away the usefulness of that familiarity with formal axioms and the insistence on (more or less) formal proofs. Points and lines are humanely undefined, but angle is defined by a pair of "rays" and things get rapidly more abstract. Papert some years ago suggested a new sort of geometry which begins by replacing this formal angle with the heuristic, "angle is a turn." The model can be a creature called a "turtle" who knows angles by turning through them. The turtle is given mobility by being able to move in a straight line (forward or backward) when he is not turning. Turtle geometry is the study of the figures constructible by such a creature and is evidently a much more active study than standard Euclidean geometry. An elementary student can easily "play turtle" to bring to bear his own intuitive knowledge about space.

To compare turtle and Euclidean geometry on common ground let us prove that the sum of angles in a triangle is 180 degrees. The usual Euclidean method involves bringing the three angles to a point to see that they sum to "opposite rays." The problem is that doing this requires the construction of a line, 1, parallel to the base of the triangle, a nonobvious activity, followed by identifying alternate interior angles.

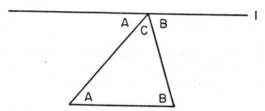

Figure 6. Proving A + B + C = 180 Degrees

To use a turtle to do the same thing, simply embed his basic angle measuring facility in a process which sums the angles: have him turn each angle in sequence in the same direction and see the result.

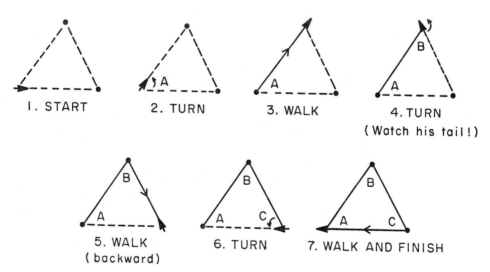

Figure 7. Measuring A + B + C

Notice the change in the turtle's heading. He's pointing exactly opposite of his initial state and thus has turned by definition 180 degrees.

Figure 7 appears much more complicated than Figure 6 on paper, but consider two important facts. First, the turtle proof as shown is rather clumsy because it is an adaptation of a representation which in essence involves motion. Here I must rely on the reader to "make a movie" of that information in his head. With less primitive technology than printing, the essential process can be much more easily presented in such a way as to take advantage of the human ability to understand and analyze motion. That ability, of course, accounts for the understandability of the proof in the first place.

Second, the control knowledge which leads to the generation of the proof is as important as the final insight for evaluating comprehensibility. The turtle proof in this respect is transparent; adding angles is merely translated to sequential performance of turns to make a proof. The motivation behind the standard proof is rather more complex.

Whether or not the second method is better than the first, it has some advantages typical of constructively defined mathematics. Among these is the important property of generalizability. The proofs of similar propositions are, in a natural sense, all the same. The second method needs no modification to work on any polygon (though it is better to measure exterior angles in case the number of sides is greater than four). Consider the turtle process to measure the sum of the *exterior* angles of a triangle shown in Figure 8.

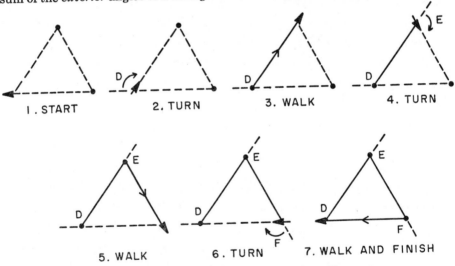

Figure 8. Measuring D + E + F

The turtle ends his trip with the same orientation as he started. He's turned 360 degrees. Try to make a construction like figure 6 to demonstrate this. If you succeed at that, notice that the turtle proof depends in no way on the figure being a triangle; it is true of any polygon! The generalization of figure 6 becomes even less transparent in such circumstances.

Many related problems of angle measurement require only slight modifications of the algorithm. Finally, the method is infalliable in the very real sense that one can add up any sequence of angles on any surface in this way, if one is careful. The Euclidean method has nothing to say about triangles on spheres, but the turtle method is a natural lead-in to nonflat geometries. I provide a glimpse of this in example 4. Incidentally, notice the fact that a turtle description is also a prescription for construction. This establishes a vital link between geometry and physics. Motion and trajectories may well be explained in turtle terms. Example 3 shows the power of uniting an active, process-oriented geometry with physics.

ORBITAL MECHANICS: A MORE COMPLEX EXAMPLE

In order to show how some of these ideas, particularly process-oriented representation, can extend beyond introductory concepts I wish to paraphrase a "deduction" of elliptical orbits from the inverse square force law. Naturally I would like to avoid the analytic language of F = ma, but again cannot take the time here to develop a complete, alternate, procedural language. Nonetheless, those who are not familiar with the analytic language should also be able to appreciate the point. The equations are markers for physicists reading this paper, and for the rest will serve to symbolize the lack of illumination a purely analytic presentation can have. Begin with F = ma

$$(A) \qquad \vec{F} = -k\hat{r}/(r^2) = m\,(d^2\vec{x}/dt^2)$$

where k is a gravitational constant, $\vec{x}$ represents position, r = radius from the sum, $\hat{r}$ is the (unit length) vector pointing radially from the sum to the planet. The first transformation is to shift emphasis from position to velocity. Forces, remember, act directly on velocity and only indirectly on position.

$$-k\hat{r}/(r^2) = d\vec{v}/dt$$

Analytically this appears to be a trivial transformation, but conceptually it is not. The procedure represented by this equation,

$$\vec{v} \leftarrow \vec{v} - (k\hat{r}/r^2 m)\,\Delta t$$

can be easily computer implemented, but one can do much better. If one shifts to looking at r as a function of ϕ rather than as a function of t, that is, look at the orbit per se rather than the time-parameterized orbit (a necessary step in any derivation unless one is willing to do elliptic integrals), one gets

$$(B) \qquad d\vec{v}/d\theta = -(k/mL)\hat{r}.$$

L is a quantity called angular momentum. Now a direct procedural translation (the act of translation is simple but is not important since in practice we would be speaking in procedural terms all along) is the following. Each change in velocity, $\Delta\vec{v}$, has a constant magnitude $(k\Delta\theta/mL)$. The change in direction of $\Delta\vec{v}$ between steps is also a constant $(\Delta\theta)$. Thus to generate the changing velocity, perform the following algorithm:

(C) (a) go forward a small amount

(b) turn by a small amount to face a new direction

(c) repeat the above.

If you think for a moment you will realize that the "procedural differential equation" is precisely the turtle geometric description of a circle.

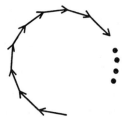

Figure 9. Representation (C), a Circle of Force Impulses (Velocity Changes)

That such an algorithm generates (a close approximation to) a circle is not a hard thing to understand; give those instructions to an elementary school child and see if he understands that they direct him to walk in a circle!

Knowing the entire sequence of *velocities*, one can go back to draw conclusions about the position orbit. The orbital problem is in principle solved, and in practice showing such facts as the orbit is an ellipse require only a few algebraic steps. Taking the velocity space solution, $\vec{v} = \vec{z} + u\hat{\theta}$ where z points to the center of the circle, u = k/mL is the circle's radius, and multiplying (cross product) by r, one gets L = r(u + z sin θ). Hence r = L/(u + z sin θ), the general equation for a conic section in polar coordinates.

Solving the problem through the intermediary of velocity is not a special ploy. Velocity *is* the thing that is changed by force. Because of this there are in fact great dividends to a velocity-oriented derivation. The effect of many perturbations becomes obvious through the intermediary of velocity. Practical and interesting problems such as guidance of an orbiting space ship become immediately accessible (Abelson, diSessa, Rudolph, 1975, and diSessa, 1975). Orbital mechanics can become a domain for student explorations rather than just more results to remember in isolation.

There is no magic in the approach. One must still solve a differential equation. But one can solve it in the form of (A) or (B)[5] which involve a great number of formal operations (check any standard derivation) or in the form of (C). There is no a priori reason to say that representation (C) is simpler than (A) or (B). In fact, it is probably as hard to solve (A) or (B) as to *prove* that (C) draws a circle. The reason that (C) succeeds in being transparent is that it is phrased in procedural terms which are very close to the knowledge store everyone must have and use to walk around this world.

Turtle Differential Geometry

One great stumbling block to doing nonflat geometry on, say, a high school level is the lack of a good definition for a "line" or more appropriately in the case of nonflat surfaces, a geodesic. The standard definition, a path of shortest or extremal distance, has some intuitive appeal but is really most appropriate in the

---

5. Actually the velocity equation (B) is already a substantial improvement over (A) even in analytic terms. In standard notation, notice $d\hat{\theta}/d\theta = -\hat{r}$, so that (B) can be written $d/d\theta\{\vec{v} - k\hat{\theta}/mL\} = 0$, hence $\vec{v} = k\hat{\theta}/mL + \vec{z}$ where $\vec{z}$ is a constant. The position space orbit can be trivially derived from this velocity space solution. Functional interpretation of $\vec{z}$ and the other term in the solution given for $\vec{v}$ provide vital links to intuition in this approach to orbital mechanics. (Abelson, diSessa, Rudolph, 1975)

analytic context of variational differential equations. A student will probably put up with the shortest distance definition but gets very uneasy when presented with a "longest distance" geodesic like a complete great circle on a sphere. In contrast to this, consider the constructive turtle definition: a geodesic is what a turtle walks if he walks straight, i.e. takes the same number and length of steps with his right and left legs.

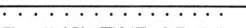

Figure 10. A Line With "Turtle Tracks"

The definition is easy to understand, and it is not hard to get high school students to make that definition *themselves* if encouraged to verbalize how *they* can know they are walking a straight line without looking; it is a simple *annotation* of some knowledge-within-process.

There are other advantages. The constructed geodesic explicitly mentions the left-right symmetry which is an essential heuristic understanding of "straight." That symmetry shows the equator is a geodesic while an 80 degree latitude cannot be. (The former divides the earth into two equal pieces and the latter does not.) The constructive definition also gives an intrinsic check on whether a path is a geodesic. Does it follow the rules for turtle construction? Can you put an equal number of equally spaced turtle tracks around the "line"?

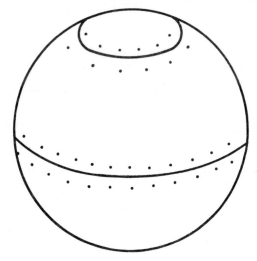

Figure 11. Equator is a "Straight" Line; Other Latitudes Are Not

Compare this process to proving that there is no shorter (or longer) path, or alternatively, constructing one.

Returning briefly to physics, students often feel quite uneasy with the variationally defined (extremal distance) geodesic. "How can a particle or light ray 'compute' its geodesic path unless it 'knows' its endpoint already?" Granted such questions are confusing, but they arise from feelings for local causality which should be encouraged rather than frustrated. This is exactly the kind of intuitive disposition one wants to take advantage of, not do battle with. Local causality is only a formal property of variational geodesics; it is manifest in the local and constructive turtle definition. (The local-global dichotomy is one of the important heuristic themes from computation which play a central role in turtle geometry and which seem quite valuable in many other areas as well.)

Along the same lines it is very easy to relate a turtle geodesic to such things in everyday experience as the path of a car with wheels straight (each wheel turning at the same speed) or a jet airplane with rudder straight and wing engines running equally fast. The formal elegance of avoiding the question of local construction through a variational definition leaves out these important experiential ties.[6]

Having bypassed formal axiomatics with appropriately active definitions, it is not hard to take this turtle quite far in nonflat geometries, (diSessa, 1975) far enough in fact to bring high school students in contact with many of the most important ideas in mathematics—the concept of transformations and invariants, continuity, the importance of topological considerations, and Stokes-like theorems.[7] It is one of the prime advantages of informal presentations that students can begin developing feelings for and even the ability to use some of these extremely valuable and broadly applicable ideas long before their formal abilities are up to very general and/or precise formulations. Planting the seeds for understanding "powerful ideas" allows time to nurture notions of purpose and use which can keep a student's head above water in the rising tide of details necessary later for true mathematical integrity.

SUMMARY

I have argued that axiomatics or other formal systems may be useful models of "good" representation of knowledge for certain purposes, but they are not sufficient as pedagogical models. Other kinds of more informal presentations have a great number of advantages in being able to take into account the specific abilities and knowledge students acquire from everyday experience.

The argument has been organized around the computational metaphor, which has two parts.

1. Human thinking and knowing is process. It is complex but exhibits organization of a type which is hardly of the logical formalist type. A particularly important outgrowth of this is a concern for knowledge which is self-directing, organizational—in short, control knowledge.

---

6. A line as an abstract entity with certain properties has been replaced by a line as the process which draws it! One may wish to generalize this comparison of turtle geometry with standard geometry by contrasting mathematics of construction to mathematics of constraint. The latter defines entities by a series of constraints (e.g. axioms), does not deal with the vagaries of models, and does not bother to tell the student either (a) that the entity of concern is a suitable generalization for everything he knows about, or (b) that there are no known examples of such a thing.

7. By the latter I mean any of the group of theorems which compute the totality of something spreading over a region by computing something else on the boundary of that region. Important examples are the fundamental theorem of calculus, the calculus of residues in complex analysis, Gauss's theorem in electrostatics and gravitation, Stokes' Theorem in electrostatics, any conservation law for flowing substances, the concept of state function in thermodynamics, and the existence of potential (e.g. energy) functions.

2. There is good reason to believe that one can produce a significantly more learnable curriculum if one augments the more traditional set of knowledge-organization schemes (such as axiomatics) with more procedurally oriented ones. This is particularly true if one can choose procedural representations with elements which match as well as possible the natural knowledge-within-process which constitutes much of the intuitive, common-sense knowledge students already have.

I conclude with an abstracted list of desiderata for pedagogical material:

1. It is a *discovery-rich* environment and is careful to organize the material with many "windows" (not just gaps or holes) for more than exercise-type individual study.

2. It discusses and develops "higher level" organizational skills such as heuristics and other control knowledge. In particular, it discusses its own organization and explains the nature of the enterprise with reference to the ultimate goals of the material. The function of ideas in problem solving etc. (qualitative knowledge) is a key part of understanding them.

3. It attempts to access and to tie closely to nonpropositional knowledge such as intuition and common sense which students have acquired about the world, whether or not the knowledge is verbally accessible. Metaphor and analogy have their place in such attempts, not only to explain a structure in approximation, but also to invoke and involve appropriate dispositions and other control structure.

4. In connection with 3 it may well be organized around procedural representations of knowledge, some characteristics of which follow:

   a. It is active and constructive rather than prescriptive or descriptive.

   b. The large-scale unifying form is process rather than deduction.

      i. The model form is:

      Initial state ——————▶ Final state
               Operation

      rather than

      Assumptions ——————▶ Conclusions
               Deduction

      ii. Important predicates and relations are: independent (as degrees of freedom in a linear system), invariants, procedural equivalence, etc. rather than true, false, follows from logical equivalence etc.

5. Presentations are from multiple viewpoints dictated by the use to which the student will put the ideas and what knowledge the student already has which can be assessed. Great care is taken to provide the kind of knowledge which interfaces well with control concerns.

6. In connection with *3* and *5* it may frequently make use of simplified schema with successive corrections and amendments. It is less concerned with the pathological special case except when this is a telling and crucial failure. It is not afraid to introduce "advanced" notions *provided* they are useful and have intuitive content: By "advanced" I mean ideas which require a large formal background for rigorous "respectable" presentation. Power to understand and to accomplish should be the first lessons of mathematical and scientific knowledge; rigor and precision are secondary.

ACKNOWLEDGEMENTS

I wish to acknowledge and thank the following for influence, reading and criticism on earlier drafts of this paper: Seymour Papert, Hal Abelson, Mimi Sinclair, Howard Peelle, Jeanne Bamberger, Don Schon, Sidney Strauss, Melinda diSessa.

REFERENCES

Abelson, H., diSessa, A., and Rudolph, L. "Velocity Space and the Geometry of Planetary Orbits." *American Journal of Physics,* 43(7), July 1975.

Courant, R. *Differential and Integral Calculus,* Vol 1. New York: Interscience Publishers, 1964.

diSessa, A. "Turtle Escapes the Plane: Some Advanced Turtle Geometry." MIT Logo Memo 21, AI Memo 348, 1975.

diSessa, A. "The Gyroscope." MIT Logo Working Paper 21, 1972.

diSessa, A. "ORBIT: A Mini-Environment for Exploring Orbital Mechanics," *Proceedings: Second IFIP World Conference on Computers and Education.* North Holland Press, 1975.

Feyerabend, P. *Against Method.* Atlantic Highlands, NJ: Humanities Press, 1975.

Kline, M. *Mathematics, A Cultural Approach.* Reading, MA: Addison-Wesley Publishing Co., Inc., 1962.

Lakatos, I. *Proofs and Refutations.* Cambridge: Cambridge University Press, 1977.

Minsky, M. "A Framework for Representing Knowledge." MIT AI Memo 306, 1974.

Moore, J. and Newell, A. "How Can Merlin Understand?" In *Knowledge and Cognition,* edited by L. Gregg, Potomac, MD: Lawrence Erlbaum Associates, 1973.

Papert, S. "Teaching Children to be Mathematicians Versus Teaching About Mathematics." *Journal of Mathematical Education in Science and Technology,* 3, pp 249-262, 1972.

Ryle, G. "Knowing How and Knowing That." In *The Concept of the Mind,* New York: Barnes and Noble, 1949.

Teaching For Cognitive Development

Robert P. Bauman, Thomas Wdowiak, and Irene Loomis

It is no secret among physics instructors that most students have difficulty with many concepts of elementary physics. One popular way of attacking this problem is to apply Piaget's model of cognitive development. But Piaget worked with children from ages one through 15, so a skeptical teacher of high school juniors and seniors or of college students must surely question whether Piaget's model really applies to high school and college teaching.

Statistics show that typically 50 to 75 percent of college freshmen are not able to cope with the kinds of problems associated with the formal operational stage. Important concepts of an introductory physics course require mental operations on abstract quantities, ratio and proportion, and an awareness of one's own thought processes. The correspondence of answers given by many college students to the answers recounted by Piaget in his interviews with children reveals a striking similarity between the thought processes of the younger students classified as concrete operational, and the physically mature students enrolled in college courses.

The more important question to be asked however is: If the Piagetian model and the numerical results of current testing programs are correct, or even partially correct, what implications does this have for our classrooms? Can we do anything to effectively move students from concrete operational or transitional stages toward formal operational thinking, so that they may understand what we and our colleagues are attempting to teach?

We believe we have found one effective means of stimulating cognitive development in college students. We have heard of other programs that also offer encouragement. Following a description of our current efforts, some of the properties of these programs will be examined in an attempt to see which elements may be important for success.

Our efforts have been directed toward the specific question of whether students can be changed, and we have therefore chosen to separate this process from our conventional physics courses. More specifically, we have devoted a one-quarter, three semester-hour course entitled Mathematical Preparation for Physics to teaching logical thinking, using mathematics as the primary medium of instruction. This course, labeled PH 10, has a typical enrollment of 50 to 60 students per quarter, most of whom plan to take a physics course but who could not pass a rather simple screening examination.

Subsequently, we have also turned attention to the marginal students who would not normally take a physics course at all, but who are in need of learning logical thought processes to survive in college. We first taught a small group of such students during a special summer program, and now offer a physical science course sequence—PHS 7-8-9, Concepts of Science and Logic—on a regular basis for disadvantaged students. The sequence employs the same materials as PH 10, plus some supplementary materials, and is spread over three quarters. We regularly offer three sections, of 30 to 35 students each, of the PHS sequence.

The premises on which the course was designed were the following:

1. The material should start at an elementary level and progress from one stage to another, without skipping stages.
2. The student should be able to move through the material at his own pace.
3. The student's mental involvement is more important than the transfer of information to the student. To ensure student involvement and allow self pacing, a form of mastery testing is employed.
4. Encouraging the student to engage in strenuous mental activity is sometimes difficult, especially for students who are convinced beforehand that physics is too hard. Our solution is to ask hard questions about easy material, and require that students express their answers clearly in writing.
5. The greatest encouragement to effort is success, so we make sure that each unit of the course concludes with a success for the student. This is accomplished through the unconventional mastery testing system where each student receives only one version of a unit exam but must get everything on that exam correct. If there is an error, the existence of the error is pointed out and the exam is returned to the student for further work.

Selection of mathematics as the medium of instruction provides face validity. Most students will accept the idea that they should understand mathematics better, whether they consider it preparatory to required science and math courses or as an alternative to more advanced science and math courses. Background experiences vary widely. Some of the PHS students have had no high school geometry, whereas roughly a third of the PH 10 students have had one or more terms of calculus.

Course material is in the form of a workbook/textbook that requires students to fill in answers and supply missing proofs as they read. It begins with problems relating to counting and one-to-one correspondence, proceeding through addition (including additivity of lengths, areas, and volumes), subtraction, multiplication and division, and counting the number of distinguishable arrangements or combinations of objects. Intermediate topics include functions, equations, and graphs; units and dimensions; elementary kinematics (displacement, time, speed, and acceleration); powers and number bases; multiplication and division of polynomials and the solution of equations; and examination of propagation of errors of uncertainties, leading to the meaning of significant figures. The final units deal with series and approximations, plane geometry, trigonometry, complex numbers, matrices and determinants, and vectors.

Units 7 and 11 introduce single-sentence symbolic logic. The first deals only with simple implication statements (*If . . . then . . .*, or $p \rightarrow q$), but requires that the students be able to translate a sentence from other English forms into the if-then format and also be able to write down all eight variations (including negatives) of a sentence and recognize which are necessarily true, which are necessarily false, and which are uncertain. Although it is only an exercise in reading comprehension, it is one of the most difficult units. Upon returning to

Table 1. Sample Problems From Workbook Units and Unit Tests

UNIT	PROBLEM
III	Mortimer wrote a check, then discovered that his bank balance was overdrawn, showing a balance of -$5.00. He didn't worry, however, because he had learned in school that 2 negatives makes a positive, so he simply wrote a check for $5.00 and cashed it to give another overdraft of $5.00. Explain, for Mortimer, how his logic went astray.
IV	One cubic foot of water has a weight of 62.4 lb. How much will 1 cubic yard of water weigh? Show how you obtained your answer. If a 10″ pizza costs $1.00, how much should a 12″ pizza cost? A cube is 3 ft on a side. How would you cut a cubic piece from it that would give you 1/64 of the total?
V	If you are given 1 bag of blue marbles, 1 bag of green marbles, 1 bag of yellow marbles, and 1 bag of red marbles, how many (distinguishable) ways could you put 1 marble into each of 5 boxes? Explain. How many different linear orders could you arrange with 1 blue, 1 green, 1 yellow, and 1 red marble? How many different distinguishable linear orders could you arrange with 3 black marbles and 1 white marble? Can you show how the second part is related to the first part?
VI	If y is proportional to x, doubling x will always double y; tripling x will always triple y. What type of equation will give this relationship? What characteristics will the graph have when a function is plotted for which y is proportional to x?
VIII	How much longer will it take to go 1 mi at 20 mi/hr than to go 1 mi at 30 mi/hr?
IX	Express the base 8 number (26/6) as a binary number. Express the base 8 number (5 + 17) as a base 12 number. Is $(n^p)^q = n^{(p+q)}$? Check your conclusion by substituting numbers.
X	Find the first 5 terms of $1/(1 + x)$ Evaluate 1/101 to 2 significant figures (without dividing!). Find approximate solution(s), by graphing, of the pair of equations: $x^2 - 4y^2 = 0$ $x^2 - 2x - 1 = y$
XIII	Estimate the value of $\sqrt{65}$ Given that ln 3 = 1.099, find ln 9 and ln 10

symbolic logic in unit 11, conjunctions (*and*) and disjunctions (*and/or*) are added and are related to implication statements. The final discovery of this unit is the meaning of an "If and only if . . ." statement, or "necessary and sufficient" conditions.

After each of the 18 units is completed in the workbook and the workbook is checked, the student is given one of several versions of the unit exam, which is answered as an open-book exam. The testing format permits us to ask quite difficult questions on the unit tests, as well as in the workbook. Some examples, drawn from different units, are shown in Table 1.

In addition to the workbook, about one week of the quarter is devoted to techniques of problem solving, employing materials prepared by Whimbey, built upon ideas suggested by Bloom and Broder. Students learn to break problems into small steps and employ aids such as sketches and analogies. Avoiding guessing, they develop a concern for accuracy and a positive attitude toward problem solving. Students work in pairs, with one (the problem solver) thinking out loud while the partner (the listener) ascertains that all steps are vocalized and the rules are followed; the two students then exchange roles. A solution, as given by a good problem solver, is available on the following page of the workbook as a means of checking the method, to provide help if required.

A closed-book examination, covering all but the last three units, is given at the end of the course. It may be retaken in alternate versions if desired, but typical performance on the rather difficult test is better than 80 percent correct.

The self-paced format of PH 10 has made it difficult to institute effective pre- and posttests, but the qualitative and anecdotal appraisals have been very encouraging. Students who struggle with early units are able to read and work expeditiously through the later units. Many students recognize that they now think differently and that it has helped them in courses in other disciplines.

More quantitative information has been obtained from the summer PHS 7 course and the academic-year PHS 7-8-9 sequence. The summer course included special instruction in English, problem solving, nonsimulation games (*EQUATIONS, WFF'N PROOF,* and *ON WORDS*), and the first third of the workbook. The students met four hours per morning, five days a week, for nine weeks. In an effort to measure general intellectual growth, as contrasted with specific course material taught, the ACT examinations were employed as pre- and posttests. Twenty-four students participated, but illness, summer jobs, and marriage limited the attendance of several of these, so that our sample was reduced to 14 who had attended at least 50 percent of the sessions. Despite the small number, the improvement in composite score was statistically significant at the .01 level, measured against national norms that incorporate a correction for normal maturation from 12th grade to the college freshman year (Table 2).

Table 2. Gains on ACT Scores of 14 Students in 9-week Summer Program

Test	Pretest Mean	Posttest Mean	Gain	t	p
Composite	10.00	11.93	1.93	2.82	<.01
English	12.07	13.50	1.43	1.99	<.05
Mathematics	7.34	9.14	1.79	1.09	n.s.
Social Studies	8.93	9.14	0.21	0.21	n.s.
Natural Sciences	11.57	15.50	3.93	3.27	<.01

Having established significance for the composite score it becomes meaningful to look at subtest scores. Clear statistical significance was observed for the natural science section and, on the basis of a one-tailed distribution, for the English section. The substantial mean improvement in mathematics scores was not statistically significant because of scatter, which could be traced to a lack of discrimination against guessing in the ACT scoring system. (Three students, who had shown substantial improvement in other areas, did not finish the mathematics test, and did not guess; random answers assigned to their unanswered questions should have improved their scores sufficiently to make the mathematics subtest mean gain for the group statistically significant.) A large drop in social science score by one student made the mean improvement of the group small and insignificant for this subtest.

A similar comparison of pre- and posttest ACT scores for 17 PHS 7-8-9 students (Table 3) showed a mean gain in composite score, from pre-admission to the spring quarter, of 1.44, statistically significant at the .005 level against national norms. Of the subtests, only mathematics was significant (mean gain 3.24, significant at .005 level); the other subtests showed smaller gains.

**Table 3. Gains on ACT Scores of 17 Students in PHS 9
Who Had Taken ACT Exams Previously**

Test	Pretest Mean	Posttest Mean	Gain	t	p
Composite	14.50	15.94	1.44	3.31	<.005
English	13.88	14.11	0.23	0.34	n.s.
Mathematics	12.64	15.88	3.24	3.67	<.005
Social Studies	11.88	13.11	1.23	1.61	n.s.
Natural Sciences	16.94	17.47	0.53	0.84	n.s.

An attempt was also made to assemble a local control group from the list of those who might have been included in the program but were not. Some of these students, and some of the students in the course sequence, had not previously taken the ACT exams, so the analysis of scores was somewhat more uncertain and statistical significance was not achieved (except for one subtest—see Table 4).

**Table 4. Differences of 31 PHS 9 Students Versus 44 Control
Students on Posttest ACT Scores**

Test	Control Mean	Subject Mean	Difference	t	p
Composite	14.23	15.48	1.25	1.08	n.s.
English	14.32	13.58	−0.74	−0.60	n.s.
Mathematics	12.00	16.00	4.00	2.94	<.005
Social Studies	13.32	14.45	1.13	0.73	n.s.
Natural Sciences	16.73	17.26	0.53	0.34	n.s.

The absence of a positive difference for the subject group in English may also have been influenced by the remedial program, available to and required of all freshmen on the basis of need, which incorporates the experience from the summer program described above.

An attempted measurement of development from concrete operational toward formal operational cannot be considered definitive because the pretest and posttest were of somewhat different form and were only indirectly calibrated against each other. Both were paper and pencil tests. The pretest included a brief interview at the conclusion of the test, primarily to clarify questions and answers. The posttest was designed by Sills. A majority of the students showed a gain of at least one step (concrete to transitional or transitional to formal), with an average gain of 0.65 step (Table 5).

Table 5. Test of Piagetian Level*

Average change: +0.65	Time Interval: 7 months
Pretest: 99 students	Posttest Differences: 32 students

Formal: 1 (+ ?)
 	
Transitional:
  Upper: 0
  Lower: 17 (+5?)
 	
Concrete:
  Upper: 23 (+ 7?)
  Middle: 32 (+ 4?)
  Lower: 8 (+ 1??)
	Up two levels
	(Concrete to Formal): 3
	Up one level
	(C to T or T to F): 15
	No change in level: 12 (+ 1 originally formal)
	Down one level
	(lower trans. to C): 1

*Numbers in parentheses indicate probable but uncertain classifications

As grades are the accepted measure of performance in the academic environment, we looked at the mean grade-point average for the PHS 7-8-9 students for the winter quarter, for all courses taken. Of the 96 students entering the program for whom grade-point averages were available for the winter quarter, seven had not actively participated at all. These students had a mean GPA of 0.53 (on 3.0 scale; Table 6), compared with 1.33 for the 89 participants. Among the participants, 28 had dropped out before the end of the winter quarter and 61 had completed both fall and winter quarters in the course. The former group had a mean GPA of 0.98, the latter a mean GPA of 1.49 for all winter quarter courses. Comparison of pretest scores revealed no significant differences among these groups of students.

The intent of these courses was to seek an answer to the question: Can we change the level of intellectual functioning of a group of students by classroom teaching? We believe we have obtained an affirmative answer to that question.

We have specifically *not* sought to answer the question of how changes in student intellectual development can be produced most efficiently. That would require different techniques, including different types of control groups. Nevertheless it may be helpful to speculate, on the basis of our observations, as to which are the critical elements, many of which are shared by other programs with similar goals.

Table 6. Grade-point Averages (All Courses) in Winter Quarter for PHS 7-8-9 Enrollees

	GPA (3.0 Scale)
All students enrolled in program (96)	1.27
Students who did not participate (7)	0.53
Students who did participate (89)	1.33
Students active through winter quarter (61)	1.49
Students who participated but dropped out (28)	0.98

Four elements appear to be particularly important. One of these is teaching the basic techniques of problem solving. Although this takes relatively little time, it influences subsequent learning experiences. A second critical aspect is strong intellectual involvement of the student in the learning process. The student is brought to a confrontation with a reasonably well-defined problem and thus is led to discover solutions to those problems. The third device that seems to be particularly effective is asking hard questions about easy material. When the student cannot dismiss the subject as too difficult, there is strong incentive to solve the problem. Finally, the motivational aspect of achieving success on every problem of every unit appears to be of great importance.

The discovery in recent years that conventional college curricula do little to improve the logical thought processes of most students has been disconcerting. It is no comfort to know that we have poured lots of facts into lots of students, in view of the well-known rapid loss of information from memory. We believe that the key to better teaching is to make students better learners. The work described here contributes toward understanding how to achieve that goal.

Tribbles, Truth and Teaching: An Approach to Instruction In the Scientific Method

Ruth Von Blum

One of the major objectives in teaching introductory college biology is to give students a working understanding of that particular approach to the world called the "scientific method." Unfortunately, most current efforts to teach students to approach problems scientifically have not been very successful. This paper discusses the design and evaluation of one instructional unit produced by Project SABLE[1] specifically to teach the basic skills involved in the scientific method.

SOME PROBLEMS IN TEACHING SCIENTIFIC REASONING

There are various ways students are taught scientific reasoning in the context of a biology course. Usually a ritual incantation of the "scientific method" in an introductory lecture is the major formal introduction to scientific methodology given to students; it is seldom included in the presentation of the myriad "scientific facts" of the discipline. Even the laboratory, which might be an ideal place for students to get first-hand experience in science, most often is concerned with simple observation of phenomena. Experiments are most often only observations, with the instructor, not the student, determining both the nature of the investigation and the conditions under which it is run. Indeed, many biology students do not begin to have a functional understanding of science until they are well into graduate school, as proven by the difficulty they have in designing and carrying out a reasonable research problem.

One approach to this problem was the initiation across the country of "investigative laboratories," following the suggestions of the Commission on Undergraduate Education in the Biological Sciences (Holt et al., 1969). In investigative laboratories, students spend a considerable part of an academic term conducting experiments of their own choosing. The tacit hypothesis behind the concept of the investigative laboratory is that, if students have received a firm conceptual background in a particular area of biology, and have learned the necessary experimental techniques and procedures, then they will be adequately prepared to undertake investigations on their own. The process of conducting the investigations will in turn effectively teach them the scientific method.

---

1. Project SABLE (Systematic Approaches to Biological Laboratory Explorations), a curriculum development project at the Lawrence Hall of Science, University of California, Berkeley, is supported in part by grants from the National Science Foundation.

There have been few reported efforts critically evaluating the effectiveness of investigative laboratories in teaching scientific reasoning, but our experience at the University of California indicates that they do not succeed in meeting their primary objective. For the past eight years students have been conducting independent investigations in Biology 1, the large course for majors which includes 600 students per quarter. We found during that time that, while students enjoyed the experience, there was evidence that they are not becoming significantly more knowledgeable about the methodology of science. An examination of the final reports handed in by students at the completion of their independent investigations, for example, revealed that a significant number neither stated nor implied the hypothesis that the experiment was designed to address. Even if an hypothesis were explicitly stated, the experiment performed often bore little relationship to it, and the results did not relate back to it. When probed, students could often state the steps in the scientific method, but give little evidence of an ability to follow these steps on their own when faced with a new problem.

MODEL FOR AN ALTERNATIVE APPROACH TO TEACHING SCIENTIFIC REASONING

This led us at Project SABLE to hypothesize that if we could explicitly teach the skills involved in the scientific method (observation, hypothesis formulation, and testing), and if mastery of these skills were coupled with an understanding of basic facts and knowledge of laboratory procedures, then students would be able to use scientific reasoning when approaching problems.

An early step in our developmental efforts was the formalization of a specific qualitative model for the scientific process that could be taught explicitly. It was our intention to make students aware of both the pervasive utility of such a general strategy, and of the techniques necessary for applying the strategy specifically in the chosen subject area. After much discussion, experimentation, and revision we derived a general model for the scientific process (Figure 1).

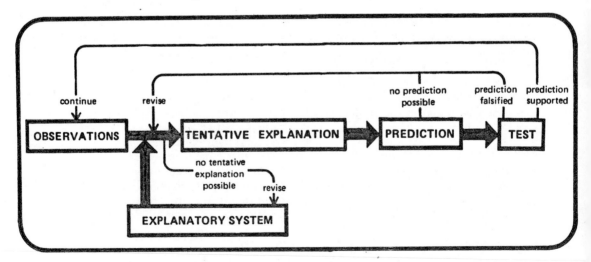

Fig. 1. Model for an Alternative Approach to Teaching Scientific Reasoning

Briefly, *observation* of natural phenomena forms the basis for all explanation. The current *explanatory system* is a formalization of pertinent, tested principles that have in the past been useful in explaining aspects of the phenomena under consideration. Using the principles and concepts in the current explanatory system, we can form a *tentative explanation* of some specific set of observations. This tentative explanation (hypothesis) is then used to form a *specific prediction* (or set of predictions) that can, in turn, be *tested* and evaluated against the initial observations. In addition to the usual scientific objective of forming testable hypotheses, one of the most significant elements of this model is that all explanations are formed from a set of concepts and principles that are explicitly stated, rather than merely implied.

We decided to introduce this model in an initial brief exercise and reinforce it by means of additional materials. The first unit of instruction would not require any previous knowledge in the subject so that the essential elements of a systematic, scientific approach could be the primary focus. The second unit would reinforce the teaching of the basic model of scientific process while building up increasingly more powerful explanations for specific biological phenomena in specific subject areas. We therefore produced one short (one to three hour) introductory exercise on scientific reasoning per se, and another considerably longer (15 to 20 hours) exercise in basic genetics. The genetics unit is described in detail in Von Blum and Hursh, 1977. Units in ecology, population genetics, and evolution are also currently being developed.

All units consist of written self-instructional texts called tutorials and computer simulations. The self-instructional written texts have the following advantages: *1.* they provide immediate feedback to the student through frequent questioning with answers supplied in the back of the text, thus helping to assure understanding of material presented, *2.* they accommodate a variety of skill levels and prior experience by means of branching, *3.* they are portable and provide a permanent record, and *4.* they are relatively inexpensive.

The computer simulation serves primarily to provide students with practice in using scientific reasoning by allowing them to test predictions generated from their tentative explanations. Two major attributes of computer simulations make them ideal for this function. First, because of the shortened time required to produce data, one experiment can be run repeatedly and many alternative explanations can be tested to determine which ones have the most predictive value. The computer also permits the student to experiment within a well-defined system corresponding to the explicitly stated principles of the current explanatory system, and free from the usual experimental variables that often obscure the results of testing on living systems.

THE INTRODUCTORY UNIT

Designing the introductory unit on the scientific method posed special problems. We wanted to produce a unit that *1.* required students to practice actually using the model of scientific process we were teaching, *2.* did not assume any prerequisite knowledge of biological concepts, *3.* could be completed in a few hours, and *4.* was enjoyable. The unit we produced, entitled "The Truth About Tribbles," was designed to meet all of these criteria.

Like other SABLE units, this exercise consists of a written tutorial and a computer simulation. The tutorial presents students with a problem, and guides them to its solution. The computer simulation simply provides the data for making observations and for forming tentative explanations and testing predictions. To eliminate the variable of background knowledge, the problem takes place on an alien planet, and the students observe alien organisms called *tribbles.*

The students are told that they are in a spaceship orbiting the planet, and that their camera probe takes photographs each day of the tribbles' location on the planet's surface. Any change in location of tribbles from day to day can be easily noted because a grid-like pattern of canals etch the surface of the planet. The computer simulates the series of photographs that follow any initial pattern of tribbles the student entered onto the terminal. Figure 2 shows the sequence of grids that are generated from the initial pattern of tribbles at C3, C4, D2, D3, E3.

On day 1 the tribbles look like this: On day 2 the tribbles look like this:

Figure 2. Initial Pattern of Tribbles

The problem presented to the student is in many ways similar to problems faced by biologists as they search for principles or relationships to explain shifts in patterns of distribution of organisms on earth. Often these shifts are due to a few simple rules with exceedingly complex interaction.

The problem presented in The Truth About Tribbles is sufficiently difficult that it is virtually impossible to determine the underlying rules governing the appearance (birth) and disappearance (death) of tribbles from day to day without the application of a systematic scientific strategy.[2] By using such a strategy, however, the complex behavior of the patterns of tribbles is reducible to a set of simple rules.

Students are guided in making detailed and systematic observations of tribbles on the computer terminal to determine whether shifts in number and position are random or follow set patterns. Next, the tutorial introduces the idea of an explanatory system. This involves making explicit the set of assumptions that influence the way the phenomenon is observed. In this case, tribbles are assumed to be like earthly organisms, but since the "organisms" are being observed from a distance, all one can do is to concentrate on certain observable aspects of tribble population.

Students are then guided in the construction of a tentative explanation for tribble growth. The tutorial provides some hints for this process: *1.* the tentative explanation should be simple and reasonable; *2.* it should lead to specific predictions that can be tested; and *3.* it should be consistent with other observations. Students are encouraged to explain simple observations first, the

2. The actual rules governing birth and death are those in John Conway's *Game of Life* (Gardiner, 1970).

procedure commonly used in science. They construct an explanation for the behavior of just one or two tribbles alone on the grid, and they test their predictions by going to the computer simulations. Their first explanation, however, quickly proves inadequate, and they must revise their explanation.

As this proves difficult, the tutorial suggests employing another common scientific procedure—breaking the problem down into its parts and seeing if progress can be made on one aspect of the problem at a time. Students first explore the rule governing death of tribbles, ignoring all births. They find that tribbles with fewer than two or more than four neighbors die. Students then undergo a similar procedure to discover an explanation for birth, and find that if an empty square is touched by exactly three neighboring squares containing tribbles, then a tribble will be born in that square.

The explanation for birth, when coupled with the one for death, predicts all of the observed complex behavior of the tribble population. In addition, it has all of the qualities of a good scientific explanation: it is simple and reasonable, it leads to specific predictions, and it is generalized to account for all observations. The student then goes on to test predictions made from this explanation against the data provided by the computer simulation. Finally, the tutorial summarizes the entire procedure that he has used, and emphasizes the relationship between biological "facts" and scientific explanation.

EVALUATION

At each step in the development of The Truth About Tribbles, trial editions were given to small numbers of students, and their comments were incorporated into subsequent versions. A summative evaluation of the unit was undertaken at American River College in Sacramento, California. The students in an introductory course for biology majors were divided into two groups, with 34 students in an experimental group working through the Tribbles unit and 28 students in a control group working through a laboratory exercise on the scientific method involving a safety pin puzzle (Riggs, 1974). All students attended the same lecture on the scientific method given by the course instructor. Evaluation consisted of personal interviews with students, an attitude questionnaire, an analysis of student performance on written questions constructed by the project staff including a set of standardized questions and a problem involving determination of the underlying rules governing the complex movement of billiard balls in a fictitious universe presented via a film loop.

The interviews and attitude questionnaire confirmed that students enjoyed the unit a great deal and could use both the written tutorial and computer simulation with little difficulty. They found the level of difficulty of the written materials appropriate, and spent from one to five hours on the unit, with an average time of 2.75 hours. For 51 percent of the students, this unit represented their first introduction to the computer. Almost all of the students recommended that the unit be kept as a part of the course, and felt that they had gained some knowledge or skills in the exercise that would be valuable to them.

There was no significant difference in pre- and posttests between experimental and control groups on the written problems constructed by the project staff; students in both groups did well. This indicates that, at some level, students in both groups had learned to *identify* important steps in the scientific method before any formal instruction was given. Students in the experimental group performed better on the standardized set of test questions (Burmester, 1953) than did the control group, but the difference was not significant. These questions emphasized the students' ability to identify statements as inference, conclusion, hypothesis, etc.

The most difficult problem presented to the students was the billiard ball problem. The skills necessary to solve this problem are the most similar to those taught in the Tribbles unit, but the underlying rules are so different that this problem can be taken as a good test for transfer. The balls follow a few simple rules, but as with Tribbles, determination of these rules requires a systematic strategy. Students must make careful observations, must break the problem down and work on it in parts, and must test out various hypotheses. Figure 3 shows the sequence of events presented in the film loop.

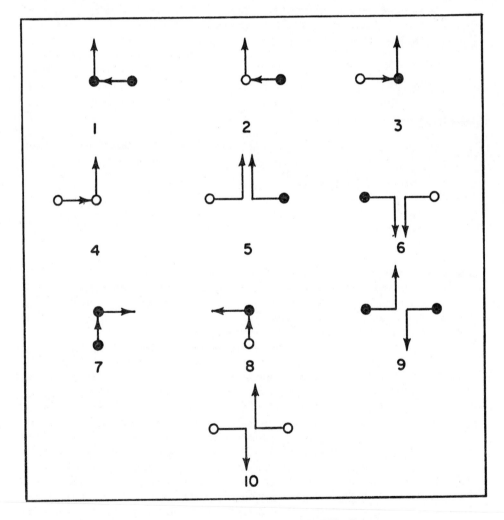

Figure 3. Events Presented in the Film Loop

After completing instruction in the scientific method, students are asked to observe the loop and protocols are collected. From analysis of the data, the SABLE staff could determine whether or not the students were able to provide an adequate explanation, to some extent the approach they used (for example, whether or not they took careful notes), and whether or not they gave evidence of checking their results against the observations.

The distribution of scores shown in Table 1 demonstrates a significant difference between the experimental and control group's performance on this task.

Table 1. Results of Billard Ball Problem (Percent)

	No Explanation Given	Restated Observations, No Generalizations Attempted	Formulated a Set of Laws, But Not Able to Account for All Observations	Formulated a Reduced Set of Laws, But Not the Best Set	Formulated the Best Set of Laws
Experimental n=34	9	21	40	15	15
Control n=28	50:	14	28	0	7

These results indicate that after working through the unit, students are able to approach the solution to a difficult and perhaps intimidating problem, whereas without such training they made far less progress. Thirty percent of the students in the experimental group came up with a reasonable to perfectly adequate solution to the problem, compared to seven percent of the control group. Indeed, a significant number of students in the control group (50 percent) made no progress at all towards the solution of the problem.

We have evidence from our evaluation that our introductory unit on the scientific method is well received by the students, can be completed in a reasonable length of time, and seems to help students approach the solution of a difficult transfer problem. We are working towards the development of more refined evaluative techniques that will elicit and unambiguously measure overt behavior indicating that students have engaged in the model scientific thinking we have taught. These techniques will evaluate this behavior by comparing it with an explicit description of expected behavior. We are looking at some recent work in concept learning, task analysis, and machine intelligence to help provide both the theoretical framework and the practical instruments for the development of such evaluation procedures.

CONCLUSION

In the design of SABLE materials we have attempted to teach a specific model of the scientific process. This instruction, delivered in the form of a written tutorial and a computer simulation, has introduced students to the scientific method by guiding them in the solution of a difficult problem. Even such a brief introduction has a measurable effect on students' performance on a transfer problem, indicating that students are using the model. If such an introductory exercise is followed by materials reinforcing the basic model of scientific process and teaching biological concepts not as facts, but as explanations for observations, there is hope that our objective of teaching the scientific method can be realized.

REFERENCES

Burmester, M.A. "The Construction and Validation of a Test to Measure Some of the Inductive Aspects of Scientific Thinking." *Science Education,* 37(2), 1953.

Gardiner, M. "The Fantastic Combinations of John Conway's New Solitaire Game 'Life.'" *Scientific American,* 223(4), pp 120-123, 1970.

Holt, C.E. Abramoff, P., Wilcox, R.V., Jr., and Abell, D.L. "Investigative Laboratory Programs in Biology." *BioScience,* 19(12), p 1104, 1969.

Riggs, J. "Safety Pins Used in Hypothesis Formation and Testing." *American Biology Teacher,* 36(8), pp 461-463, 1974.

Von Blum, R. and Hursh, T.M. "Mastering Genetics, With a Little Help from GENIE." *American Biology Teacher,* 39(8), pp 468-472, 1977.

A More Personal Approach to Teaching a Technical Course

R. E. Sparks

The lecture system is bad. We all agree to that. But I have not seen it change in my years of educational experience, except through the addition of practical experience in a laboratory and problem-solving recitation sessions. Of course, the lecture system is very difficult to get away from in technical courses because of the enormous amount of factual information and analytical techniques involved. Most of us teach such courses as if it were not possible to think until all these facts and techniques were mastered. Our whole course tends to be devoted to these and we end the course hoping the student will somehow to able to think properly about all we have given him.

After such a sweeping condemnation of the lecture system, I would like to be specific about its disadvantages, because this will help us be specific about the changes we ought to make. If we stay on the sweeping condemnation level, we will only be able to wring our hands in despair.

Any of us who has spent much time talking to students know that the poorest way to help someone learn something is to tell him. And, of course, there is a good reason for this—his mind is likely to be passive. For most of us, listening is usually an inert activity, conducive to daydreaming and general mental laziness, unless the subject is inordinately interesting. During a lecture the only mind working at reasonable efficiency is that of the lecturer. The only time such a situation can be highly educational is when the minds of the audience are also working. This will occur when the audience has strong interest in the subject matter. A good example is a seminar being given by a technical man to a group of his peers. Somehow we must arrange for the students' minds to be working too.

Another problem is that it is well-nigh impossible to have a good, inclusive discussion in a class of 30 students. If the instructor asks questions, a few of the sharper and more vocal students may start a discussion but the rest of the students are quite willing to let these few carry the burden. They feel little need or desire to participate in the discussion.

Very few students in a class of 30 will take time to ask a question that is troubling them unless they know it is troubling a great many other people too. And yet, what would be the most valuable input that an instructor could give a student? Obviously to answer his particular questions or to address the subjects that particularly interest him at the time. What is most relevant to the student is what is on his mind. No lecture, then, could possibly be very relevant to many of the students in a class.

The instructor can only lecture at a pace and level which, in his judgement, will get through to most of the students. But this means that many of the sharper students will be bored by the pace and level. It also means that some of the slower students will not be able to understand well enough to benefit.

AN ATTEMPT

I would now like to describe how I tried to meet these goals in my junior-level chemical engineering course on Heat, Mass and Momentum Transfer. Clearly this is a course loaded with sophisticated technical content. I began by asking the class what they would like to do, but they had little experience with anything other than the lecture system and there were few original ideas. Some useful comments were made about such things as the use of homework and the desire to do almost anything different.

I could not have let the class go completely on its own, for then I would have been shirking my faculty responsibility. In addition, the few such experiments by my fellow faculty members had shown results varying from disappointing through disastrous. I could have divided the class into groups around tables, letting the students help each other, occasionally coming by to offer encouragement or straighten out a difficulty. Again, the experiences of some of my fellow faculty showed that this was not ideal, that the groups tended to be inefficient and frequently floundered when the professor was not available. Even the best groups can do little more than explore the surface of the material. No students had sufficient experience or perspective to recognize good questions. Another practical disadvantage is that very few students have the ability to pace themselves properly, particularly when several other courses are making rigid demands on their time. Finally, this was an engineering course containing a great deal of material in which they had to be proficient by the end of the semester. Only someone with a broad idea of their possible future professional duties would be able to set a suitable pace or to set priorities in the subject matter. Hence I decided that one of my arbitrary decisions had to be what the subject matter would be and the pace at which it would be covered.

We were fortunate to have available a fairly good textbook and we decided to let the textbook be the main teacher of facts, basic principles and calculation techniques. The students would have to read the text critically and use it as the source of raw materials for our thinking, since they would not get it all from my lectures.

My prime informational input was two lectures each week to present relevant material which was not presented in the text, or to place in heightened perspective some portion of the subject matter. I did not lecture on any material presented with reasonable adequacy in the text. This freed me from the time-consuming and inefficient task of organizing all the subject matter for presentation in a complete set of lectures. As a result I became free to think about the subject matter and to think about how best to lead the students to understand and think about it. This eliminated a great deal of the tedium from the course for me and, I hope, for the students also. The remainder of the class work was done in groups of five to seven students. There were 36 students in the class so we divided into six groups. Each group met with me at an assigned time once each week. Since I lectured on Monday and Wednesday, I arranged their schedules so they could meet with me on Thursday or Friday. This gave them a few days to digest the lectures as well as to complete the homework assignment for the week.

Our meetings were held in a conference room furnished with a coffee urn. The main aim of these groups was to work toward a familiarity, or "gut feeling," for what was really in the subject matter. It didn't take long for the students to see that there was a great difference between understanding the subject matter and exhibiting a surface facility in making the calculations required for the homework. In these small groups we were striving almost entirely for the former. I felt that if the proper images and understanding of the physics of the processes could be developed, the students would be able to think intelligently about the subject long after they forgot the equations, the professor, and the homework problems.

In the small groups we tried to construct a framework for the subject, even at the expense of some content. This is preferable to throwing a lot of content onto a very flimsy framework. Putting it another way, we established a basis from which the students could learn.

One difficulty with asking questions in a large class is that the questions catch the students cold. Homework helps prevent this from occurring. One of the purposes of the homework is to give everybody a common ground of thought from which to start. In the normal problem only a few minutes are taken to discuss calculational techniques, normal sources of error, and the meaning of solution. In many classes this would be considered sufficient treatment of the homework. However, having gotten the majority of the students to think seriously about a particular problem with its set of limitations, we started a critical and extended discussion of the assumptions built into the problem, reasons for the assumptions, and alternative ways of working the problem. From there we moved into a general discussion of the basic concept behind the problem and what its value might be.

As an example, consider one problem where the students calculated the effect of a threefold increase in molecular weight on the liquid-phase diffusion coefficient of a solute. A sufficient answer to this problem can be obtained using the normal liquid-phase diffusion correlations with the proper correction factors and constants. However, having made this correct calculation, the student has learned very little. At this point in our discussion we began to look carefully at the diffusion correlations and at what were the important variables in determining the diffusion coefficient. The students saw the temperature written explicitly in the equation, and it was obvious from the form of the equation that the diffusion coefficient is a stronger function of temperature than it is of solute molecular weight. However, after sufficient discussion of the effects of temperature on the correlation, some of the students noted that the diffusion coefficient is inversely proportional to the viscosity, and recalled correctly from the facts in the book that the viscosity of a liquid is a strong function of temperature. Upon looking back at this functional relationship they began to see that the viscosity is possibly the most important parameter in the correlation since it can be varied greatly in many cases. We then discussed techniques for varying the viscosity, the obvious one being the change in temperature which led us to this part of the problem in the first place. However, after sufficient discussion the group came upon the idea of the addition of small amounts of a second substance to change the viscosity of the basic substance. For example, a small amount of the low viscosity solvent will greatly reduce the viscosity of a highly viscous material. In contrast, the addition of a small amount of gelling agent or high-molecular weight thickener will drastically raise the viscosity of a low-viscosity material.

Particularly for chemical engineers, this is a critical stage, because they have gone a step beyond determining a mass transfer coefficient. That step, so important to the engineer, is the recognition of the important variables and how one controls them. If you were to ask nearly any of these students in what major way one can control mass transfer, I am sure most of them would answer "by controlling the viscosity of the field fluid." This type of understanding is not given in the text, but is basic to understanding the nature of mass transfer, its interrelation to the other transport phenomena, and the engineering use and control of mass transfer processes.

It's obvious from the above discussion that one of the main functions of the homework was simply to prepare ground from which we could launch a discussion of the important points in the material. From this point of view, hard work on the homework problem is much more important than obtaining the correct answer. To encourage the student to work on the problems rather than simply turn in an answer, we removed the grade credit from the homework assignments. When the homework came in, the only recording was a check signifying that the student had made a reasonable effort to solve the problems. In order to give him ample time to do the homework and to allow him to fit the homework into his schedule, homework assignments were made about five days before they were due.

It is up to the student to check the details of his work; we did not waste time in the small groups or in the lecture going through the detailed solution of any homework problem. Taking care of this part of the course was the main function of the graduate assistant. His main job was to do the homework problems conscientiously, in great detail, and sufficiently ahead of time so that copies of his solutions could be passed out to the students in their small groups. In the small groups we had three sets of solutions to compare: those of the graduate student, the instructor, and the students.

SOME OBSERVATIONS AND EVALUATION

This proved to be an unusually personal way of conducting a technical class. By the second meeting I knew almost every student by his first name. What a difference it makes to talk to only six people! In a large lecture it is difficult to have eye contact with, or to talk to, a single student for more than a few seconds at a time. In a group of six, we can talk to each other person-to-person. With that kind of conversation going on, nobody sleeps. I don't think anybody even daydreams very long. If a student has a small question, which would never come out in lecture, it's likely to come out in a group of six or seven people where the conversation is informal and his instructor is sipping coffee and has his feet on a chair.

A word of warning: any kind of put-down in such a small group should be absolutely forbidden. In such close mental quarters, a put-down would be a deadly weapon and could turn the small groups into torture sessions for the students instead of what they should be—experiences in learning to think about the material, and to ask questions which will lead them into new insights into the material.

At the end of the fifth week I asked the students to take five minutes to write down opinions on how the course was going, what they thought of various aspects of it, and how it would be possible to improve the course. The opinions were of course unsigned. Out of 29 responses only two indicated that we should have more lectures. The rest of the general comments about the nature and structure of the course varied from mildly positive to enthusiastic. Seven people said words to the effect that "the course is set up well and is working very well." Somewhat more informative comments were the following: "If we had met for the normal lectures, I wouldn't have learned anything;" "I am enjoying the course as a good educational experience—this is probably due to the course structure and attitude, because many times the material is somewhat dry;" "I like the stressing of concepts behind the equations;" "I like getting away from 'these equations work and these don't.'" There were also some highly positive responses such as "This is the most enjoyable and stimulating technical course I've taken;" "If we are covering all that is essential, then this is by far the best-taught course at the university." However, my favorite comment was made by one of our good students to his faculty advisor about two weeks after the course started. He said he thought the course was "the greatest thing since peanut butter." A few comments such as these give one a great deal of encouragement to keep trying.

Most of the students commented about the small groups. All but one were positive, and that one was tolerant. Many simply said they liked the idea of the small groups. Somehow, the message was getting through to the students, as judged by the following comments: "The small groups are good because one has to think;" "The small groups offer much communication;" "During small groups it is much easier to understand what is going on, and to learn something;" "The small groups are very good when one is prepared, but the motivation instilled in the small groups doesn't carry over to the next time;" "They are more effective than any lecture to the entire group could be;" "I really enjoy the small session and seem to be getting more information and a better feel for what's happening from them, than from lecturing or reading," "The idea of the small groups is excellent; interaction at the group level is far superior to that in the lecture system," "I like the idea that you talk about problems which we have worked on for awhile. That helps to make the time spent worthwhile." "I feel the small group sessions are very worthwhile, especially since although the homework is used as a basis to get the discussion started, the point of these meetings is not just to see if one was able to grind out the right answer or not. Instead, the significance of the problem or principle is being considered which I believe is much more interesting and worthwhile." Four people thought the small group sessions were too short, and one person brought me up short with the response, "I don't seem to be able to ask too many questions. You seem to be directing the class."

The comments regarding homework procedure varied from "good" to one student who thought it was better handled than in any other course he had taken. One student simply wanted to spend more time discussing the homework problems.

Now for a few of my personal observations. Having been freed of the task of organizing a set of inclusive lectures on the subject matter, I found that I could spend my preparation time thinking about what was really important in the material and how, through questions and comments and a general discussion, to get the students to gradually see and understand the basic phenomena we were studying. This was an enjoyable experience for me. In addition, in working through the homework problems, making several different sets of assumptions and looking at the homework as something to learn from rather than simply problems to do, I found out that I was learning also. As a matter of fact, most of the concepts which we have discussed in detail in this course were not so sharply etched on my own mind before as they are now. By way of a practical comment, I should mention that we were using a new textbook, so all homework problems had to be worked out from scratch. If I subtracted a fair fraction of this extra time spent on the homework, then the time required for the course would be 12 to 14 hours per week. This is approximately the time I had spent on the course when I taught it as a normal lecture course.

My judgment of how the course has gone is affected by many things, but is determined primarily by what I see in the eyes of my students, and what I read on their faces as I talk with them. If I permit myself some loose judgments, I would say that previously in a good lecture I would have perhaps ten to twenty percent of the students really thinking with me. By contrast, in my small groups, I would guess that 70 to 80 percent of these students are seriously trying to turn their minds over with us and grapple with the problems of understanding the material. I have begun to think of teaching now as leading people to see, and helping them to learn how to lead themselves to see.

I have always considered myself a reasonably good lecturer, and have always told people, in truth, that I liked teaching. However, for the first time since I joined the university I am thoroughly *enjoying* teaching a course. In fact, I can hardly wait to get to school on mornings when I have my small groups. Another response which has been growing, but has only been recognized recently in some surprise and disbelief is the feeling "I am a teacher." It's an exhilarating feeling.

Mathematics and Learning: Roots of Epistemological Status

James J. Kaput

INTRODUCTION

This paper is concerned with certain vital aspects of the platonism-constructivism issue and how they are reflected in our everyday work of teaching and learning. It is suggested that we are confronted with a pair of universes: one being the stark, atemporal, formal universe of ideal knowledge; the other being the organic, interior, processual universe of human knowing. The former, Plato's, has monopolized status and power, and is responsible for the fundamental dominance of product over process, the values by which legitimacy of knowledge is conferred, and the rules, particularly linguistic rules, under which any inquiry occurs. This paper illustrates how the manifold consequences of this status dominance permeate our academic lives at all levels, concentrating mainly on its exclusionary function in mathematics and mathematics knowing/learning.

The illustrations include discussion of the following phenomena:

1. The symbol system of mathematics denies the reality or importance of the knower/learner. Our use of mathematical equality systematically denies the process/product distinction, a distinction that is fundamental and real in the universe of human knowing. It also denies the various and distinct heuristic/linguistic functions of equality.

2. The processual meaning of mathematical operations is achieved through an essential, yet covert and unacknowledged, act of anthropomorphism, a projection from our internal cognitive experience onto the timeless, abstract-structural mathematical operations. (This includes, for example, operations in arithmetic, algebra and calculus.) In some respects this anthropomorphism acts as a metaphorical "structure preserving mapping," a morphism.

3. The basic, irreducible and essential metaphoric nature of human thinking has only an accidental, unacknowledged, and denigrated role in mathematics. As it is in any circumstance, the metaphorizing process in mathematics is our primary means for creating and, especially, transferring meaning from one universe to the other. However in mathematics this process is forced into a smuggling and bootlegging role, and never acknowledged for its crucial function. For example, virtually all of basic calculus (the study of change) achieves its primary meaning through *an absolutely essential collection of motion metaphors.* These metaphors control the notation. Hence we write limit statements using arrows and use image-laden words such as "diverge," "converge," "increasing," "constant," and "transform." However, the formal mathematical definitions associated with these notations, being atemporal, are not connected to motion.

In all three illustrations we see the active human contribution to knowing denied. The notion of "perfect knowledge" apart from such essential human activities as anthropomorphizing and metaphorizing (to say nothing of the twin operations of generalization and instantiation, which are not discussed in this paper) is fundamentally dishonest and destructive.

Our failure to acknowledge the acts of knowing and learning is analogous to the Victorian attitude toward sex. One cannot develop mathematical conceptions without engaging in the torrid act of learning. There is no such thing as immaculate conception!

In teaching, especially in the teaching of those who will not become mathematicians, this Victorianism becomes fundamentally unhealthy and, ultimately, destructive. In curriculum design and in teaching at all levels we witness a prudish attention to the knowledge *products* that leaves the underlying *processes* discreetly behind closed doors, presumably to happen spontaneously and naturally. However, mathematical knowing does *not* happen spontaneously and naturally, at least not the sustained and sophisticated knowing we expect from today's students. If necessary, we should remind ourselves of *1.* how few civilizations on earth developed mathematical cultures, *2.* how few of these— namely one—used the logico-deductive mode for organizing this culture, *3.* how many centuries were needed to develop our current culture, and *4.* the dimensions of the intellects responsible for most of its development.

The aim of this paper is to redefine, or at least enlarge our view of, the transactional process between the subject and the knower in such a way as to give genuine epistemological status to certain absolutely essential aspects of this process, aspects that currently are denied as real, important, or even respectable—particularly when held up against the ideal of "perfect understanding of pure mathematics." The paper closes with a brief discussion of the role of mathematics in describing human knowing and the problems associated with a constructivism-structuralism that takes its mathematical models too literally— or at least too uncritically.

MULTIPLE USES OF EQUALITY

My discussion of where and how the universe of human knowing and the universe of formal knowledge fail to fit will turn first to where these universes necessarily must come into contact—in symbolism, its function and meaning. In this section I will illustrate how the symbol system of mathematics misrepresents, and especially, underrepresents, the cognitive processes related to its learning and use.

The Process-Product Disacknowledgement

It is no secret (except perhaps to students) that we use the mathematical equal sign "=" in many different ways. Nonetheless, we are told, the equality can be reduced eventually to equality of sets. Furthermore, mathematics states absolutely and unequivocally that equality is symmetric: a = b if and only if b = a.

Let us now consider the equation

$$2 + 3 = 5$$

The left side is inevitably read as an operation, a process, something we do. The right side is the result, a number. Wittgenstein correctly stated that in mathematics the process and the result are the same. But for human beings they are *never* the same! Hence the traditional mathematical representation of a process and its result can never reflect, account for, or accurately represent the cognitive activity associated with it. An equation such as the above is inevitably telling a kind of lie, a half truth. The equation

$$2 + 3 = 3 + 2$$

tells a slightly different sort of lie. Cognitively, $2 + 3$ and $3 + 2$ are two different activites. But the equation ignores this while at the same time agreeing, in fact asserting, that the result of each process is the same. A more serious problem occurs with unary operations: $\sqrt{9}$ is both a process *and* a product, a fact students find deeply confusing in the case of, say $\sqrt{2}$. The confusion is inevitable in this case because there is no equation to help separate the two. In fact, we should point out that if we use a slightly different symbol which stands for "gives," or some such phrase, then the "equation" accurately represents what is occurring on the human end: $2 + 3 \Rightarrow 5$.

Similar representation problems occur when we write fractions: $5/2$ is either a division statement or a number. As usual, the *mathematical* distinction is of no consequence. Writing $5/2 = 2\frac{1}{2}$ we have either the result of the division process (a quotient of two plus a remainder of one-half—with the plus sign omitted in this case), or two different names for the same number. This problem of the double meaning of fractions confuses students even more in the context of differentials, where dy/dx is either a quotient or a derivative.

These representational difficulties could be shrugged off as unimportant epiphenomena of little relevance were it not for the undeniable fact that such a large proportion of the population fails to learn mathematics.

Let us continue illustrating the uses of the "=" sign. Consider the cognitive meaning of the next two equations.

$$x^2 + 5x + 6 = (x + 2)(x + 3)$$
$$(x + 2)(x + 3) = x^2 + 5x + 6$$

The first represents the factoring process while the second represents the multiplication process. Cognitively, these processes surely are not symmetric, but the mathematical representation does not respect this difference. Similar "directional" uses of equality occur at virtually all levels. Consider

$$D_x(x^3 + 5x) = 3x^2 + 5$$

The reverse process of differentiation yields a further difficulty:

$$\int (3x^2 + 5)dx = x^3 + 5x + C.$$

The left side, when regarded as an operation, is now equal to an infinite class of functions—although the use of C is tricky here: it is a constant, but by convention it is also a variable whose domain is the set of all real numbers. (All through mathematics, context-shifts change constants to variables and vice-versa.) It is usually the case with process-product equations that the cognitive processes involved in the two directions are very different, with one much more difficult than the other.
Consider:

$$\frac{2}{x + 3} + \frac{5}{x - 2} + \frac{3}{x^2 + 1} = \frac{7x^3 + 14x^2 + 10x = 7}{x^4 + x^3 - 5x^2 + x - 6}$$

$$\frac{7x^3 + 14x^2 + 10x - 7}{x^4 + x^3 - 5x^2 + x - 6} = \frac{2}{x + 3} + \frac{5}{x - 2} + \frac{3}{x^2 + 1}$$

The first equation is associated with a basic high school combining-fractions problem, the second with a partial fraction decomposition.

Alterations in Usage and Meaning

Robert B. Davis (1975) points out that we often use algebraic equality in two ways, often in close proximity. In one way it acts as a requirement or condition that an unknown number x must satisfy. This is a *semantic* use, as in the equation

$$\frac{6}{3x + 1} = \frac{2x - 3}{x}$$

On the other hand, in solving this equation, we multiply both sides by $3x + 1$ and write

$$\frac{6}{3x + 1}(3x + 1) = \frac{2x - 3}{x}(3x + 1) = \frac{6x^2 - 7x - 3}{x}$$

The second equality (on the right) is merely a connector, a *syntactic* use of "=" in an identity. (Note also the *indicated* multiplication on the far left and the "real" multiplication on the right.) As anyone who has tried to teach algebraic activity such as the above to a neophyte will attest, the various uses of the same symbol are no trivial matter to be brushed aside. Because they are given no attention in formal mathematical statements, they are given only anecdotal attention in pedagogy—if that.

The Fundamental Anthropomorphism

What *is* the relationship of cognitive activity to the process-product identification in our formal mathematical statements? Formally, the equality of sets behind

$$2 + 3 = 5$$

states (given a collection of assumptions) that the cardinality of the union of a certain disjoint pair of sets equals the cardinality of a certain set. When learning how to add, a child pulls a pair of sets together and counts their elements. Somehow, with repetition over a variety of sets and situations, this process is "internalized." However, the formal structure of that which is internalized turns out to be precisely the formal mathematics mentioned above. (Note: this does not necessarily imply, as many constructivists once suggested, that there is then a formal mathematical structure thus embedded within the child!) The crucial factor to recognize, however, is that the formal mathematical structure itself is timeless, it is merely a structure, and is knowable (other than as a set of marks on paper) *only* by being invested with the meaning projected from our prior knowledge and experience on which this structure is modeled. The abstract mathematical formalism (sets, union, cardinality, etc.) gets its meaning via an *anthropomorphism*. We project or superimpose on the mathematical formalism, our own internal cognitive experience.

Indeed, the mathematical-morphism analogy holds up—at least in this case—by *construction*! We can think of the anthropomorphism as a "structure preserving map" from the universe of human experience into the formal, abstract world of mathematical structure. It is no accident that this map preserves this particular structure—that is, it is no accident that the mathematical structure resembles the structure internalized from experience. However, without the anthropomorphizing projection we could see no meaning in the formalism. It gives virtually all the *operations* of mathematics their initial or primary meaning. After all, as an abstract formal ideal, abstract operations could not be knowable otherwise. By anthropomorphizing, we simultaneously impose on and draw from the mathematics its meaning (this includes the imposition of a time-sequence). Thus, although formal mathematics is atemporal, and cannot recognize a process, we impose our own cognitive processes on it. This is one aspect of what it means to "know" mathematics. On the other hand, when we *use* formal mathematics, we actually use the mathematics enriched with the content from cognitive experience.

Much more could be said in this linguistic framework regarding the "quality" of understanding and the "faithfulness" of the anthropomorphism. The anthropomorphism is weakly acknowledged in symbolism for most finitary operations because of the separation provided by the equal sign. For many operations (recall $\sqrt{2}$, for example) this is not the case and confusion is inevitable. For infinitary operations involving limits, such as infinite series, the anthropomorphism is dysfunctional. The imposition of a time sequence gets in the way of the limit process. The notion of an infinite process—with no last step—taking place in a finite amount of time is experientially incompatible with the limit-definition of infinite series. This is an excellent illustration of a situation requiring an improved experental base for the anthropomorphism so that the infinite process can yield a result in a finite amount of time.

It is perhaps worth pointing out that the anthropomorphism, as a mapping into the universe of formal mathematics, necessarily forgets much of the psychic activity of its domain because the range universe is without elements or notation to receive this activity. For example, it forgets *intention*, so it cannot easily preserve heuristic structure. As a result it does not reflect the heuristically distinct uses of equality in the algebraic equation-solving illustration described earlier.

A closer look at this metaphorical anthropomorphism mapping suggests qualities of the structure-semantics functors from classical category theory, and is tempting to posit an inverse map bearing meaning. Unfortunately, as is often the case when dealing with the cognitive process, there is a reflexivity problem. The anthropomorphism is both a means and an end of our understanding: we project and view in the same mirror simultaneously.

It is also tempting to use the currently popular word "representation" in an effort to clarify this situation. However, for this author this word has a number of overlapping meanings yet to be satisfactorily disentangled. For example, science as a collective public enterprise is said to "represent" reality (in its public record). On the other hand, individuals are said to "represent" reality cognitively, with the word "reality" variously interpreted as the source or result of a private constructive process. Furthermore, cognitive science aims to "represent" (in the former sense) this "representing" (in the latter sense) in some formal way. It takes as its data the incomplete clues provided by combinations of *1.* the close observation of individuals and *2.* the scrutiny of their concrete "representations" of cognitive "representations."

Lastly, certain cognitive scientists, interested in science and mathematics learning, concern themselves with individual "representations" of scientific and mathematical "representations" and ask how these interplay with and build upon naturally—or experientially—based "representations." Since this paper deals with precisely those areas where the usages of "representation" blur into one another, the more idiosyncratic term "anthropomorphism" is used.

Notice that the anthropomorphizing act gives the implication arrows of formal logic their primary meaning. We invest these arrows with our own cognitive experience of logical deduction. However, since the meaning-creating role of the anthropomorphism is not commonly acknowledged, mathematics and philosophy departments around the world attempt to teach "logical thinking" via the manipulation of the formal symbols of statement or predicate calculi. Part of the confusion relates to the persistent metonymical use of key words across the two universes. A clean illustration is provided by the three meanings of the word "sum:" We use, respectively, the verb "sum" and the noun "sum" for the process of adding and then the result of that process—in the human cognitive universe. The third meaning is the formal "sum" of two numbers in the universe of formal mathematics. The first two meanings are seldom distinguished from the third.

Other illustrations are provided by the word "logic" and the particularly troublesome "abstract." The word "abstract" is used in one universe as mathematical abstraction, i.e. formal generalization reducible to set inclusion; it is used in the other universe as cognitive-perceptual abstraction, i.e. a complex human activity with many subtle aspects (kinesthetic, hormonal, neurological, etc.)

SILENT METAPHORS, SECRET METAPHORS

Metaphorizing is a central process by which we transfer meaning. We regard the creation and application of metaphors as an irreducible element of all human thought at all levels.

> That impulse towards formation of metaphors, that fundamental impulse of man, which we cannot reason away for one moment—for thereby we should reason away man himself—is in truth not defeated nor even subdued by the fact that out of its evaporated products, the ideas, a regular and rigid new world has been built as a stronghold for it. (Nietzsche, 1971, p 86)

We will examine a few metaphors that underlie, and in fact are responsible for, our understanding of classical calculus. Since calculus is the study of change, it should come as no surprise that at the roots of some of its fundamental concepts is a collection of motion metaphors. Perhaps the most obvious is the motion metaphor that gives

$$\lim_{x \to \alpha} f(x) = L$$

its primary meaning. "As x moves towards α, f(x) moves towards L". This is frequently written:

$$\text{as } x \to \alpha, f(x) \to L.$$

Nowhere is the schizoid relationship of mathematics to metaphor more blatant than here. On the one hand the role and responsibility of the metaphor are openly acknowledged in the symbols, especially in the use of the arrow, which symbolically carries the meaning of the motion metaphor. It is more important than Cupid's arrow. On the other hand, for very good reasons, generations of brilliant mathematicians struggled to attain logical control over and clarification of the meaning of this metaphor. They finally succeeded in squeezing out the metaphor, leaving us with the familiar (timeless) $\epsilon - \delta$ conditional statement as the ultimate definition. They also left us with all the arrows used to denote it. In view of the importance of motion (both physical and more abstract) in the creation of the calculus, it is less surprising that it took so long to disentangle the logical, formal definition from motion than that it was done at all. In particular, the attempt to wring out from the calculus its motion-content *1.* was historically very irrelevant to the stupendous success of calculus, *2.* was historically very difficult to achieve, and *3.* is today disastrous in the teaching of calculus. (In fact the transformation of calculus in the 19th century was the development of an entirely new subject in the Kuhnian sense.)

When we try to squeeze the motion metaphor from an undergraduate's understanding of limit and replace it with ϵ's and δ's, then, of course, the ϵ's start moving (toward zero, naturally). If we can stop the ϵ's, then the δ's start moving. If finally, through coersion and threats, we are able to stop the ϵ's and δ's, everything stops, especially thinking. The ϵ-δ definition was designed to bar monsters that cannot fit through such an undergraduates' conceptual door.

The nonstandard approach to calculus (which, of course, was the *original* approach) replaces the experientially-based motion metaphor with the experientially-based notion of "closeness." The hyperreal numbers and the idea of "infinitely close" are reflected in a notation which describes more honestly the underlying metaphor that carries the meaning. The apparent success of the nonstandard approach as a conceptual framework for learning calculus relates to the diminished dissonance between the universe of human knowing and the universe of formal mathematics, and the integrity of the notation that enables students to bridge these two universes.

There is a rather dramatic difference between physics and mathematics in the amount of respect each shows for its pre-twentieth century knowledge. In physics, pre-twentieth century knowledge is still taught as valid, although in a limited domain; in mathematics this knowledge is regarded only as "motivation" and is ultimately considered to be *wrong*. This is a consequence of the strict adherence of mathematics to maximal universality as dictated by its internal logic and in its systematic disacknowledgement of any other universe.

Again Nietzsche (1971, p 81):

> Everything which makes man stand out in bold relief against the animal depends on this faculty of volatilizing the concrete metaphors into a schema, and therefore resolving a perception into an idea. For within the range of those schemata something becomes possible that never could succeed under the first perceptual impressions: to build up a pyramidal order with castes and grades, to create a new world of laws, privileges, sub-orders, delimitations, which now stands opposite the other perceptual world of first impressions and assumes the appearance of being the more fixed, general, known, human of the two and therefore the regulating and imperative one. Whereas every metaphor of percep-

tion is individual and without its equal, and therefore knows how to escape all elements to classify it, the great edifice of ideas shows the rigid regularity of a Roman Columbarium and in logic breathes forth the sternness and coolness which we find in mathematics. He who has been breathed upon by this coolness will scarcely believe, that the idea too, bony and hexahedral, and permutable as a die, remains however only as the *residuum of a metaphor* . . .

Formal mathematics says that a function is a purely static matter—a set of ordered pairs. But of course the motion metaphor again provides the meaning and allows us to think of it as a rule for transforming, or even moving, numbers. Even in its quietest state, as a relation between variables, the variables vary— one independently, the other dependently. Furthermore, we use such words as *increasing, decreasing, converge, diverge, bounded* and *unbounded*. We even feel it necessary, in the midst of all this traffic, to proclaim that F(x) = 4 is a "constant function"—"it *takes* all real numbers to 4." Where would we be without that metaphor?

It even seems that in this case at least, the metaphor rather than logic dominates the symbolism. Contrast

$$\lim_{x \to \alpha} f(x) = L, \qquad \lim_{\|P\| \to 0} \left(\frac{\Sigma\, f(x_i)\, \Delta x_i}{P} \right), \qquad \lim_{x \to +\infty} f(x) = L$$

The meanings are very different (although an excursion into topology could connect the first and last) yet the symbol "lim" is the same for all three.

Not all mathematical metaphors are based on motion. When we write

$$\lim_{x \to \alpha} f(x) = +\infty$$

we are using "$+\infty$" as a metaphorical number, and, of all things, "=" as a metaphorical equality! Of course some teachers, rightly wary of the dangers of that metaphor, prefer to be "more mathematical," avoid the itchy word "infinity," and say instead that (the set of ordered pairs) "f grows without bound."

Western culture's tradition of excluding the poet from the laboratory is both strong and long. Note, for example, Bochner (1966, p 19):

> Poetry is also profoundly involved with symbolization, in its distinctiveness perhaps as much as mathematics; but there seems to be a major difference between mathematics and poetry nonetheless. Since the beginning of history, rationality has been urging that science and other knowledge be articulated through mathematics; but Aristotle maintains in his *Poetics*, Chapter 1, sentence 11, that genuine poetry is not the proper medium for a discourse on "prosaic" science. And Aristotle is apparently right; . . .

Traditionally, the inclination in mathematics, science, and especially philosophy, has been to regard analogy and metaphor as a kind of unreliable, inferior knowledge, to be replaced as soon as possible by purer, more substantive knowledge. In the next section we shall examine this conviction in a broader context.

IDEAL VERSUS HUMAN KNOWLEDGE: EPISTEMOLOGICAL STATUS

In the previous three sections we have looked at some of the ways that mathematics and humans interact, pointing first to the bias of the symbol system against acknowledging the reality or importance of cognitive and heuristic functions; second we noted that pure atemporal mathematical operations get their primary meaning via an anthropomorphism based on our internal cognitive experience; third, we illustrated our tacit use of metaphors in giving mathematics its raw, prelogical meanings by which we communicate with, or out of which we construct its architecture, its structure. We now wish to relate these observations to the functional yet impossible ideal of mathematics as pure, abstract knowledge. One expression of this long-standing thematic ideal was provided by Laplace and recently addressed by Polanyi (1975, p 29):

> Laplace's ideal of embodying all knowledge of the universe in an exact topography of all its atoms remains at the heart of the fallacies flowing from science today. Laplace affirmed that if we knew at one moment of time the exact positions and velocities of every particle of matter in the universe, as well as the forces acting between the particles, we could compute the positions and velocities of the same particles at any other date, whether past or future. To a mind thus equipped, all things to come and all things past would stand equally revealed. Such is the complete knowledge of the universe as conceived by Laplace.
>
> This ideal of universal knowledge would have to be transposed into quantum-mechanical terms today, but this is immaterial. The real fault in the kind of universal knowledge defined by Laplace is that it would tell us absolutely nothing that we are interested in. Take any question to which you want to know the answer.

What is the distance between Boston and New York City? What is Boston? What is New York? Without knowledge from experience, the Laplacean knowledge is meaningless. In fact, perfect Laplacean knowledge amounts to perfect ignorance. The brilliant physicist and mathematician Hermann Weyl points to the meaning problem in another way (1973, p 35):

> When Bertrand Russell and others tried to resolve mathematics into pure logic, there was still a remnant of meaning in the form of simple logical concepts; but in the formalism of Hilbert, this remnant disappeared. On the other hand, we need signs, real signs, as written with chalk on the blackboard or with pen on paper. We must understand what it means to place one stroke after the other. It would be putting matters upside down to reduce this naively and grossly understood ordering of signs in space to some purified spatial conception and structure, such as that which is expressed in Euclidean geometry. Rather, we must support ourselves here on the natural understanding in handling things in our natural world around us. Not pure ideas in pure consciousness, but concrete signs lie at the base, signs which are for us recognizable and reproducible despite small variations in detailed execution, signs which by and large we know how to handle.
>
> As scientists, we might be tempted to argue thus: "As we know" the chalk mark on the blackboard consists of molecules, and these are made up of charged and uncharged elementary particles, electrons, neutrons, etc. But when we analyzed what theoretical physics means by such terms, we saw that these physical things dissolved into a symbolism that can be handled according to

some rules. The symbols, however, are in the end again concrete signs, written with chalk on the blackboard. You notice the ridiculous circle.

On the other hand, given the history of science, especially in the last 500 years, it is surely unnecessary to argue the extraordinary potency of formal, abstract mathematics as a *tool* for understanding and dealing with the world. Mathematics serves our intentions individually and collectively at all levels. Minutes ago I witnessed my four-year-old son rank-ordering his chocolate-chip cookies according to the number of chocolate chips contained therein. He then traded the least valuable cookie with his two-year-old brother. (In breaking apart the cookies to discover hidden chocolate chips, the four-year-old also had his first taste of the Heisenberg principle.) Neither is it necessary to argue that mathematics is itself a source of meaning in the understanding or generation of further mathematics. Whether applied to the external world or to itself, the functional dependability and wide utility of mathematics is, of course, based on its independence of empirical phenomena and its independence of the subjective knower. (Presumably, the same is true of any abstraction, as Neitzsche pointed out.) Hence on one hand a human contribution is absolutely necessary before abstract mathematical knowledge can have any meaning whatsoever, and on the other, the functional value of mathematics depends on its abstract universality.

Where are we then? Where we always are: between heaven and earth. Meaning is our transaction between them.

But while this active transaction, which is also called "knowing," works both ways, the history of our culture has given most of the status to the formal, timeless, abstract—the heavenly—side of the transaction. It is not the intention here to discuss the wider social meanings of this fundamental status problem. Our narrower interest is in its consequences in education, particularly in the context developed earlier: the learning, teaching, knowing, using of mathematics.

In the United States at least, the status problem did not become publically visible until well into the 20th century. Not until then had there been an attempt to teach mathematics beyond arithmetic to a population which would not "live" mathematics as did the mathematically elite of earlier centuries. The problem was further complicated by the hard-won triumphs of the axiomatic method during the latter half of the 19th century in providing the first semblance of logical foundations. The logical efficiencies of axiomatic organization of formal structures, coupled with the simultaneous recognition that mathematical truth had a fact-free foundation in such axiomatic organization, led mathematicians and teachers alike to shift emphasis and attention accordingly. Despite early and persistent warnings that the essence of mathematics should not be identified with this logico-hierarchical organization, and that these organizing principles are entirely inappropriate and inadequate instruments of curriculum organization, this organization nonetheless became the template for construction of curriculum. The best critical statement is perhaps given by Morris Kline (1977). Further, the shift toward fact-free, abstract mathematics provided an additional epistemological status pull, away from experiential knowledge and toward logico-deductive abstract knowledge. But what is meant here by epistemological "status?"

"Status" is a value-term and can therefore be expected to embody certain subtleties and complexities, so rather than invest verbiage in a general discussion, we shall attempt briefly to illustrate the function of status in controlling the ways we allocate our attention, and structure or value our various teaching acts.

Basically, the greatest epistemological status rides with the formal, abstract, universal, atemporal, ideal knowledge regarded as an independent entity, somehow (as Neitzsche knew) felt to be superior in a vague sense to the knower or learner. Of much lower epistemological status is, as Polanyi put it (1975, p 17) *"the part which we ourselves necessarily contribute* in shaping such knowledge." (his emphasis) Earlier in this paper we have shown the *existence* and *necessity* of such knowledge (although completeness was not attempted regarding existence—in particular, the twin activities of categorical generalization and exemplification were relatively ignored). In doing this we also illustrated the second-class status of this knowledge in several ways:

1. It is systematically ignored or denied by the symbol system—especially regarding the process/product distinction and heuristic content.
2. The secret anthropomorphisms by which operational mathematics gets its primary meaning, though crucial, is unacknowledged.
3. The secret metaphors that convey meaning from common experience terms taken from natural language, such as *limit, decreasing, divergent,* and in denotative symbols, such as $\lim_{x \to \alpha} f(x)$[1] have an entirely tacit role, and are ignored in the logical structure of the subject. (Notice that in order to *prove* that $\lim_{x \to \alpha} f(x) = L$, one must *first find L,* presumably from an informal, global knowledge of the function—this fact is virtually never acknowledged in the calculus culture.)

In practical terms, we see epistemological status functioning as a value all around, and even within us, every day. It organizes our teaching and our textbooks. Fifty percent or more of the mathematics presented to students at any given time is cognitively inappropriate. We attempt proofs instead of explanation. Close examination reveals that most upperclass mathematics majors neither understand *nor believe* what they have heard about limits. It is not merely the ϵ-δ definition that has rolled off, but the whole notion that limits exist in the sense that $.999 \ldots = 1$. Furthermore, few mathematics majors, when pressed, seem to have any confidence in the real number system, especially regarding its "continuity." When discussing convergence of geometric series, for example, it becomes evident that these students' basic understanding is at about the level of the ancient Greeks. (The same is true of geometry, by the way, and as for philosophy, they are pre-Socratics.) My colleague Bob McCabe and I use the bouncing of an ideal basketball to illustrate the point that an infinite process—one with no last step—can in principle be completed in a finite amount of time. (Initially, most students believe that the ball will bounce forever.) After talking through all the sidewalk physics (quantum mechanics is the standard reservoir of ambiguity that students draw on whenever their beliefs are questioned by a physical model), some students *are* convinced that—despite the anthropomorphism-temporizing problem mentioned earlier—the ball

---

1. Presumably these are cultural traces remaining from earlier times.

ideally does stop. However, I have noticed in retrospect that we *never actually computed* the time involved. How much stronger the explanation would be if the time were computed! But the fact that we did not allocate our own time and energy for this task illustrates the epistemological status problem at work in both of us.

The status issue gets reflected in the increasing respect granted to instructors, courses, and students as the level of mathematical abstraction increases. The words "pure" as opposed to "applied" mathematics tell us something as well. Note, incidentally, that there are two levels at which mathematics interacts with the rest of experience: the private and the public. So far, we have discussed only the "private" kind, relating to cognitive experience. The public-level interaction (applied mathematics—in the sense of deliberate modeling) can be seen to have higher status on the epistemological totem pole—although not as high as "pure" mathematics. Status issues also, of course, have impact on reward structures in academia: one need only check salary schedules, language used in evaluations by peers, the number of teachers fired for poor teaching as opposed to the number fired for poor research, the allocation of resources within any college or university, definitions of fundable research, etc.

Similar questions of worth come up at the level of philosophy and history of science. The style and thematic assumptions underlying choices of what is "fit" to be studied reflect parallel epistemological biases. As Gerald Holton points out in *The Thematic Origins of Scientific Thought,* the process/product cleavage is evident also in the private/public separation of knowledge. This relates to the widely-held and useful analogy between the process of original research and discovery on the one hand, and the process of learning (rediscovery) on the other. In each case, the processual activity gets less attention than the *products* of the processual activity. (The activity referred to here is at the individual rather than collective level. Kuhn, Lakatos and others often overlooked the two-dimensional nature of the above analogy which is often used as a bridge between the collective-aggregate process of discipline-development and the individual development of a learner.) Holton (1973, p 35) asks for a fuller discussion of the discovery phase in science:

> Einstein himself pointed frequently to both the interest and the difficulty in any such discussion. For example, he wrote, "Science as an existing, finished (corpus of knowledge) is the most objective, most unpersonal (thing) human beings know, (but) science as something coming into being, as aim, is just as subjective and psychologically conditioned as any other of man's efforts," and its study is what one should "permit oneself also." Elsewhere, Einstein used the suggestive phrase "the personal struggle" to describe what seemed to him to deserve central attention in the analysis of scientific development.
>
> This advice is, of course, exactly counter to that of many other scientists, historians, and philosophers of science. Among the last, Hans Reichenbach's dictum is typical: "The philosopher of science is not much interested in the thought processes which lead to scientific discoveries . . . that is, he is interested not in the context of discovery, but in the context of justification." Historians of science also have not paid much attention to the nascent phase because, as one of them put it, the reasons why scientists embrace their guiding ideas in individual cases "lie outside the apparent sphere of science entirely;" therefore they are all too easy to dismiss as not leading to certain knowledge. Those few philosophers of science who have looked at such problems have tended to label them as "metascience problems," hidden at the basis of science, and not really

part of it.

Scientists themselves, by and large, have traditionally helped to derogate or avoid discussions of the personal context of discovery in favor of the context of justification.

Once again we see the exclusionary metaphysical ground rules at work. The underlying positivist-leaning epistemology has as much respect for the processual universe as classical behaviorism had for the psyche—it is not a legitimate object of study. The wider message for students of cognitive processes is clear. *Ours is not the coin of the realm!* An appropriate response has yet to be formulated, although the straight behaviorist-positivist currency switch has already been rejected. On the other hand, counterfeiting of sorts is currently underway in certain developmental circles. The subterfuge goes as follows. Mental constructs are posited and these constructs are assumed to have a logico-mathematical structure. Simultaneously, these structures are represented using formal mathematics as the language. The fundamental legerdemain consists of *identifying the mental structure with the formal mathematics used to represent it.* Carol Fleisher Feldman and Stephen Toulmin (1975, p 417) address this problem directly in the following remarks:

> There are, of course, special reasons why it is particularly hard to keep these two layers of reality—the representation and that which it represents—distinct in cognitive psychology. Roughly speaking: the formal systems we use to express our theories are themselves among the most striking products of human cognition, and so exemplify those very human capacities that cognitive psychologists have been most interested in studying. When we study cognition, logic is thus both the subject studied and also the medium of theoretical representation; so we are especially prone to reify the logical structures of our formal representations. For instance, when we study a child's knowledge of formal logic and express our results in terms drawn from formal logic, we must be clear that we are concerned with two ontologically distinct levels, on one of which the theoretical psychologist employs logic to give a formal representation of the child's logic existing on the other. If we slide unthinkingly from one level to the other, we may end by losing track of our own theoretical procedures and attributing to the child's mind formal characteristics that belong more properly only to our own theoretical representation.
>
> What, then, does it tell us about actual empirical phenomena that they are susceptible of being represented in formal terms? Not as much (perhaps) as we are sometimes tempted to suppose. We select a particular formal model as an *instrument* which permits us to express ourselves clearly, precisely, and communicably but that is all. The model will be chosen as expressing certain aspects of our theory in a particular elegant way, but it will always have certain formal features that are irrelevant—even mismatched—to the particular theoretical application in question. Those superfluous features cannot, then, tell us anything about our own theoretical ideas, still less about the aspects of reality we are using the formal model to explain. We learn about reality by checking the consequences of our theoretical ideas against our experience.

These category problems occur in both arenas of interest to the classical developmentalists, the study of cognitive activity within an individual at a particular stage or time, and the study of the developmental-maturation process through an individual's lifetime. (As Feldman and Toulmin point out, this latter study is best formulated and carried out in a probabilistic populational context rather than an individual context.)

In either case, the taking of the mathematical metaphor as literal truth represents a subtle attempt to use status-bearing currency. It satisfied the twin needs, both to *use* mathematics and to *create a legitimate object of study*. But the words of Alexander Pope thunder out of the 18th century:

Presume not God to scan,
The proper study of Mankind, is Man.

Surely the appropriate language for this study will need to be more expressive than mathematics. We have already seen that at one level, mathematics systematically denies the knower. Can we now expect that at another level it becomes the best tool for understanding the knower? Northrop Frye, (1957, p 352) among the 20th century's leading theoreticians of literary criticism, reminds us:

> Mathematics is at first a form of understanding an objective world regarded as its content, but in the end it conceives of the content as being itself mathematical in form, and when a conception of a mathematical universe is reached, form and content become the same thing. Mathematics relates itself indirectly to the common field of experience, then, not to avoid it, but with the ultimate design of swallowing it. It appears to be a kind of informing or constructive principle in the natural sciences; it continually gives shape and coherence to them without being itself dependent on external proof or evidence, and yet finally the physical or quantitative universe appears to be contained by mathematics.

My deep prejudice is now evident. We all agree that the problems of teaching and learning mathematics are not merely technical problems to be overcome by a better delivery system. I am suggesting, further, that these problems are inextricably tied to the larger issues of what knowledge is and, ultimately, to what a knower is. And, most crucially, we are to beware of any system of understanding and language, such as mathematics, that not only systematically denies the reality of that to which it is being applied, but further, *functionally depends on this denial for its effectiveness as a means to truth.*

I suspect that understanding cognitive process—even as it relates only to the learning of mathematics—will require mixing the precision of mathematical language with the richness of natural language. Further, we should be neither surprised nor dismayed if more than one web of truths presents itself to us. Frye (1957, p 354) puts it this way:

> . . . a language in itself represents no truth, though it may provide the means for expressing any number of them. But poets and critics alike have always believed in some kind of imaginative truth, and perhaps the justification for the belief is in the containment by the language of what it can express. The mathematical and the verbal universes are doubtless different ways of conceiving the same universe. The objective world affords a provisional means of unifying experience, and it is natural to infer a higher unity, a sort of beatification of common sense. But it is not easy to find any language capable of expressing the unity of this higher intellectual universe. Metaphysics, theology, history, law, have all been used, but all are verbal constructs, and the further we take them, the more clearly their metaphorical and mythical outlines show through. Whenever we construct a system of thought to unite earth with heaven, the story of the Tower of Babel recurs: we discover that after all we can't quite make it, and that what we have in the meantime is a plurality of languages.

REFERENCES

Bochner, S. "Myth, Mathematics, Knowledge." In *The Role of Mathematics in the Rise of Modern Science,* Princeton: Princeton University Press, pp 19-20, 1966.

Davis, R. B. "Cognitive Processes Involved in Solving Simple Algebraic Equations." *The Journal of Children's Mathematical Behavior,* 1(3) pp 7-35, 1975.

Feldman, C. F. and Toulmin, S. "Logic and the Theory of Mind." In *Proceedings of the 1975 Nebraska Symposium on Motivation,* edited by W. J. Arnold, Lincoln: University of Nebraska Press, 1976, p 417.

Frye, N. *Anatomy of Criticism.* Princeton: Princeton University Press, 1957, p 352.

Holton, G. *The Thematic Origins of Scientific Thought.* Cambridge: Harvard University Press, 1973, p 35.

Kline, M. "Logic and Pedagogy." *American Mathematical Monthly,* 77(3), pp 264-282.

Nietzsche, F. "On Truth and Falsity in their Ultramoral Sense." (M. A. Mügge, trans) In *The Existentialist Tradition,* edited by N. Langiulli, New York: Anchor Books, 1971.

Polanyi, M. and Prosch, H. *Meaning.* Chicago: University of Chicago Press, 1975, p 29.

Pope, A. "Essay on Man." *Pope Poetical Works,* edited by H. Davis, London: Oxford University Press, 1966.

Weyl, H. "Wissenschaft als Symbolische Konstruction des Menschens." *Eranos Jahrbuck,* Zurich: Rhein-Verlag, 1949 pp 382, 427-8. (quoted in Holton, *The Thematic Origins of Scientific Thought*).

The Feeling of Knowing When One Has Solved a Problem

Herb Koplowitz

In order to be a good problem solver, one must be able to determine *when* a problem has indeed been solved. In working with undergraduates who have difficulty in solving problems, I have come to believe that a major aspect of the difficulty for some students is that they do not have an appropriate sense of when they have solved a problem. Two examples may serve to illustrate the difficulty. They are both taken from a Math Learning Skills course I taught which was designed to improve undergraduates' problem-solving abilities and to develop other skills needed in order to learn mathematics.

A RATIO WORD PROBLEM

I gave the following problem to a group of students:

> I go to a certain place at 40 miles per hour and it takes me 20 minutes to get there. I return at 50 miles per hour. How long does the return trip take?

The following ways of reasoning were offered by various students:

1. It takes 20 minutes to go 40 miles per hour. So you're going 10 miles every 5 minutes. (The student has blurred the distinction between miles and miles per hour.) So to go 50 miles, it'll take an extra 5 minutes. So, it'll take *25 minutes.*

2. It takes 20 minutes at 40 miles per hour, and 40 miles per hour 20 minutes; it would take 10 minutes at 80 miles per hour, so 50 miles per hour 17½ minutes; it would take 15 minutes at 60 miles per hour, 60 miles per hour 15 minutes, which is half way between 40 and 80 miles per hour, 80 miles per hour 10 minutes. So it would take *17½ minutes* at 50 miles per hour which is half way between 40 and 60 miles per hour.

3. It takes 20 minutes to go 40 miles per hour. (Again, the distinction between miles and miles per hour is blurred.) So you're going 10 miles per hour for every 5 minutes. But if you're going faster, like 50 miles per hour is faster than 40 miles per hour, it'll have to take less time. So you subtract 5 minutes. So it takes *15 minutes* to get back.

4. You set up a formula, see: 20/40 = X/50. So it takes *25 minutes* to get back.

Each student was confident of his solution as he offered it, but registered no surprise when other students arrived at different solutions. I was able to lead the group to see the correct answer to the problem. My way of solving the problem did not, however, appear to seem more logical to the students than their own ways, I think they saw my method as one more way of coming to an answer to the problem, but a way which had the stamp of the authority's approval.

A GEOMETRIC RATIO

I gave some students sets of wooden circles, asking them to measure the circumference and diameter of each circle and to make a chart listing C, D, C+D, C−D, CxD, and C/D for each circle. Eventually, they discovered that the circumference of any circle is a little more than three times its diameter. The students also came to understand this as the meaning of the formula they had long ago learned but never understood: $C = \pi D = 3.14D$. I then attempted to explain that the formula $C = 3.14D$ contains two pieces of information: *1.* The ratio between the circumference and the diameter is the same for any circle, and *2.* that constant ratio is 3.14. In order to illustrate the first point, I drew a picture of a square on the blackboard with a horizontal line bisecting the square. I asked, "If we consider the distance around the square to be its circumference and the line across the middle to be the diameter, what is the ratio between C and D for the square?" The students all agreed that C/D was 3.14 for the square because it was 3.14 for the circle. In 15 minutes of probing, confronting, and nudging with these students I was able to get them to say that C/D was four for the square even though C/D is 3.14 for the circle. I am not sure, however, if they really believed me.

DISCUSSION

What intrigues me about these examples is not that the students devised incorrect answers for the questions, but that they were confident in those answers. It is not epistemology—the nature of the students' knowledge—that interests me here but phenomenology—the nature of their experience. In particular, I am interested in the experience or feeling that one has solved a problem.

This is an important area for the simple reason that students stop working on a problem not when they have in fact solved it but when they feel they have solved it. The student who figured that in Problem 1 it would take 17½ minutes to return was just as confident of his answer as I am of mine. In fact, I believe he had the same experience that I had on deriving my answer, the feeling of "O.K., I've solved this one. Let's see what the next problem is." And it is that feeling rather than the fact of having solved the problem that leads the student to stop working on the problem. A major difficulty the students I have described suffer from is that they have that satisfied feeling inappropriately. What is needed here is *1.* a description of what leads good problem solvers to feel that they have solved a problem and *2.* a way of training poor problem solvers to have a more appropriate sense of when they have solved a problem.

This is also an important area because it sheds light on these students' understanding of what it is to solve a problem and on their conception of such analytic disciplines as mathematics. The students I was working with had no way of comparing one solution with another and no student was surprised that another student would devise a solution contradictory to his own. I believe that, for these students, "solving a problem" meant "deriving a number from a formula which uses the data in the problem." They feel, therefore, that their problem-solving task is completed when they have managed to derive a number from the data in the problem by means of a formula. They do not have a sense of the appropriateness of a given formula to a given problem. When I questioned the student who had derived his answer from the formula $20/40 = X/50$ as to why he chose that formula, he was unable to defend his choice of formula. In fact, he seemed to think it peculiar that such a choice would have to be defended. A student who does not understand what it is for a formula to be appropriate for a particular problem does not really understand formulas. What is needed here is *1.* a description of what it is for a formula to be appropriate for a particular problem and *2.* a way of training students to have a better sense of the appropriateness of formulas for particular problems.

A student who cannot compare one solution with another and who is not surprised at another student's having derived a contradictory solution to his or her own will also not have a sense of *what might be the answer to the question* different from a sense of *the answer to the question.* Any number which can be derived from the problem by a formula seems to be just as good as any other number. Such a student will also not be able to make appropriate use of guessing strategies by which the student first derives a possible answer by means of a formula which might be appropriate—but of which the student is unsure—and, second, tests the derived answer to see if it makes sense. These students have no way of testing whether an answer makes sense.

These students will clearly have great difficulty in mathematics and in analytic thinking in general. In a sense, they are in a position similar to that of children in Piaget's preoperational stage who believe that the taller of two glasses must be the larger and also that the wider of two glasses must be the larger. If glass A is both wider and shorter than glass B, such a child will say that A is bigger than B if he or she happens to focus on width first, but will say that B is bigger than A if height is focused on first. Piaget says of such a conception of size that it is "over-determined"; the answer to the question "Which glass is bigger?" is influenced by a number of factors (height and width), and the factors are not coordinated. The child does not have a unique way of deriving an answer from the given factors, and is not bothered by the contradictions deriveable from such reasoning. We might say of the students described above that all of mathematics is overdetermined for them. They do not have unique ways of organizing the factors affecting the answer to a given problem. If, for example, they focus first on the similarity in the phrases "C/D for a circle" and "C/D for a square," they will say the latter is 3.14. If, on the other hand, they focus first on the square's having four sides each equal to the "diameter", they will say C/D = 4 for the square. Like the preoperational child, they seem unconcerned by the contradictions which are derived from their approach to problem solving.

CONCLUSION

Many poor problem solvers are inappropriately satisfied with their problem-solving attempts and have inappropriate understandings of what it is to solve a problem or, indeed, of what mathematics is. If we are to help these students become better problem solvers we must first help them develop a better sense of when they have solved a problem and a better sense of when a particular formula is appropriate to use in a given problem. At the present time we really do not understand how these feelings arise in expert problem solvers or what role they play in controlling the expert's problem-solving strategy. Research on these questions could yield important insights with immediate instructional implications.

Teaching Analytical Reasoning
In Mathematics

Arthur Whimbey

INTRODUCTION

The acquisition of basic mathematical skills and competence in systematic sequential problem-solving techniques is important to success in a variety of fields of study, including numerous "non-mathematical" college disciplines such as economics, psychology, and philosophy. However many students, particularly many of the non-traditional students, do not have a firm foundation in these skills. They have no schema for sequential, step-by-step analysis of problems, and thus cannot solve mathematical problems involving even very low levels of abstraction. They tend to believe, particularly when dealing with word problems, that they either know the answer to a problem or do not, and they lack the skill to break a problem into steps, and then proceed in an orderly fashion toward a solution. Or they may fail to identify relevant information and important variables.

At Bowling Green State University we have been experimenting with devices for teaching analytical reasoning in primary mathematics to educationally disadvantaged students. Evidence of the need for such instruction comes from a recent analysis of the mathematics placement test scores earned by 200 freshmen from low-income families. At least 63 percent of these students were in need of basic pre-algebra math. Other evidence comes from the steady increase in enrollment in a noncredit algebra course offered at the University. Thirty-one percent of the students who take the University math placement test are advised to enroll in this course because they are seriously deficient in math skills.

The experimental program at Bowling Green State University is designed to teach analytical reasoning to educationally disadvantaged students. The program is based on principles derived from Piaget and other investigators of human thought processes. This research recognizes that effective thinkers in any area engage in mental activities which are different from the activities of novices. The cognitive-skills approach to teaching which we employed begins by identifying the mental activities used by successful thinkers as they solve problems and master ideas. Other students who have not yet demonstrated these competencies are then taught the techniques used by high aptitude thinkers. Mastery of such skills is facilitated initially by providing guidelines that lead the student through all the necessary steps or operations and then by providing enough drill and variety to facilitate refinement and generalization of the new skills. This training process is not haphazard or random. Rather, all the necessary steps for solving a problem are clearly specified. Increasingly complex problems and exercises become the basis for the practice necessary for reinforcing the component skills. The student is led systematically through the necessary sequences in a variety of problems until he learns to apply the new skills to new situations.

INSTRUCTION IN PROBLEM SOLVING

The 10-week instructional program which we have developed includes teaching students to solve problems of increasing complexity by thinking out loud. In vocalized problem solving, students are asked to verbalize as much of their mental processing as they can, including counting, reading, arithmetic operations and selection of equations. Some instructors (Lochhead, 1976; Whimbey, 1977) have the students work in pairs, taking turns as problem solver or vocalizer and as listener. They learn each other's strategies of problem solving, and continually check for errors. In addition, students read transcribed protocols of good problem solvers working similar problems. A variation, used by DeLeeuw in Stanford's fundamental math course, schedules students to meet one session a week for verbalized problem solving with a tutor, and other sessions for individual silent problem solving.

Students who are weak in analytical reasoning frequently tend to be ineffective readers and careless in arithmetic. Vocalization brings them face to face with their laxness in carrying out operations and making interpretations.

Besides stimulating a concern for precise thought, vocalized problem solving allows communication of thinking strategies which are normally hidden from view. Students learn from each other and from tutors, seeing how and why particular operations are performed in certain problems as well as the ordering of these operations. And with advanced topics, where even professors may become temporarily stumped, students see the fumbling trials and false starts which deflate the myth that real problem solving proceeds in the smooth, elegant steps of the final solutions which are printed as textbook examples.

Consider the following problem:

If 3 days before tomorrow is Thursday, what is 4 days after yesterday?[1]

This exercise has been particularly useful for teaching sequential analysis, since it is suitable for expansion to more abstract and more complex forms. It can be solved by a very direct process of step-by-step reasoning. Other advantages are:

- It is a math "word problem," the hurdle most feared by nonanalytical students, yet it involves only the simplest mathematical operation—counting.
- The "given" information is easily separable from the question asked—a first step in the right direction for nonanalytical thinkers.
- It can be systematically varied in difficulty without drawing upon additional mathematical knowledge by varying the number of steps required to process the given information and/or the question asked.
- The "given" information is in a form requiring an inverse transformation which is basic to much mathematical comprehension.

---

1. In these problems assume "one day after" means 24 hours later. Thus, for example, one day after Tuesday is Wednesday.

- The steps in the solution to a problem can be readily diagrammed, which is a great pedagogical asset. A sample diagram for a similar problem is shown below.

If Thursday is 2 days after tomorrow, what is 1 day before yesterday?					
SA	SU	M	TU	W	TH
1 day before yesterday	yesterday	today	tomorrow	1 day after tomorrow	2 days after tomorrow
answer					*information given*

This "days of the week" problem demonstrates the schemata of sequential analysis quite vividly. A student must be able to separate the problem into parts and solve each component in turn. Research by Bloom and Broder (1950), Bereiter and Engleman (1969), and Whimbey (1978) have shown that low-achieving students do not possess skill in breaking problems into parts. Instead their problem solving is often characterized by random, unsystematic efforts and unsupported assumptions.

TRAINING

One week of the basic 10-week math course is devoted to teaching students to analyze "days of the week" problems. The three classes of problems used are outlined below:

Class 1. Complex first half, simple second half
Example:
Sunday is 2 days after 6 days before tomorrow. What is today?

Class 2. Simple first half, complex second half
Example:
Tomorrow is Sunday. What is 2 days after 3 days before yesterday?

Class 3. Complex first half and second half
Examples:
Pure Operations: If 6 days after tomorrow is Tuesday, what is the day before 3 days after tomorrow?
Novel Operations: If 3 days after yesterday is Saturday, how many letters before Z in the alphabet is the first letter of 4 days before tomorrow?

Analyses of individual errors have shown that Class 1 problems are more difficult than those from Class 2. Low-achieving students, seeing the word "after" in a problem, tend to count automatically to the right of the day given. But in the Class 1 example, the word "after" in the first half of the problem requires counting to the left because Sunday is already "after" the day sought. Reversing (or inversing) an operation this way is necessary in much mathematical problem solving, but is extremely difficult for low-achieving—in Piaget's terms, concrete operational—students.

Students begin by working through a series of progressively difficult Class 1 problems. They work in pairs and take turns solving the problems, explaining their thoughts and representing their moves on a diagram showing the days of the week. Later on they work in a similar way—with vocalized thinking and diagrams—through a series of Class 2 problems and finally Class 3 problems. Some of the Class 3 problems include novel operations, requiring decisions about facts and relationships other than days of the week. Research has shown that non-analytical students initially have difficulty with the novel operations introduced, but that once they develop the skill to analyze pure "days of the week" problems, they can handle the novel operations with little if any additional guidance. The novel operations were included to make the exercises more interesting and to allow students to practice their newly acquired analytical strategies on different terrain.

NEW MATH SKILLS

Training in "days of the week" problems served as a prototype of sequential analysis to which instructors and students referred in later units dealing with negative numbers, fractions and elementary algebra word problems. Sometimes instructors and tutors reminded students to break new problems into steps as they had done in "days of the week" problems. Often students did this spontaneously. Several students reported that the "days of the week" problems not only taught them to work math problems in steps, but also to read all their textbooks with greater attention to meaning. The results of 39 students answering the representative math problems before (pretest) and after (posttest) taking the course is shown in Table 1. Different questions were used for the pre- and posttests.

The results indicate that training in analytical reasoning can be effective in strengthening ability to solve both computational and word problems. It is especially interesting that even students who had long histories of failure in mathematics were able to build math skills as long as developmental experiences were provided for growth. In fact, Larkin's research (1977) indicates that increasingly complex schemata may be acquired even later through professionally active careers in the sciences.

These findings, along with the successes of large-scale compensatory education programs in Israel (Peleg & Adler, 1977) suggest that much can be done through education to teach people the thinking patterns used by experts in various academic areas.

Table 1. Results of 39 Students Taking Pre- and Posttests

correct responses on *Pretest*	correct responses on *Posttest*	Problem
8	31	If 3 days before tomorrow is Thursday, what is 4 days after yesterday? Show your steps on a labeled diagram.
6	30	A company bought 20 chairs at $6.00 each and 5 tables for $10.00 each. What fraction of the total bill was spent for tables?
11	38	A tree grew 5/7 of an inch every day for 11 days. How much did it grow in total? Express your answer as both (a) a mixed number and (b) an improper fraction.
13	38	Two-fifths of some number equals 16. What is that number?
2	32	A car traveled some number of hours at 30 mph and 6 less hours at 40 mph for a total of 460 miles. For how many hours did it travel at 30 mph? For how many at 40 mph?
4	28	A thermometer dropped from −3 to −11. a. Show the change on a thermometer diagram. b. Compute the absolute change using the formula: a.c. = high temp. − low temp. c. Compute the directed change using the formula: d.c. = final temp. − initial temp.

REFERENCES

Bloom, B.S. and Broder, L.J. *Problem-Solving Processes of College Students.* Chicago: University of Chicago Press, 1950.

Bereiter, C. and Engelman, S. *Teaching Disadvantaged Children in the Preschool.* Englewood Cliffs, NJ: Prentice-Hall, Inc., 1966.

Inhelder, B. and Piaget, J. *The Growth of Logical Thinking from Childhood to Adolescence.* New York: Basic Books, Inc., 1958.

Larkin, J.H. "Processing Information for Effective Problem Solving." University of California, Berkeley, 1977.

Lochhead, J. "The Heuristics Laboratory: Dialogue Groups." University of Massachusetts at Amherst, 1976.

Peleg, R. and Adler, C. "Compensatory Education in Israel: Conceptions, Attitudes, and Trends." *American Psychologist,* 32, pp 945-958, 1977.

Whimbey, A. (with Whimbey, L.S.) *Intelligence Can Be Taught.* Stamford, CN: Innovative Sciences, 1978.

Whimbey, A. "Teaching Sequential Thought: The Cognitive-skills Approach." *Phi Delta Kappan,* 59, Dec 1977, pp 255-259.

Can Heuristics Be Taught?

*The Elements of a Theory and a Report on the Teaching
of General Mathematical Problem-Solving Skills*[1]

Alan H. Schoenfeld

INTRODUCTION

Can students be taught general strategies that truly enhance their abilities to solve mathematical problems? Or are the heuristics described by Polya and others merely a *description* of the actions of accomplished problem solvers? Are they essentially valueless as *prescriptions* for problem solving? While many mathematicians are convinced that they employ heuristics, there is little evidence that general problem-solving skills can be taught.

I offered a course based on the applications of heuristics to mathematics majors at the University of California, Berkeley. This article presents the rationale for heuristics and notes some questions about their effectiveness in the teaching of problem solving. I offer some suggestions regarding these questions, and describe the course I used to implement these suggestions. I discuss what we can and cannot expect students to assimilate—heuristics they can learn to use and obstacles that prevent them from employing others effectively.

SECTION 1. Problem Solving in Perspective: Theory and Practice

George Polya's *How to Solve It* was published in 1945. That and his subsequent work laid the foundations for the study of general strategies for problem solving in mathematics, focusing on the broad strategies he called "heuristics." Definitions vary, but the following is compatible with Polya's usage:

> A *heuristic* is a general suggestion or strategy, independent of subject matter, that helps problem solvers approach, understand, and/or efficiently marshal their resources in solving problems.

Examples of heuristics are: "draw a diagram if possible," "try to establish subgoals," and "exploit analogous problems"; a more complete list is given in Section 3. A rationale for the study and teaching of heuristics is the following:

1. Through the course of his career, a problem solver develops an idiosyncratic style and method of problem solving. A systematic use of these strategies may take years to develop fully.

2. In spite of these idiosyncracies, there is a surprising degree of homogeneity in the approaches of expert problem solvers.

---

1. Revised version of a paper to appear in *Applied Problem Solving,* edited by R. Lesh, ERIC, 1979.

3. One could possibly extract a global problem-solving strategy, using first the introspections of the experts and then incorporating the systematic techniques of artificial intelligence.

4. This strategy can serve as a guide to the problem-solving process. Students instructed according to this plan could shorten the long and arduous process of arriving at these general principles themselves.

Most mathematicians who have read Polya readily accede to the first three points in the rationale. Certainly my personal experience was compatible with them. My problem-solving techniques when I left graduate school were substantially different from those I used as a college freshman and even as a first-year graduate student. To quote Polya's "traditional mathematics professor," "a method is a device which you use twice." If it succeeds twice, you remember using it successfully, and you think of using it when given another similar problem. A method becomes a *strategy*. Over a period of time the student remembers some of these strategies; some he does not. A personal, idiosyncratic, and relatively stable approach to problem solving evolves.

Although problem-solving methods are generally idiosyncratic, there are a number of general principles involved in arriving at solutions. Consider a problem that is within the grasp of freshmen but sufficiently unusual even for experts. Ask both students and colleagues individually to solve the problem *aloud,* and observe the *process* of solution. The experts will most likely engage in some form of systematic exploration to "get at the heart of the problem." The approaches of the novices will seem comparatively unstructured—even when they succeed in solving the problems.

Polya recognized this. His description of the general problem-solving strategies used by mathematicians is excellent. I first saw *How To Solve It* after completing my dissertation in pure mathematics and was amazed at Polya's accuracy. Page after page I nodded my head in agreement with his words. My response was similar to that of most mathematicians, and I doubt that many would seriously dispute claims *1* through *3* above. Number *4*, however, is another matter. I discovered that surprisingly few mathematics professors actually use Polya's work in any substantive way when giving instruction in problem solving. A colleague who has very successfully coached his university's team for competition in the nationwide W. L. Putnam Mathematics Competition told me that his students did not find Polya's works useful. Though they enjoyed the books a great deal, they neither seemed to solve problems more effectively, nor did they perceive themselves as learning more useful techniques for solving problems than before reading them. The coach of the team that won the Putnam Competition that year told me much the same thing.

Those entrusted with training students to solve problems have generally followed this pragmatic but successful rule: *One learns to solve problems by solving a large number of problems.* In practice, the formats of their problem-solving courses are remarkably consistent. The students are given a set of problems to solve for the next class meeting. When the group next meets, solutions are presented, various approaches to the problems are discussed, and comments about particular techniques to solve similar problems are made. This approach might be considered as an attempt to accelerate the process described in *1* by means of a concentrated dose of problem solving with direct feedback.

For at least some students (generally the most talented ones), this method is highly successful. Yet there are a number of drawbacks. Chief among them is that it is still up to each student to synthesize the material he has seen: to ingest it, place it into context, organize it, and have it readily accessible. The product of many hours of work and much thought can be irretrievably lost if the connections between those ideas are weak. A student may deal at length with a certain problem and understand it completely at the time. Yet this does not guarantee that he will retain either the "lesson" to be learned from it, or even the solution. Unable to make the proper connections, the problem solver may later be totally frustrated, knowing that he has solved the same problem before but is now unable to recall even the general approach.

Surely this has happened to the reader more than once. The most dramatic evidence I have of this kind of phenomenon was given to me by the first colleague mentioned above. One of the problems on the Putnam examination had been assigned in his problem-solving seminar before the examination was given. The three members of the university team and the three alternates had all seen the problem solved in detail, yet none could solve it on the examination.

The instructor of a course in problem solving is caught between two extremes. On the one hand, he has available an attractive theoretical structure that describes the expert's approach to solving problems but that has not been shown effective as a prescription for problem solving. On the other hand there is a practical process that does accelerate the growth of problem-solving ability in some students. This approach is inefficient, suitable only for a minority of students, and lacking in theoretical coherence.

The solution, of course, is to attempt a synthesis of the two: a judicious selection of problems presented within the context of an overall problem-solving strategy. There are some obvious and some subtle reasons that efforts to teach problem solving via heuristics have not succeeded to date. These are discussed in Section 2. Instruction in heuristics alone will almost always prove insufficient: Students need to be trained in a means for selecting the appropriate strategies for problem solving and for allocating their resources wisely. The means for doing this, which I call a *managerial strategy*, are described in Section 3. The framework, consisting of a managerial strategy combined with instruction in individual heuristics, was the foundation for my course in problem solving at the University of California, Berkeley, in the fall of 1976. The course provided clear evidence that students can be taught to use certain heuristics effectively. These "successes" are discussed in Section 4. In Section 5, I describe what students *cannot* be expected to pick up, and why. Section 6 provides a discussion of both practical and theoretical concerns for those interested in teaching problem solving, and some ideas for future work.

SECTION 2. The Major Obstacles

For a student to succeed in solving a problem using a heuristic strategy, at least three things must happen:

1. The student must have a "general understanding" of what it means to apply the heuristic.
2. The student must have a sufficient grasp of the subject matter at hand that he can apply the heuristic correctly.
3. The student must think to apply the heuristic!

Points *1* and *2* are certainly substantive and worthy of attention. The bulk of instruction on heuristics until now has focused on *1*. The subject-matter competence described in *2* is an obvious *sine qua non* for the specific application of any heuristic. While *3* is apparently trivial, insufficient attention to it may account for the failure of many attempts to teach problem solving via heuristics. We will discuss this further after some elaboration of the first two points.

Understanding the Heuristic

It is easy to underestimate the degree of sophistication required to understand and use even the simpler heuristics. To illustrate this, consider one heuristic and a series of problems to which it can be applied. First, the heuristic:

> Exemplify the problem by considering various special cases. This may suggest the direction of, or perhaps the plausibility of, a solution.

Now consider its application in each of these problems:

1. Determine a formula in closed form for the series

$$\sum_{i=1}^{n} \frac{1}{(i)(i+1)}$$

2. Let $P(x)$ and $Q(x)$ be polynomials whose coefficients are the same but in opposite orders:

$$P(x) = a_0 + a_1 x + a_2 x^2 + \ldots + a_n x^n \text{ and}$$
$$Q(x) = a_n + a_{n-1} x + a_{n-2} x^2 + \ldots + a_0 x^n.$$

What is the relationship between the roots of $P(x)$ and those of $Q(x)$? Prove your answer.

3. Given the real numbers a_0 and a_1, define the sequence $\{a_n\}$ by $a_n = \frac{1}{2}(a_{n-2} + a_{n-1})$ for each $n \geq 2$. Prove that $\lim_{n \to \infty}(a_n)$ exists and determine its value.

4. Two squares, each s on a side, are placed such that the corner of one square lies on the center of the other. Describe, in terms of s, the range of possible areas representing the intersections of the two squares.

5. Of all triangles with fixed perimeter P, determine the triangle that has the largest area.

Problem 1 is most often encountered in courses on infinite series, where the clever use of partial fractions reveals the sum to be a "telescoping series." This is entirely unnecessary, however; computation of the partial sums for n = 1, 2, 3, 4, . . ., yields an obvious pattern that can be trivially verified by induction. In problem 2, one employs the heuristic somewhat differently. In the linear case the relation between the roots of $P(x) = a_0 + a_1 x$ and $Q(x) = a_1 + a_0 x$ is clear. This is obscure when $P(x) = a_0 + a_1 x + a_2 x^2$ and $Q(x) = a_2 + a_1 x + a_0 x^2$, however. A choice of conveniently factorable polynomials such as $P(x) = x^2 + 3x + 2$ and $Q(x) = 1 + 3x + 2x^2$ makes things much more apparent, and one can suspect the answer. (Proving it is another matter.)

In problem 3 the computations for a rapidly become complex. By setting $a_0 = 0$ and $a_1 = 1$, it is possible to compute a easily, and (especially if one draws a picture) to generalize back to the original problem. In problem 4, one should try a variety of positions where the area is easily calculated; this suggests the desired conclusion. And in problem 5, fixing the perimeter at some convenient value and then calculating the areas for various triangles—including extreme cases, isoceles right triangles, and the equilateral triangle—strongly suggests the result, which can be verified analytically.

For a relatively inexperienced problem solver, deriving these five distinct types of actions from the strategy given is by no means trivial. Most often the statement of a heuristic is quite broad and contains few clues as to its actual use. The heuristic is not in itself precise enough to allow for unambiguous interpretation. Rather it is (for the expert) a label attached to a related family of specific strategies. "Using the heuristic" in a particular instance means sorting through the family of specific strategies and selecting the appropriate one.

The more general the statement of the heuristic or the more difficult it is to apply, the worse these difficulties become. Consider the following heuristic, taken from *How To Solve It:*

> If you cannot solve the proposed problem, try to solve first some related problem. Could you imagine a more accessible related problem?

One can construct many *types* of related problems, but even this is only the beginning. To apply this heuristic successfully, the student must be able to:

i) determine an appropriate, related problem that is more accessible,

ii) solve the related problem, and

iii) exploit *something* from the solution—perhaps the answer or the method employed.

One can see that learning to use a particular heuristic, even under ideal conditions, is far from simple. What for the expert is a label that serves as access to a variety of specific techniques is to the naive student a vague and almost useless suggestion. Even illustrating the use of the heuristic in one or two examples is insufficient; the student must see it interpreted and applied in many contexts and then be given training in, and feedback on, his use of it. In brief, we must be as serious about instruction in heuristics as we are about any other mathematical techniques; with any less than that degree of classroom attention, we cannot realistically expect students to learn to use heuristic strategies. For a more detailed description of pedagogic "necessities" and a sample classroom hour, see Schoenfeld (in press (b)).

Prerequisite Subject-Matter Competence

Clearly, the student must have some grasp of the subject matter at hand in order to solve a problem. But equally important in the application of heuristics is some kind of perspective that enables the student to sort out the essential details. With respect to the second heuristic described above, the student's ability to perform (i) and (iii) may depend on possession of this perspective. One example taken from elementary physics will suffice here. A student had difficulty with the following problem:

A block weighing 12 lbs moves on a frictionless plane inclined at 30° and is connected by a light cord passing over a small frictionless pulley to a second hanging block weighing 8 lbs. (Figure 1). What is the acceleration of the system?

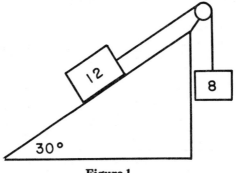

Figure 1

The student was baffled when asked to consider a more accessible, related problem. From the teacher's point of view, some aspects of the problem were critical and others merely technical complications; he could construct and examine a problem that retained the essential elements of the example. The student found the problem as a whole so confusing that the heuristic was valueless. This kind of "cognitive overload" is extreme, but one can encounter such difficulties. Subtle factors can also have adverse effects, as we shall see in Section 5.

You Can't Solve It If You Don't Think of It

The obvious is not always insignificant, but it is too often ignored. This is the case with the truism above. It may reflect the most important reason that attempts to teach problem solving with heuristics have failed in the past. A student may learn to employ a series of individual heuristics, but this does not guarantee that he can solve a heterogeneous collection of problems. He may demonstrate a subject-matter competence and still lack an efficient means of sorting through the heuristics at his disposal and of selecting the appropriate one. He needs a means of assessing and allocating his resources; he needs a *managerial strategy*. Without this, the student may lose the possible benefits of his heuristic resources.

To illustrate this point let us consider the problem of teaching students in a first-year calculus class to perform indefinite integrals. Here, the various techniques of integration—substitution, parts, partial fraction, and other methods—function as the heuristic strategies in general problem solving. I indicate elsewhere (Schoenfeld, 1978) that the difficulty students have in learning to integrate effectively arises not so much from *applying* a particular technique as from *selecting* the appropriate one when the problems appear out of context. If you doubt this, try giving two versions of the same test on integration to a class. Let the problems on the two tests be identical, but on the second test suggest the most appropriate method of solving each problem. The differences in student performance will be dramatic.

In an experiment described in my book (Schoenfeld, in press (a)), a group of students provided with a managerial strategy for approaching integrals easily outperformed a control group, though they spent less time per student preparing for the exam. This was because students in the experimental group apparently could use their resources more effectively than the controls, not because they were more competent in applying particular techniques—they had had no extra practice in that.

When we consider general mathematical problem solving, we find that the role of a managerial strategy is even greater. Even when problem-solving techniques are nearly algorithmic (as in the case of integration) and domain of problem solving small enough to expect students to formulate strategies on their own, they do not. Providing them with replacement strategies however, yields significant results. The techniques in general mathematical problem solving are heuristic, often subtle and difficult to apply. The problem solver can be lured down mathematical "blind alleys" by using the inappropriate heuristics, so the need for guidance in their use is great. Since the domain of problem solving is immense, we cannot expect students to develop an efficient strategy on their own, especially when they are still learning to employ the heuristics themselves. We must provide them with the perspective.

The point to keep in mind here is that the managerial strategy must be *prescriptive* rather than *descriptive* and sufficiently detailed that students can learn to apply it. Remember when reading about strategies that the important question is not: "Does this reflect the problem solving behavior of experts?" but "Is this strategy sufficiently explicit that we can expect students to use it reliably?"

SECTION 3. The Prescribed Remedy

We may take the points discussed in Section 2 as a rough description of the behavior we would like students to exhibit after completing a course in problem solving. One way to guarantee (or at least raise the probability) that students will have subject-matter competence is to require junior standing as a mathematics major for admission to the course. The students will have seen a reasonable amount of mathematics, and we will have some latitude in choosing our examples.[2] It is up to the instructor then to provide training in the heuristics and a strategy for applying them.

Polya (1945) and Wickelgren (1974) together present just about all the heuristics one could expect to use. Thus the major problem I faced before offering the course was to design a workable managerial strategy. Assuming that the students could be taught to employ certain heuristics effectively, I considered the students as "information processors" with fairly well-defined attributes. In terms of Artificial Intelligence (AI), the strategy I sought to design was an "executive program" for the information processor. The systematic observation/distillation/modeling cycle typical of AI[3] ultimately yielded a strategy that served as the foundation for the course. Even in its nascent form, however, it

---

2. By their junior year in mathematics, students are certainly ready for the techniques. If the students in my course were at all representative, we cannot assume that they have developed either the heuristics or managerial strategies on their own. Some arguments for and against easing the requirements for admission to the class are discussed in Section 6.
3. See Newell & Simon (1972) for the most thorough description and Schoenfeld (in press (a)) for a brief discussion of the managerial strategy for integration.

would have overwhelmed the students. So I provided them with a flow chart (Figure 2) and promised to elaborate upon it during the course of the quarter.

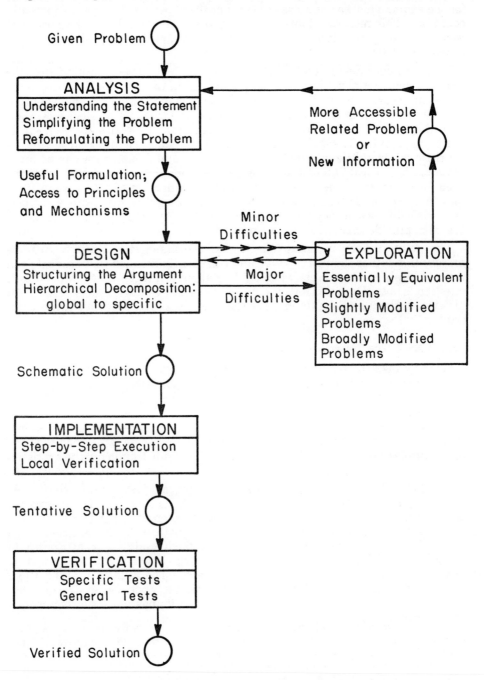

Figure 2. Schematic Outline of the Problem-Solving Strategy

Of course the diagram in Figure 2 was too general to be useful to the students when I first gave it to them. It served rather as an indication of things to come and as a frame of reference. At the appropriate time, each individual box in the flow chart was discussed in detail. This included a listing of the heuristic strategies applicable to that stage of problem solving and instruction in the individual heuristics. The class hour(s) devoted to the strategy "examine special cases" would, for instance, be spent in having students work the problems I gave in Section 2. Since the emphasis in this approach to problem solving is on *process* rather than *product*, classroom dynamics must reflect this. Class meetings were true "discussion sections" where each problem was examined in detail. Students were encouraged to develop solutions on their own, essentially without my help. After solving a problem in as many ways as possible, I would show the way in which the heuristic was applied and how it might work in a related problem. We tried to work within the context of a global strategy, to reinforce that as well as the application of the particular heuristic(s).

In the discussion that follows, the reader is reminded that the strategy is meant to serve only as a guide. Each sentence should be read as if it were followed by the phrase "with all other factors being equal." Some circumstances that tend to make certain factors *not* equal will be discussed later.

A LIST OF MAJOR HEURISTICS

Analysis

1. Draw a diagram if at all possible.

2. Examine special cases.
   a. Choose special values to exemplify the problem.
   b. Examine limiting cases to explore the range of possibilities.
   c. Set any integer parameters to 1, 2, 3 . . ., in sequence, and look for an inductive pattern.

3. Try to simplify the problem.
   a. Exploit symmetry.
   b. Use "without loss of generality" arguments (including scaling).

Exploration

1. Consider essentially equivalent problems.
   a. Replace conditions by equivalent ones.
   b. Recombine the elements of the problem in different ways.
   c. Introduce auxiliary elements.
   d. Reformulate the problem by
      i) changing perspective or notation
      ii) considering argument by contradiction or contrapositive
      iii) assuming you have a solution, and determining its properties

2. Consider slightly modified problems.
   a. Choose subgoals (obtain partial fulfillment of the conditions).
   b. Relax a condition and then try to reimpose it.
   c. Decompose the domain of the problem and work on it case by case.

3. Consider broadly modified problems.
   a. Construct an analogous problem with fewer variables.
   b. Hold all but one variable fixed to determine that variable's impact.
   c. Try to exploit both the result and method of solution in any related problems which have similar form, "givens," and conclusions.

Verification

1. Does your solution pass these specific tests?
   a. Does it use all the pertinent data?
   b. Does it conform to reasonable estimates or predictions?
   c. Does it withstand tests of symmetry, dimension analysis, or scaling?

2. Does it pass these general tests?
   a. Can it be obtained differently?
   b. Can it be substantiated by special cases?
   c. Can it be reduced to known results?
   d. Can it be used to generate something you know?

The first stage of the problem-solving process is *Analysis*. It begins, of course, with the reading of the problem, and is complete when the problem solver has formulated a convenient representation of the problem and has found a suitable mathematical context. After *Analysis* one should be able to sense the nature of the problem. The teacher should stress the importance of categorizing and establishing a context for a problem, since accurate classification often indicates immediately a set of procedures to deal with it. For example, if one recognizes a "maximization" problem as such, one will find an appropriate analytic representation of the quantity of interest and use calculus to find its maximum. Another important aspect of *Analysis* is selecting the proper representation for a problem. For instance, the game "number scrabble" is difficult unless one knows that it is isomorphic to tick-tack-toe (see Newell and Simon's *Human Problem Solving*, 1972), and that "connection lists" are also difficult without a visual representation for them (see Hayes' chapter in Kleinmuntz, 1966). We can impress students with this idea simply by asking them to multiply the Roman numerals MMCDLVII and MMMDCCCLXXXIV.

I have presented a list of the particular heuristic strategies that operate during *Analysis*. In Section 2 we saw how complex even one of those strategies, "examining special cases," can be. Perhaps the best way to demonstrate *Analysis* is to work an example through that stage of the strategy. Remember that the heuristics listed within any stage are not given as an ordered set. Any of them may be used at any time during *Analysis* (or, less often, at another stage).

Example: Find the area of the largest triangle that can be inscribed in a circle of radius R.

Analysis: As one reads the problem, one begins a process of orientation and categorization and assumes that calculus will be involved. One should make a diagram and search for an analytic representation. One might examine the unit circle (scaling) and assume (without loss of generality) that the base of the triangle is horizontal. Using a few more diagrams and special cases, one realizes that, for any particular horizontal base, the triangle with the greatest height (and thus the largest area) is isoceles. At this point, the problem is finding the base of the isoceles triangle in a unit circle that yields the largest area. This is essentially a "standard," one-variable, maximization problem. With the appropriate analytic representation, the problem solver now has a ready-made plan for solution.

One proceeds from *Analysis* to *Design*. At the level of straightforward or routine problem solving, *Design* consists mainly of the intelligent ordering and structuring of an argument. The problem solver should have an overview of the solution process; he should be able to say, at any point in the process, what is being done and why, and how that action relates to the solution. He should proceed hierarchically and avoid intricate calculations pertaining to one part of a solution while global aspects of another phase remain unsolved. (We have all suffered the discomfort of solving a difficult equation, discovering later that it didn't have to be solved in the first place.)

For more complex problems, however, *Design* takes on more significant dimensions. It differs from the other phases of problem solving and, in a sense, pervades them all. *Design* is a "master control," monitoring the whole process and allocating problem-solving resources efficiently. It keeps track of alternatives so that if a chosen approach proves more difficult than expected, a better one can be selected. If there is difficulty in making a straightforward plan, *Design* sends the problem solver into *Exploration*. Problems resolved fairly easily are returned to *Design* and elaboration of the plan continues. However, if *Exploration* provides new insights into the problem or the solution, the "master control" may return the problem, with the new information, to *Analysis*.

Clearly, then, *Design* is the most difficult to prescribe of all the stages in the problem-solving strategy. In the classroom the teacher must be aware of, and open to, individual problem-solving styles. He must not say what is "right," but rather help the student decide what is "right" for him.

It is in *Exploration* that the majority of heuristics generally come into play. From the list of heuristics, we see that *Exploration* is divided into three stages. The suggestions in the first stage are more useful and easier to apply than those in the second stage. This is also true for the relation between the second and third stages. All other factors being equal, the problem solver begins *Exploration* by considering the heuristics in stage 1 and selecting any which seem appropriate for trial. If the strategies in stage 1 prove insufficient, he considers those in stage 2 and, if needed, the strategies in stage 3. Whenever substantial progress is made, the problem solver may either return to *Design* to plan the balance of the solution, or reenter *Analysis*, using the insights gained in *Exploration* to make the problem more accessible.

Implementation is self-explanatory. As with *Design*, it should be hierarchical and used in the last stages of the solution. *Verification* is mentioned only because it is often neglected. Certainly, checking over one's solution allows one to catch silly mistakes. In general, by reviewing the solution process, one can often find alternate approaches, see connections to related subject matter, and incorporate some aspects of the solution process into one's global strategy.

Finally, I should return to my earlier comment that the strategy is intended to provide students with a useful framework for approaching problems and is not meant to be a rigidly followed algorithm, with the students as human automata. A well-defined strategy is neither alien nor constricting to students, but is meant to be incorporated into their own framework and modified accordingly. Further, the phrase "with all other factors equal" should serve as a means to personal alteration of the strategy. For example, one problem solver may be able to consider in *Analysis* what another might reach only in *Exploration*. Also, experience may allow for bypassing some of the strategy; one might start a problem in vector analysis by decomposing the vector, though "breaking into parts" is formally a stage 2 operation. Similarly, certain cues may lead the problem solver to try a particular strategy "ahead of its time." This should and does happen with expert problem solvers. For example, five mathematicians considered the problem:

> Let a, b, c, and d be given real numbers between 0 and 1. Prove that
> (1-a) (1-b) (1-c) (1-d) > 1-a-b-c-d.

All five individually decided, within one minute, to use the approach of examining the two-variable problem and extrapolating the result. The presence of too many variables with similar roles prompted the action, as it should. At least, it should prompt consideration of strategy. The "master control" in *Design* should decide if and when its use is appropriate.

Ultimately, a refined version of the strategy might take such cues into account. Perhaps the "condition-action" language of the production systems will prove a good vehicle for describing such behavior (see Larkin's "Hi-Plan" model: Larkin, in press).

SECTION 4. What Impact Can We Expect Heuristics to Have?

The following are brief statements concerning my students' abilities to learn problem solving via heuristics:

1. Individual heuristics can have a dramatic effect on students' abilities to solve particular problems. Often the mere mention of a particular heuristic can point out an alternate approach, resulting in a solution that was previously inaccessible. With proper training, students can learn to apply heuristics in rather sophisticated ways.

2. Students *can* learn the essential ingredients of a managerial strategy. Most importantly, they can develop skills in determining the appropriate heuristic for dealing with a wide range of problems. This implies that we can hope to have impact on students' problem-solving behavior *outside* of the classroom.

One must be cautious, however. The class was experimental, with an enrollment of eight students. Extrapolating from a small sample, especially when the students received such individual attention, is risky, though the small enrollment did permit a detailed monitoring of students' performance and growth. The numbers that I offer in the sequel are an attempt to substantiate these qualitative statements, and not a statistical argument based on eight students.

Sometimes, just the mention of a heuristic can access a solution process. We saw an example of this in Section 2 where students were given the problem:

1. Find an expression in closed form for

$$\sum_{i=1}^{n} \frac{1}{(i) (i+1)}$$

Of the eight students in the class, two had succeeded in solving this as a homework problem. One recalled having seen it as a telescoping series, and the second remembered performing some algebraic manipulation; he later saw the partial fractions decomposition for the denominator. The other six students said that they had tried but simply could not solve the problem. I then presented them (for the first time[4]) with the heuristic "examine special cases." The students worked by themselves for a while. Within four minutes all saw the pattern, and all had verified it. Virtually everyone (except the mathematicians who know the telescoping series) is initially stumped by this problem but finds the solution apparent once the heuristic is suggested.

Similar results occurred with the problem:

2. For what values of (a) does the set of simultaneous equations

$$\left\{ \begin{array}{l} x^2-y^2 = 0 \\ (x\text{-}a)^2+y^2 = 1 \end{array} \right\}$$

have either 0, 1, 2, 3, or 4 solutions?

This problem can be handled algebraically, although the "bookkeeping" can get sufficiently complex that students make careless mistakes. The problem is number 173 from the USSR *Olympiad Problem Book*, (Shklarsky, 1962) and the algebraic solution is given on pp. 276-277 there. Only three of my eight students handed in error-free solutions.

Now consider approaching this problem using the heuristic *Draw a diagram if at all possible*. The first equation becomes two straight lines intersecting and at 45° with the origin; the second equation a circle of radius 1 with center at (a,0). (see Figure 3) With the graphic interpretation the problem is trivial. Again seven students out of eight solved it completely, all claiming the problem was easier. (Two had used the heuristic before.)

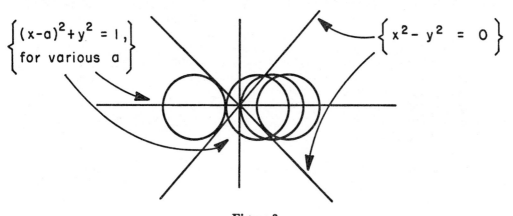

Figure 3

4. Students themselves must be convinced that a strategy is valuable in order to apply it. For that reason I occasionally gave students a problem like the one above to work on and *later* gave them the appropriate heuristic. The effect is impressive and helps to convince them of the heuristic's utility. Remember, though, that if students are given too many inaccessible problems, they can become frustrated, and the approach counterproductive.

Another case in which a heuristic produced drastic results was in the problem:

    3. Let n be an integer. Prove that if $(2^n - 1)$ is a prime, then n is a prime.

Two students out of eight solved the problem within 10 minutes, while the rest made little progress. Unfortunately the heuristic employed for question 1 yields the sequence $(3, 7, 15, 31, 63, 127, \ldots)$, which is not especially suggestive; at this point the students were left without any ideas. I then suggested trying the "reformulating" strategy and all that it includes. Argument by contradiction was an appropriate choice as the negative of "prime" is "composite" and the problem becomes:

    Let a and b be integers greater than 1. Prove that $2^{ab} - 1$ is composite.

The problem was easy for the four of the remaining six students who realized that $2^{ab} - 1 = (2^a)^b - 1$ has a factor $2^a - 1$.

As an example of the efficacy of particular heuristics consider the problem:

    4. Show it is impossible to find numbers a, b, c, d, e, A, B, C, D, E, such that

      $x^2 + y^2 + z^2 + r^2 + s^2 = (ax + by + cz + dr + es)(Ax + By + Cz + Dr + Es)$

The problem is quite complex, and the only three students to solve it on their own did so by employing the "fewer variables" heuristic that I later offered to the whole class. With that suggestion, another three students succeeded in solving it.

These examples are atypical. There are relatively few problems that are inaccessible to students when presented out of context, and completely solvable when a heuristic is suggested. Yet the examples are significant. They indicate that heuristics can drastically affect the solution rates on problems, if only because they focus the problem solver's attention on approaches they might have used but did not think of. The impact of heuristics goes beyond this, however. In the classroom discussion of problem 4, I discovered that only one of the students had ever heard the heuristic explicitly mentioned before. It is safe to say that the problem is too complex for them to solve without recourse to the strategy. Thus training in heuristics adds to the student's problem-solving repertoire and gives greater access to some techniques already known. But this claim is empty unless we can deal with the difficulties raised in Section 2. While particular heuristics can help in solving some problems, students should be able to determine independently their proper applications. At least they must have a strategy for approaching problems and employing the heuristics. Otherwise the heuristics are useless.

Part of the final examination in my course was designed to see if students could select the appropriate means of approaching a variety of problems. The students were given one hour to examine 12 problems and were told to answer the following questions for each of them:

    Which of the techniques studied in class do you think are helpful in understanding the problem and in finding a solution?

    How would you approach this problem if you had an hour to work on it, and why? Mention specific heuristics.

The students were then asked if they had seen that problem or a similar one before; and if so, how much of the solution they remembered. Since they were assured that their answers to the question would in no way affect the grading of their papers, I am fairly confident they answered honestly. A sample of the questions and the responses follows:

1. Let S be any nonempty finite set. We define $E(S)$ to be the number of subsets of S which have an *even* number of elements, including the null set and possibly S. Determine $E(S)$ in closed form for any finite set S, and prove your answer.

Seven of the eight students indicated that they had not seen the problem before. All of them indicated that they would examine sets of 0, 1, 2, 3, . . . elements to see if a pattern emerged; if it did, they would prove it by induction. The one student who had seen a solution, outlined a combinatorial argument.

The next problem has appeared in a variety of contexts. We saw it as an example of "expert" response to "cues" in Section 3. It was also used as a pretest problem in an experiment where seven students with backgrounds similar to those in my course were asked to solve it; only two students of seven considered the "fewer variables" strategy.

2. Let a, b, c, and d be real numbers between 0 and 1. Prove that

$$(1\text{-}a)\,(1\text{-}b)\,(1\text{-}c)\,(1\text{-}d) \geqq 1\text{-}a\text{-}b\text{-}c\text{-}d$$

This problem was new to all eight of the students. Seven of them indicated that they would go about solving it by first examining the one and two variable cases,

$$(1\text{-}a) \geqq 1\text{-}a$$
and
$$(1\text{-}a)\,(1\text{-}b) \geqq 1\text{-}a\text{-}b$$

The eighth student left the problem untouched.

3. Let C_1 and C_2 be two smooth non-intersecting closed curves in the plane. Prove that the shortest line segment which connects a point of C_1 to a point of C_2 is perpendicular to both C_1 and C_2.

None of the students had seen this problem before, though we had discussed some aspects of the isoperimetric problem. Not surprisingly, six of the students mentioned drawing a diagram, and the other two suggested trying examples. But four of the students thought of an argument by contradiction, taking the line segment L as given and proving there would be a shorter one if L were not perpendicular to both C_1 and C_2. Three of the remaining four students noticed the symmetry of the problem statement and that it was sufficient to prove L perpendicular to C_1; one of these further noted that it is sufficient to examine the degenerate case where C_2 is a point. (The eighth suggested examining two circles or perhaps two ellipses.)

4. Let f be a function whose domain is ordered pairs of polynomials and whose

range is the set of all polynomials:
f[p(x), q(x)] = r(x), where p, q, and r are polynomials.

We define I(x) to be an *identity polynomial under f* if
p(x) = f[I(x),p(x)] = f[p(x), I(x)] for all polynomials p(x).

Prove the identity polynomial is unique.

The phrasing of the problem was deliberately vague for I was curious to see how the students would manage their choice of strategy. Two left the question blank. The other six assumed the existence of two identities and proved them equal, one student commenting "We do this any time we want to prove uniqueness."

These results are far from conclusive, but they are suggestive. Judging from the quality of their homework papers at the start of the quarter, the students' ability to make such determinations improved substantially during the course. But an ability to select the proper approach to a problem does not guarantee that a student will be able to solve it. Finding an appropriate heuristic may not be a sufficient condition for success, but it may be necessary. If the students had not used a heuristic approach, they probably could not have solved the problems above.

SECTION 5. What Should We Not Expect of Heuristics?

The examples in Section 4 show that students can learn to apply a variety of heuristics, using a managerial strategy to select them. Yet clearly there are limits to this approach to problem solving. In this section we examine some of the reasons for this.

The statements of heuristics are often very general, and leave much interpretation to their users. Consider the heuristic "examine an easier analogous problem with fewer variables" when applied to the following:

    1. Prove that for all real numbers a, b, and c,
$$a^2 + b^2 + c^2 = ab + bc + ba$$
implies that a = b = c.

I asked the students to construct an easier, similar problem. One suggested setting c = 0 to obtain the simpler problem:

    Prove that $a^2 + b^2 = ab$ implies a = b

The class agreed. That choice is inappropriate, of course. The correct statement obtained from setting c = 0 (which is *not* analogous) is:

    Prove $a^2 + b^2 = ab$ implies a = b = 0

Considering the cyclic nature of the terms on the right-hand side of the original problem, the correct analogous statement is:

    Prove $a^2 + b^2 = ab + ba (= 2ab)$ implies a = b.

The solution is easier and generalizable to the original problem.

How successfully one employs a heuristic depends significantly on the way he encodes information and the perspective he brings to the subject matter. As an example, the last lines of the proof to problem 1 read:

Since $(a - b)^2 + (b - c)^2 + (c - a)^2 = 0$, we have a = b, b = c, and c = a, as desired.

I left the solution to this problem on the blackboard and gave the following problem:

2. Let $\{a_1, a_2, \ldots, a_n\}$ and $\{b_1, b_2, \ldots b_n\}$ be given sets of real numbers. Determine necessary and sufficient conditions on $\{a_{l_i}\}$ and $\{b_i\}$ such that there are real constants A and B with the property that:

$(a_1x + b_1)^2 + (a_2x + b_2)^2 + \ldots + (a_nx + b_n)^2 = (Ax + B)^2$
for all values of x.

I told the students that the two problems were related, and that they should try to exploit the first in solving the second. I gave them 15 minutes and asked them to work individually. None of the students made any progress for they did not see the structural similarity between the two problems.

This is not surprising. The second equation is, after all, a morass of symbols none of which are quite comparable to those in the first: the number of terms is different, the quantities are polynomials instead of numbers, there are subscripted variables, and the right-hand side is a quadratic polynomial with undetermined coefficients. Nonetheless when I read problem 2, I was reminded of problem 1, although an hour had passed since I selected problem 1 for discussion. In solving problem 1, I had been impressed by the fact that much information is contained in an equation where a sum of squares equals zero. When I read problem 2, I saw a sum of squares equal to *something*. Thus, I would gain information if *something* were replaced by *zero*. The compact form of encoding "sum of squares equals . . ." enabled me to do this. Without this concise yet powerful means of summarizing, the structural similarity between the two equations remains obscure.

Similarly the perspective one brings to a problem (largely a function of one's experience) may determine what is seen in that problem and thus the success in solving it.

Consider the following problem:

3. Let T be a given triangle of area A. Using a ruler and compass, construct two lines parallel to the base of T such that the three resulting areas (see Figure 4) are all equal.

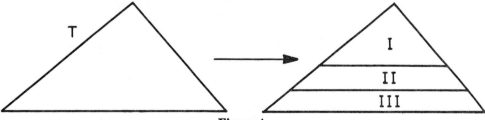

Figure 4

A graduate student with a background in physics had difficulty with this problem because he was only able to perceive it as illustrated in Figure 5. He found it nearly impossible to equate the area of the upper triangle with that of each of the two trapezoids.

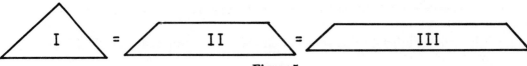

Figure 5

For myself, for a colleague who had taught high school geometry, and for four of the eight students confronted with this problem on my final examination, Figure 6 is a reasonable representation of our perception of Figure 4.

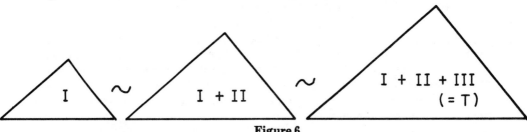

Figure 6

We now see three triangles of areas A/3, 2A/3, and A, respectively. Since the areas of similar plane figures are proportional to the squares of their sides, the problem reduces to that of constructing $1/\sqrt{3}$ and $\sqrt{2}/\sqrt{3}$ with ruler and compass. But one must see the similar triangles in Figure 4, and no amount of training in heuristics can compensate for not seeing them.

The fundamental assumption in teaching heuristics is that once students have been shown how to apply a heuristic in a variety of ways, they will be able to apply it themselves. To a certain degree this was true in my course. By the end of the quarter students were substituting n = 1, 2, 3, . . ., for integer parameters, even when the parameters were only implicit; they were analyzing the impact of particular variables on problems by holding all but that variable fixed and letting that one vary; and employing other techniques. But when the class encountered something that was beyond their range of experience, they ran into trouble. For example, the following was a homework problem:

4. Let N be any integer. Find D(N), the number of distinct integer divisors of N, including 1 and N.

The table of values for D(N) does not, at first suggest any pattern.

N:	1	2	3	4	5	6	7	8	9	10	11	12	13	14	15	16	17...
D(N):	1	2	2	3	2	4	2	4	3	4	2	6	2	4	4	5	2...

At this point the class was stuck. It did not occur to them to ask which values of N give particular values of D(N). Once I pointed that out, it became apparent to them that whenever D(N) was 2, N was prime, and that N was a prime square when D(N) was 3. They were now on their way to solving the problem. But this application of "looking for a pattern" was beyond their range of experience and thus not useful to them. This particular method of seeking a pattern may later be incorporated into their interpretation of the heuristic. But this is no guarantee that they will be successful the next time an unusual application of the heuristic is required.

This difficulty becomes more apparent when the successful solution of a problem depends on the concurrent or synthetic use of two heuristics.

5. Let P be a polygon drawn in the plane whose vertices are all points with integer coordinates (lattice points). Find a simple formula for the area of P, which depends on the number of lattice points in the interior, and on the boundary of P.

The answer (known as Pick's Theorem) is that the area is $\frac{1}{2}(2I + B - 2)$, where I and B are the number of interior and boundary lattice points of P, respectively. This formula is sufficiently complex that most students are unlikely to think of it with a random selection of special cases. The best approach to the problem is to combine two familiar heuristics. First fix one variable (say, I = 0) and then take sequential values B = 3, 4, 5, 6; then fix I = 1 and repeat the process. Do the same for I = 2. The formula should now be accessible. But this kind of synthetic approach was not at all apparent to the students even though they had in the past kept variables fixed and had certainly looked for inductive patterns. We cannot really expect students to combine the use of other heuristics they have learned individually.

SECTION 6. Conclusions

The teaching of general mathematical problem solving is still in a primitive state, though Polya's notion of "modern heuristics" has been with us for more than 30 years. The modern theory still does not provide a usable prescriptive model for problem solving; current practice is largely in the hands of isolated individuals who proceed on the basis of personal experience. This paper has dealt with various aspects of a prescriptive theory of problem solving. Practical and theoretical issues have been mixed for they complement each other. In this concluding section they are separated. In the next section I deal with some practical concerns for those who might consider offering a course in problem solving.

Practical Concerns for the Teacher

The success of teaching a course in problem solving depends on more than the compilation of the strategies which serve as its theoretical foundation. For example, the role of affective considerations, which I have barely mentioned in this paper, is crucial. In a situation where the confidence of the individual problem solver may determine whether or not he solves any particular problem, we cannot really ignore such concerns. However, this is not the place for a discussion of teacher-student relationships; see chapter 14 of Polya (1965) and Schoenfeld (in press (b)).

In preparing my course, I found two things to be crucial. The first is the philosophical framework, summarized in the next section, and the second is the choice of examples for discussion. Selecting valuable and instructive problems may be the hardest task a teacher of problem solving faces. The purpose of the examples is to illustrate, in a variety of contexts, how each general heuristic can prove valuable. Problems requiring only a single heuristic or those that *dramatically* demonstrate the heuristic's utility are very uncommon. Nonetheless they are absolutely necessary in the process of instruction; they serve to convince the students of the importance of heuristics and to illustrate their use.

Only when the student has mastered the use of heuristics in simple problems can he apply them in more complex situations. Some of the problem sources I found most useful are listed in the bibliography. The examples in this paper are a fair sample of those that meet my criteria.

There are some types of problems that I feel should be avoided, such as those for which the solution depends on "divine revelation." Consider the following, taken from Shklarsky (1962):

1. Prove that $n^2 + 3n + 5$ is never divisible by 121 for any integer n.

The argument in the book begins:

We shall use the identity $n^2 + 3n + 5 = (n+7)(n-4) + 33$.

I cannot see how anyone except the author of the problem can be expected to arrive at the identity himself, and I see no instructional value here.

Other problems to avoid are those selected just for subject-matter theme. It is easy to pick a series of problems from a text or a collection like the Olympiad book that deals with a particular subject in mathematics. Such problems do appear frequently in problem-solving competitions, and a student who wishes to compete should be familiar with these subject areas. However, the techniques used are often unique to that area and of little help in developing general problem-solving skills. An effective presentation of heuristics as general strategies should not include, at least in the beginning, this area-by-area approach. Students can be given problems united by subject-matter theme after they have learned to employ a variety of heuristics.

Finally, I should comment on the level of the problems we can use in a course on problem solving. Surprisingly, they need not be so advanced as we might expect. Often a problem that is easy within a particular context can be quite difficult for students if they are out of practice. For example, consider the following problem. It is routinely solved in high school geometry classes.

2. Prove that in any circle, the central angle which subtends a given arc is twice as large as any inscribed angle which subtends the same arc. (Figure 8a)

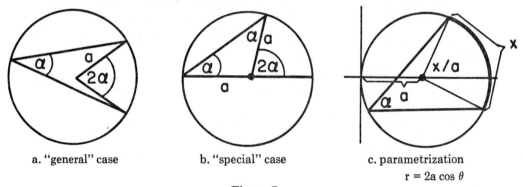

a. "general" case   b. "special" case   c. parametrization
                                                    r = 2a cos θ

Figure 7

When I assigned this problem, I had intended that the students consider a special case in which one side of the inscribed angle is a diameter of the circle

(Figure 7b). The general case can be obtained by adding (or subtracting, if the inscribed angle does not include the center of the circle) two special cases. To my surprise, the problem caused my students some difficulty. One person whose problem solving was particularly idiosyncratic found the solution by parametrizing the circle and performing an arc length integral in polar coordinates. I pointed out that he might have suspected a simpler proof.

Thus advanced mathematical problems are not always necessary. But it was good that I gave the problem as homework and not as an in-class example. My students may have been insulted by such an elementary problem. Purely as an instructional strategy, we must convince them that they can learn from such "sample" problems. This is best done by letting the students discover that the problems are not so trivial as they think.

Since one can learn the "object lessons" of the heuristics from some problems at a fairly elementary level, one could possibly ease the admissions requirements for the course. The course I offered could probably be tailored to students who have completed one year of calculus, but I wonder about the benefits of such a course. It might be more appropriate to offer them basic instruction in mathematical logic. But the lowest level at which the strategies of heuristics can be presented, with benefit to students, is an empirical question. That it can be done at all has been established.

Theoretical Issues

The following assumptions will provide a context for this discussion:

- Although there is both idiosyncracy and individualism in problem-solving behavior, experts demonstrate discernible patterns in their approach to mathematical problems. They often use certain global strategies.

- These global strategies can be described. Polya has done this for the most important heuristics, with Wickelgren covering whatever Polya has not. Polya has also offered insight into the effective aspects of teaching problem solving.

- Many mathematicians, and scientists in general, accept Polya's descriptions of heuristics. They claim to have used them and wish their students would.

- There is still no real evidence to show that one can substantially improve students' abilities to solve problems by teaching them heuristics.

If we accept these four assumptions, we may draw either of these conclusions:

1. Heuristics may well serve as a description of expert problem-solving behavior. They fail, however, as prescriptions or guides to problem solving for nonexperts. Unfortunately, an individual must develop his own personal approach to problem solving.

2. Some element is lacking in the theory of problem solving and/or in the means that have been used to teach it. If we can extend or modify the theory to incorporate the missing element, we may be able to teach problem solving with heuristics successfully.

I personally reject the first conclusion. In this paper I have discussed why I believe the present theory is incomplete. I have offered a means for providing the elements that it seems to lack, and given some evidence that these ideas can be implemented in a practical way. Let me summarize.

Even if we succeed in training students to apply each in a series of individual heuristics, we still cannot expect improved performance when students are tested on a wide range of problems, requiring a variety of heuristics. Students must be able to select the appropriate means of approach to each problem. If they lack this ability, they may not benefit from their heuristic resources. It is our job, then, to train students in this selection process.

The means I propose is a "managerial strategy," discussed in Section 3. The strategy was designed to simulate the way that experts would approach problems that were new to them. If the heuristics are taught within the context of this managerial strategy, the difficulties mentioned above might be overcome.

Although my class was small, the results, described in Section 4, are definitely suggestive. They imply that students can learn to employ individual heuristics, and they can indeed learn to quickly select the proper means of approach to a variety of problems. In short, I believe that a managerial strategy is the missing element in the established theory of problem solving. Including it may be the key to successful teaching of problem solving via heuristics. Remember, however, that my class was an experiment using human subjects being trained in the most complex and least-understood thought processes known to man. There were necessarily many uncontrolled variables. Experiments providing irrefutable evidence of the success of heuristic strategies simply do not yet exist. Progress is being made, however. The experiment described in Schoenfeld (in press (a)) shows that, under certain conditions, we can demonstrate the impact of heuristics. It also provides a mechanism for further study of heuristic behaviors. Using protocol analyses to investigate cognitive processes and production systems to form models may give both information about problem-solving strategies and a language for discussing them. More generally, the field of cognitive science may provide useful insights into the nature of human thought. There is indeed reason to believe that we can better understand productive human thinking, and that we can use this understanding to the benefit of our students.

REFERENCES

Bloom, B.S. *Problem Solving Processes of College Students.* Chicago: University of Chicago Press, 1950.

*Burkill, J.C. and Cundy, H.M. *Mathematical Scholarship Problems.* New York: Cambridge University Press, 1961.

*Bryant, S.J., et al. *Non-Routine Problems in Algebra, Geometry, and Trigonometry.* New York: McGraw Hill, Inc., 1965.

D'Amour, G. and Wales, C. "Improving Problem-Solving Skills Through a Course in Guided Design." *Engineering Education,* 67(5), 1977.

Daniels, P. *Strategies to Facilitate Problem Solving,* cooperative research project No. 1810. Brigham Young University, Provo, Utah, 1964.
_____
*Problem Sources

Dodson, J. *Characteristics of Successful Insightful Problem Solvers.* Ann Arbor, MI: University Microfilms International, 1970, No. 71-13, 048.

Duncker, K. "On Problem Solving." *Psychological Monographs,* 58(5) Washington, DC: American Psychological Association, 1945.

*Dynkin, E.B., et al. *Mathematical Problems: An Anthology.* New York: Gordon & Breach, Science Publishers, Inc., 1969.

*Eves, H. and Starke, E.P. *The Otto Dunkel Memorial Problem Book.* Washington: Mathematical Association of America, 1957.

Flaherty, E. *Cognitive Processes Used in Solving Mathematical Problems.* Ann Arbor, MI: University Microfilms International, 1973, No. 73-23, 562.

Goldberg, D. *The Effects of Training in Heuristic Methods on the Ability to Write Proofs in Number Theory.* Ann Arbor, MI: University Microfilms International, 1975, No. 75-07, 836.

Hadamard, J.S. *Essay on the Psychology of Invention in the Mathematics Field.* New York: Dover Publications, 1954.

Hayes, J.R. "Memory, Goals, and Problem Solving." In *Problem Solving: Research, Method, and Theory,* edited by B. Kleinmuntz, New York: John Wiley and Sons, 1966.

Kilpatrick, J. "Problem Solving in Mathematics." *Review of Educational Research,* 1969.

Kilpatrick, J. *Research in the Teaching and Learning of Mathematics.* Presented at MAA, Dallas, 1973.

Kilpatrick, J. "Variables and Methodologies in Research in Problem Solving." University of Georgia, 1975.

Lakatos, I. *Proofs and Refutations.* New York: Cambridge University Press, 1976, revised 1977.

Landa, L.N. *Algorithmization.* Englewood, NJ: Educational Technology Publications, 1974.

Landa, L.N. "The Ability to Think: How Can It Be Taught?" *Soviet Education,* 18(5), March 1976.

Larkin, J. "Skilled Problem Solving in Physics: A Hierarchical Planning Model." *Journal of Structural Learning,* in press.

*Lidsky, D., et al. *Problems in Elementary Mathematics.* Translated by V. Volosov, Moscow: MIR Publishers, 1963.

Lucas, J. *An Exploratory Study on the Diagnostic Teaching of Heuristic Problem-Solving Strategies in Calculus.* Ann Arbor, MI: University Microfilms International, 1972, No. 72-15, 368.

Newell, A. and Simon, J. *Human Problem Solving.* Englewood Cliffs, NJ: Prentice-Hall, Inc., 1972.

Polya, G. *How to Solve It.* Princeton: Princeton University Press, 1945.

Polya, G. *Mathematical Discovery,* Volumes 1 and 2. New York: John Wiley & Sons, 1962 (Vol. 1) and 1965 (Vol. 2).

*Polya, G. and Kilpatrick, J. *The Stanford Mathematics Problem Book.* New York: Teacher's College Press, 1974.

*Rappaport, E., Trans. The Hungarian Problem Book. Washington: Mathematical Association of America, 1963.

Schoenfeld, A. "Presenting a Strategy for Indefinite Integration." *American Mathematical Monthly,* 85 (8), Oct 1978, pp 673-678.

Schoenfeld, A. "Heuristics DO Make a Difference." *Journal for Research in Mathematics Education,* in press (a).

Schoenfeld, A. "About Heuristics." *1980 Yearbook of the National Council of Teachers of Mathematics,* in press (b).

*Shklarsky, D.O., et al. *The USSR Olympiad Problem Book.* Washington: Mathematical Association of America, 1963.

Skinner, B.F. "Teaching Thinking." In *The Technology of Teaching,* New York: Appleton-Century-Crofts, 1968.

Smith, P. *The Effect of General Versus Specific Heuristics in Mathematical Problem Solving Tasks.* Ann Arbor, MI: University Microfilms International, 1973, No. 73-26, 367.

*Trigg, C.W. *Mathematical Quickies.* New York: McGraw-Hill, Inc. 1967.

Turner, R. "Design Problems in Research on Teaching Strategies in Mathematics." In *Teaching Strategies: Papers from a Research Workshop,* edited by T. Cooney, Columbus, OH: ERIC Clearinghouse for Science, Mathematics, and Environmental Education, 1976.

Wickelgren, W. *How to Solve Problems.* San Francisco: W.H. Freeman & Company Publishers, 1974.

Woods, D.R. "Teaching Problem Solving Skills," Annals of Engineering Education, December 1975.

*Yaglom, A.M. and Yaglom, I.M. *Challenging Mathematical Problems with Elementary Solutions.* Edited by B. Gordon, translated by J. McCawley, Jr., San Francisco: Holden-Day, 1967.

Participants,
Conference on
COGNITIVE PROCESS INSTRUCTION

University of Massachusetts
Amherst

June 19-21, 1978

Hal Abelson
Virginia Adams
Arnold Arons
Robert Bauman
John Seely Brown
Frederick W. Byron Jr.
Susan Chipman
Norman Chonacky
John J. Clement
Gene D'Amour
Andrea A. diSessa
Jack Easley
Bat-Sheva Eylon
Donald Finkel
Gerhard Fischer
Colin F. Gauld
Ira Goldstein
Robert L. Gray
George Hanson
John R. Hayes
Richard Hendrix
Julia Hough
James J. Kaput
William Kilmer
Herb Koplowitz
Jill H. Larkin
Herbert Lin

Joseph H. Lipson
Jack Lochhead
Donald McGuire
Erik D. McWilliams
Irwin Marin
Edwina Michener
Melton Miller
George S. Monk
Sondra Perl
Alexander Pollatsek
Frederick Reif
Alan H. Schoenfeld
Dorothea Simon
Elliot Soloway
Robert E. Sparks
Mel Steinberg
Ting Wei Tang
Richard Thompson
Ruth Von Blum
Ernst von Glasersfeld
Arnold Well
Richard Wertime
Keith T. Wescourt
Arthur Whimbey
Sheldon H. White
Donald R. Woods
Richard A. Yanikoski
Joseph L. Young